CYBERSPIES

By the same author

The Art of Betrayal
Shopping for Bombs

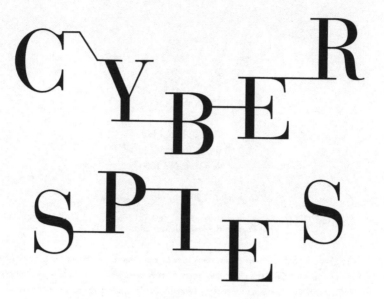

CYBER SPIES

THE SECRET HISTORY OF SURVEILLANCE, HACKING, AND DIGITAL ESPIONAGE

GORDON CORERA

PEGASUS BOOKS
NEW YORK LONDON

CYBERSPIES

Pegasus Books Ltd.
80 Broad Street, 5th Floor
New York, NY 10004

First Pegasus Books hardcover edition July 2016

ISBN: 978-1-68177-154-0

10 9 8 7 6 5 4 3 2 1

Printed in the United States of America
Distributed by W. W. Norton & Company, Inc.

For Jane, Joseph and Samuel

The computer was born to spy. The first computer was created in secret to aid intelligence work, but all computers (and especially networked computers) are uniquely useful for – and vulnerable to – espionage. The speed and ingenuity of technological innovation has often blinded us to understanding this historical truth and its implications. Many books have been written about the emergence of cyber-espionage in recent years, but this book is about something broader. It aims to weave together related stories like encryption and code-breaking, the rise of the computer industry and its complex relationship with the secret world, and relate these to the emergence of what was first called computer security (now cyber security) and with it the rise of cyber-espionage. Much of what people today discuss as if it is new – big data, cyber security and 'going dark' thanks to encryption to take three examples – is, in fact, decades old and only an understanding of the history can illuminate what really is novel.

This is a book that explores that history in the context of the important relationship between Britain and America. Britain played a key role in pioneering computing – thanks to Alan Turing and others in the thirties and forties – but also in the use of bulk intelligence collection even earlier. The precedents for the kind of activity Edward Snowden revealed can be found in British as much as American history. The book also provides the most detailed account yet of the birth at Bletchley Park of the alliance between the two nations' code-breaking services. It explores how America eclipsed its partner and the rise of Silicon Valley (one reason was the huge state subsidies fed into the private sector thanks to work for the secret state) and the relationship of convenience between spies and companies that lasted until recently exposed.

The focus on the international story should also aid the reader in seeing the history in a global context. In many ways being the most advanced nation in computing has imbued the US with a peculiar mix of insecurity and assertiveness. It understood better than anyone how computers could be exploited for espionage and sabotage because it pioneered many of the techniques. It also realised how its role leading the way in computing left it dependent and vulnerable. Many nations have long worried about their dependence on American hardware and software. Now the US is also worrying about the rise of global players like Chinese telecoms giant Huawei – a company whose importance is explored in this book based on extensive, original reporting. The view from China and other countries also helps put some of the issues of great concern in the US – such as intellectual property theft – into a broader perspective.

Events in the last year have thrown many of the issues in the book into sharp relief, none more so than the extent to which the group called ISIS in Iraq and Syria and its supporters in the West have used technology. Five years ago, Western spy agencies faced a threat from a group (Al Qaeda) which used one major communications tool (email) and which was, in the most part, relatively unsophisticated. This meant it could be tracked relatively easily. Today's members of ISIS are from a new generation who have grown up with the latest technology and make use of the latest tools, which emerge at dizzying speed, driven by the desire and ingenuity of developers and the public's appetite for the new. Some ISIS operatives (such as the Briton Mohammed Emwazi, also known as 'Jihadi John,' who killed American and British hostages on video) proved adept at using advanced technology to propagandise and operate online but also hide their tracks to make it harder to find them (it took more than a year to track Emwazi resulting in his death in a US drone strike in November 2015).

Attacks in San Bernadino and Paris – as well as the extent to which ISIS uses technology more broadly to propagandise, recruit and organise as well as perhaps even to carry out nascent cyber attacks – has heightened the tensions between Silicon Valley and the spies and compounded a relationship already fractured by the Edward Snowden revelations. At the same time as ISIS's use of social media has increased the state's desire to get more from companies, the companies are focusing more on privacy. ISIS and others are using American tech companies as their platforms to reach around the world, including into America itself to encourage and support terrorism. The early signs of individuals being radicalised are also often displayed on social media. So how far should companies be responsible for the platforms they create and the content they carry? Is monitoring social media content for threats a matter of good corporate social responsi-bility for companies, or is it the outsourcing of intelligence-gathering to the private sector in which they spy on their users? And how do you deal with such issues when American companies increasingly operate globally?

Sometimes a thorny knot of issues is presented as simply a case of requiring us to 'balance' security with privacy, but the reality is far more complex and multi-layered. Encryption might at once offer individuals more privacy and security but also pose challenges for law enforcement and the state in providing collective security. This book traces the story of encryption from the days of Bletchley Park to its central role today in political debate – with a crucial meeting at Stanford in the 1970s, the pivot point from the secret world into the public eye. Does Apple's pro-vision of end-to-end encryption for its phones help make individuals safer by protecting us from cyber criminals or does it make the world more dangerous by giving jihadists a way to communicate out of reach of the state? The row between Apple and the FBI over the phone of the San Bernadino shooter is only the latest chapter in a longer story. One thing that the history shows is that the 'going dark' debate in the US is not new – Britain faced it with an upgraded Enigma machine on the eve of war, Britain and America faced it in the early Cold War when the Soviets improved their systems, and they feared it again in the 70s and 90s with the spread of encryption and fibre-optics. The history suggests that intelligence agencies (more so than law enforcement) are highly inventive in finding ways round obstacles which at first seem insurmountable (sometimes through high technology and sometimes through low cunning).

Spies are also struggling to adapt to the new world. In recent months, the CIA has radically reorganised itself with a new directorate to focus on bringing digital intelligence into the heart of human espionage. Computers certainly offer new ways to steal data but it also poses real challenges to spies. You could forgive the chief of MI6 – known as C – a slight shudder as he watched the last Bond movie *Spectre*. Not the scene in which MI6 HQ is blown up but the more worrisome plot device that his Secret Service was going to be swallowed up by a new data driven super-agency. Spies know that just as they can use technology to uncover secrets so the same technology can be used to strip away the secrecy that they consider vital to doing their job. Few outside the intelligence world understand the extent to which spies in the US and Britain perceive technology as an existential threat to their work. An arms race is on between spy services to exploit technology. Only those who adapt will survive.

★

The fear of what could be done with data was evident in Washington's neuralgic reaction to the cyber-intrusion into the federal government's Office of Personnel Management in which the personal details and vetting material on 21 million people – many of them government workers – were stolen (allegedly by the Chinese). The material could be used to spot vulnerabilities and approach an official for blackmail. A smart intelligence service could also correlate who at an embassy, say in Beijing, was on the database and, by elimination, work out that anyone not on the database might be an undercover intelligence officer. 'It was a great intelligence operation,' one top American spy says, 'and given the chance we would have done the same thing.'

Recent events have also suggested the NATO alliance of western countries is behind Russia in integrating cyber into information operations and into 'irregular' or 'hybrid' warfare, as witnessed in Ukraine. Soon there will be no such thing as 'cyberspies' as it will be a given that spying will be cyber enabled and the same is true for cyber-warfare – it will become an intrinsic part of all military doctrine. Russia appears willing to push the boundaries in 'live testing' cyber weapons. This is one part of a notable recent proliferation and escalation of cyber-attack (as well as espionage) capability. The signs are that the long-standing concerns over the use of cyber techniques to cause real world effects rather than just pilfer data are now turning into reality – whether it knocking out power stations or interfering with live TV broadcasts. The nightmares of people maliciously manipulating health and banking data or carrying out real cyber terrorism (rather than just vandalism) may be about to be come true.

What are the implications of the rising tide of cyber espionage for all of us? What is clear is that it is becoming harder and harder to keep secrets in the modern world. Just ask the people who signed up to Ashley Madison and then had their details hacked. Or the people whose credit card information is sold on the Dark Web. Or the teenagers who realise their social media trail of embarrassing pictures is harder to erase than they thought. Or the twenty one million people affected by the OPM breach. Or the NSA after Edward Snowden. It is now easier than ever for information to be stolen and leaked. An ever-increasing dependence on inherently insecure technology will only accelerate this trend as cars, watches, fridges and an array of everyday items start to get hacked by a range of malicious actors.

The rising toll of cyber security breakdowns begs the question of whether a tipping point is approaching when the insecurity under-mines confidence in the much-heralded future of big data and the internet of things. If we do not lose our trust in technology, we may instead be forced (partly by our own commercial choices) to refashion our notions of privacy and secrecy. Companies are keen to point their finger at the state for compromising privacy but less keen to talk about the implications of their own practises in exploiting data, while spies like to point their fingers at the companies, ignoring some of the differences. In reality, a complex symbiotic relationship exists between corporate data collection and the intelligence activities of the state – the existence of pools of data gathered for profit acts as both a tempting target for spies but also a precedent for their actions. Europe and America are also at odds over the power of American technology companies and the implications of their dominance.

Should we simply shrug our shoulders and enjoy the latest app – which sucks up our data – and accept the intrusions of the state as necessary in the fight against terrorism? Or should we instead rage, as some would put it, against the dying of the light of privacy and internet freedom. This contest for our future has only just begun. But it will become an increasingly central struggle in the social and polit-ical discourse within and between nations. The hope is that the history in these pages will help inform that debate.

CONTENTS

An enigmatic telegram arrived at the port of Dover just past midnight. It was in code so its meaning would have been lost on anyone save for its intended recipient, an officer named Superintendent Bourdeaux. 'WHERE COULD SECOND LETTER FIND YOU,' the telegram read. It was 5 August 1914 and Bourdeaux began urgent preparations as he waited for a second message to confirm his orders. 'We were taking considerable risk,' he noted in a melodramatic handwritten report buried deep in the archives of Britain's Post Office. At 1.52 a.m. Bourdeaux was on board a ship, the *Alert*, as it set sail. The bulk of the crew were ignorant as to their mission. The weather was still good. War had been declared with Germany just minutes before the telegram had been sent. The *Alert* was about to undertake the first offensive act of the conflict.

The *Alert* arrived at its first destination at 3.15 a.m., lowered its hook to the seabed and began to dredge. After twenty minutes it had hauled up the end of a cable. Bourdeaux had been instructed to bring it to the surface to check it was the right one. Then he was to cut it. By the time a fourth cable was reached, it was six in the morning. Heavy rain and wind were battering the ship, making the grappling much harder. At this moment, a foreign ship was spotted for the first time. With relief, Bourdeaux realised it was from Britain's ally, France. '*Alert*: What are you doing?' the curious French captain asked as he pulled up alongside. 'Cutting German cables,' Bourdeaux replied with disarming honesty for a man on a secret mission. 'He and his crew then gave us a splendid round of cheers which we heartily replied,' Bourdeaux writes. By ten in the morning, the *Alert* was heading back to Dover.

The *Alert*'s order had come by telegram – the near-instantaneous form of communication that had transformed everyday life in the previous half-century by sending an electrical signal down a wire to deliver a message. The telegraph had been the driving force in the first wave of globalisation. It allowed businessmen, diplomats and ordinary people to communicate with each other across vast distances in a matter of moments rather than weeks or months. It had brought the world so close together that optimists believed it had banished the possibility of conflict. But Europe was now descending into the abyss of the First World War. And, in carrying out the order sent by telegram, Bourdeaux and the *Alert* were undertaking one of the first strategic acts of information warfare in the modern world, an act whose consequences would ripple out in unforeseen ways through the war years and beyond, leading to the birth of modern communications intelligence. The *Alert* had cut off almost all of Germany's communications with the outside world.[1] It had hit the kill switch.

On 4 August, just before the *Alert* set sail, a man arrived at the cable station at Porthcurno in Cornwall. On the secluded, sandy beach, telegraph cables carrying traffic across the Atlantic came ashore and the messages they conveyed would then be relayed from inside the building. The man's job title was not one that we associate with espionage. He was a 'censor'. In the office of the Eastern Telegraph Company in the British colony of Hong Kong, another 'secret censor' walked into his new office. A similar figure did the same in every far-flung corner of Empire, from Malta to Singapore. Once the censors were in position, instructions had told them to send a message to London reading 'Fixity London, Fixed'. The system was up. At midnight on 2 August 1914 a global system of interception had been instituted. Known as 'censorship', its aim was 'to prevent intelligence being conveyed to the enemy and to cut off the enemy's correspondence with his agents'. Britain was taking advantage of its control of much of the international telegraph infrastructure to create the first global communications surveillance system, from Cairo to Cape Town, from Gibraltar to Zanzibar. Fifty thousand messages would pass through the hands of 180 censors at UK offices alone every single day. Another 400 worked in 120 stations overseas. In all, 80 million messages would be subject to censorship during the war.[2]

Spy fever had gripped the country as war approached. This had been fuelled by lurid tales from novelists and newspapers suggesting German agents were lurking undercover in every port and village, secretly conveying vital information back to the Fatherland that might cost Britain victory. The fears led to the creation of MI5 to catch German spies at home and MI6 to make sure the favour was being returned. It was hoped that the 'censorship' of telegraph cables would help catch those (largely non-existent) agents. And some men would meet their deaths – blindfolded and shot in the moat of the Tower of London – based on the letters and telegrams that were intercepted. The impetus of war and the fear of the unknown would lead to intelligence collection on an industrial scale.

An army of vans pulled up outside Strand House in London and unloaded bags of international mail onto trolleys. Bulk intercept in the pre-computer age was low-tech and laborious. The building was damp, cold, poorly ventilated and cramped, declassified files record with some sadness, and a full-time nurse was on duty to limit the high sickness and absence rates. The task of reading other people's mail was so vast that it spilled out of Strand House and into even the new Science Museum in South Kensington. At its peak, 375,517 letters were 'censored' in a single day. At the end of the war, it was reckoned that 630 million letters had passed through the system, with 1.3 million of them stopped for further checks.

How were these selected? Letters to and from certain addresses and names on a blacklist were intercepted, the most sensitive being sent straight to MI5. There was mission creep as the system expanded from looking for German agents to keeping track of anarchists, finding obscene literature and even 'information on political matters from well-informed private individuals'. Members of Parliament could have their post stopped, since discontented soldiers at the Front frequently communicated their concerns (and supposedly sensitive information) by letter. This was all about international mail. There were trials of blanket interception of domestic letters as part of the pursuit of spy rings, but installing censors at every local post office would be too obvious and generate too much material to process.[3] The technology did not yet exist to facilitate such an ambition. The latest piece of communications equipment – the telephone – added a new complication. Phoning Europe required permission from the Cable Censor's office. In naval towns like Sheerness, Devonport and Harwich, all domestic calls were listened to and transcribed in the hunt for enemy agents. A single line in the official report summarises the results: 'Nothing of any value was discovered.'[4]

Intercepting mail with Germany was fair game. But what about neutral countries? In May 1915, a warrant was signed for the interception of all mail coming from or heading to America (a whole new office in Liverpool was opened to cope with the volume), and by the end of the war America accounted for ten times as much material as any other neutral country, and well over half the total.

Britain was learning that reading someone's messages to glean intelligence might be more useful than stopping them communicating. Just as modern commerce depends on the internet, international trade then depended on the telegraph. Clever detective work could help expose German attempts to evade the economic blockade. Was a company in Amsterdam seeking to buy metal from America really a front for the German firm Krupps? If so, it would be placed on the 'Secret Blacklist'. The next step was using this intelligence as a weapon. When the Irving National Bank of New York was found to be conducting a little too much business with Germany, its cables were delayed until it changed its behaviour. The same trick was used on all commercial traffic to the Netherlands because the Dutch were allowing sand, gravel and cement for the German war effort to transit. This, a British official remarked, inflicted an 'infinity of harm' until they agreed to stop.

When the *Alert* severed telegraph cables, it did not cut Germany off completely. There was the new medium of radio. Germany had invested in a high-powered kilometre-long radio antenna at Nauen. For the German and British navies radio was changing warfare, allowing ship-to-shore communication for the first time and centralised command and control. And German radio would be the crucial stimulus to the conception of Britain's communications intelligence machine. A decade before the war, the Eastern Telegraph Company had set up a radio receiver on a small headland above Porthcurno beach, still known as Wireless Point. It was there to spy on the transmissions of Guglielmo Marconi, the flamboyant businessman and inventor who had just sent a radio message across the Atlantic. The Eastern Telegraph Company feared this new upstart technology would destroy their business. But radio had a problem – it was a broadcast medium – anyone with a receiver could listen in. Surely that meant it was insecure for any sensitive communication? Marconi had claimed it would be possible to transmit on a narrowly defined wavelength to stop people 'tapping' a message. To prove him wrong, the Eastern Telegraph Company turned to a magician. At a June 1903 lecture at the prestigious Royal Institution, Marconi's team was supposed to pick up a secure message from Cornwall in front of a packed crowd. But just before the agreed time, an assistant started to hear a rogue signal. He realised with horror that the word 'rats' was being spelt out in Morse code.[5] A rhyme followed:

There was a young fellow of Italy,
Who diddled the public quite prettily

Word soon seeped out that Marconi had been – to use the modern term – 'hacked'. The culprit was Nevil Maskelyne, the magician in the pay of the Eastern Telegraph Company, who had interfered with Marconi's communication to show it was not secure. The lesson was that radio was subject to interception. That meant security was going to have to be integrated into the message itself rather than relying on the mode of transmission. That required secret codes. And where there were codes, there would be code-breakers.

On the first day of the war, the Director of Naval Intelligence took Sir Alfred Ewing, a man with shaggy eyebrows and an academic bent, to lunch. A Marconi company engineer was delivering messages intercepted from Germany on a specially chartered train to Liverpool Street Station in London. Some were in code and a team was needed to decipher them. Ewing was asked to take charge. 'These were the decipherers and a rummier set of fellows I never came across in all my born days,' one observer said of the men who would assemble under him in Room 40 of the Admiralty building.[6]

The magnificently named Fleet Paymaster Charles Rotter – a naturalised German – was the first to tackle the gobbledygook of encoded messages. The Germans, as did the British, used code books, which were like a dictionary shared between sender and recipient with a list of common words and phrases and a corresponding code to substitute for them. Rotter and his team had got hold of all three main German code books and began to decipher German communications. The First Lord of the Admiralty, Winston Churchill, was a man who loved intrigue and intelligence and he immediately grasped the enormous potential. In November 1914 he issued a charter marked 'exclusively secret'. Room 40 would study messages past and present 'in order to penetrate the German mind'. Churchill would have decoded messages rushed in to him as he was having his morning bath; he would then grip them with a 'dripping hand'.[7]

Spies had pilfered letters for kings and generals for hundreds of years, but Churchill's charter marked the emergence of something new. This was communications intelligence as a formalised discipline – the systematic gathering, classifying, processing and decoding of messages. When used correctly it was a powerful tool for both the soldier and the diplomat. A decoded German message – known as the Zimmermann Telegram – would help bring America into the war by revealing Berlin's plans to offer chunks of US territory to Mexico.

British code-breakers also learnt that sometimes you did not need to actually break a code to extract useful intelligence. German Zeppelins were bringing the war to the Home Front by dropping bombs on British towns. But it emerged that when an airship took off, messages would be sent to German anti-aircraft guns so that they would not fire on it. The Zeppelin would also use regular call signs to report back as it cruised over the North Sea and asked for bearings. You might not be able to break those codes, but if you systematically analysed this behaviour over time you could begin to predict when a Zeppelin was taking off and what its path might be – an early form of geo-location based on communications data. That would then allow you to alert your anti-aircraft guns at the right positions to get ready. This type of intelligence – studying the externals of a message for value – became known as 'traffic analysis'. Once inside your enemy's communications you could also deceive: during one raid false bearings were signalled to send Zeppelins off-course, allowing them to be shot down over France.

Just as the war was ending, the first sign was emerging of something new that would supplant the centuries-old practice of code books. On 23 February 1918 an engineer named Arthur Scherbius, whose other inventions included an electric pillow, filed patent number 416129 in Berlin. The German military could not yet see the potential of his new device but after the war Scherbius would sell it to businesses with the slogan, 'One secret, well protected, may pay the entire cost of the machine'.[8] Confident of its ability to create a mystery that other people could not understand, Scherbius christened the electro-mechanical device Enigma. The era of machines was arriving.

In the East End of London, a boy stood in his front garden during the war. He could see a searchlight scanning the dark skies above. He was a working-class bricklayer's son not quite in his teens who loved to tinker with toy steam engines and build things with Meccano sets. As he gazed up into the night sky he saw the menacing form of a dark, bomb-laden Zeppelin suddenly illuminated by a searchlight, perhaps spotted with the help of the code-breakers. Guns opened up from the ground and the hulking mass was brought down over London in a flaming wreck. By the time of the next war, young Tommy Flowers would build his own machine, a creature made up of hot valves and paper tape – a machine that would not only help win that war but take the world into the computer age.

In 1929, a computer arrived at Britain's War Office. 'The work of a Computer, Class II, is mainly in connection with the calculation of the trajectories of projectiles in artillery fire,' reads the official description of its task. The computer's name was Kathleen Marion Lewis.[1] This was not one of those pet names that engineers like to give to their creations. Kathleen was a person, not a machine. When the job had been advertised, it had been made clear that applicants needed to be British and have a maths qualification. The advert also said that women would be paid £100 a year – £20 less than men – and they would have to resign if they got married. That did not put Kathleen off. Her job was to perform calculations because a computer in those days was someone who computed.

Out in the foothills of northern India in the 1920s, another member of the War Office called John Tiltman had the title 'Signal Computor', as he worked with pen and paper. His job was to break the codes of Russian 'signals' or communications, a task he would labour on for another half-century. Even though he resolutely remained old-school in his methods, his work would contribute to the arrival of the first computer – a machine called Colossus – designed to break German coded communication in the Second World War. Tiltman went on to work for Britain's Government Communications Headquarters (GCHQ) and America's National Security Agency (NSA), only retiring in the 1980s.[2] That decade saw the arrival at the NSA of a black box, roughly the height of a man, named Frostburg. With its red blinking lights (which were mainly for show), it could perform 65 million calculations in a second, far more than Tiltman could have scribbled.[3] During the intervening decades, computers went from being a tool for espionage – by breaking codes and collating data – to a target of espionage – because they held valuable information – and finally to being the means of espionage itself: because they could talk to each other, one machine could steal another's secrets. This is the story of how that happened, what came next and why it matters.

There is a saying, attributed (perhaps falsely) to the information theorist Marshall McLuhan: 'We shape our tools and then our tools shape us.' The birth of the computer and the contours of its youth were shaped by the demands of code-breaking and espionage. But in time, the power of computing would in turn transform the spying business. And as computers become increasingly all-pervasive in our modern society, the intertwining forces of computers and espionage are now reshaping the entire world – from the rise of China to the phone in our pockets – making what was once the preserve of a few intelligence agencies something that has implications for all of us.

This was originally intended to be a narrower book about 'cyber espionage' – the stealing of information over computer networks. But as I tried to understand what it was and where it came from, it soon became clear there was a more interesting story to tell. Every time I pulled at a thread about today's world it drew me further into the past; I wanted to understand the history of computers and spies, how it began and how it has shaped the present. It drew me into how spies first developed computers for a specific task – code-breaking – and how the battle over encryption and protecting or exploiting communications is today more fiercely contested than ever. It drew me into understanding how spies were the first to appreciate the power of computers to unlock the value hidden in large amounts of data – now a staple of the private sector and part of all our future. And finally, it drew me into the story of how computers and communications merged with the creation of the internet and the emergence of 'hacking' to exploit vulnerabilities, which in turn has changed the age-old practice of spying. The revelations from Edward Snowden that came after I had started the book added extra spice to the tale. What I found was a story that was far more interesting and mattered far more than I had expected.

What is spying? At its simplest it is finding out secrets. Since time immemorial, that has involved establishing the intentions of another state, such as its plans and capabilities for waging war. Those secrets may also be the identities of people – such as those who want to remain hidden, like enemy spies operating in your country – or terrorists planning to attack. This takes spying into the trickier domestic domain, where it can also be used to root out dissent or as a form of social control, in the way the Stasi deployed surveillance in East Germany during the Cold War. The secrets you are after might be discovered through human spies and agents – the subject of my last book on MI6 – but they can also be unearthed by intercepting the communications of others. In England the practice of the state clandestinely reading letters dates back to at least the fourteenth century. The need for a single Post Office with a monopoly over mail was justified in 1657 by the state requiring access to communications in order 'to discover and prevent many dangerous and wicked designs against the Commonwealth'. It was an era rife with fears of plotting at home guided by the hidden hand of those abroad. The fears and desires that have driven spying have changed little over the centuries. But the technology has. Our story begins a century ago when Britain built a formidable machine to search for the wicked designs of the enemy and its agents in war.

The First World War may seem a slightly odd place to start a history of computers and spies since it pre-dates the electronic world. But starting here serves two purposes. Firstly, it is the time when 'signals intelligence'– studying your adversary's signals (mainly, although not exclusively, communications) – came into being. The *Alert*'s cutting of cables marked the beginnings of information warfare and the establishment of an industrial-scale global intelligence operation. This gathered material in bulk from a commercial cable infrastructure that Britain dominated. Computers were a long way off. But the concepts that would underlie their use in modern espionage – from mass sifting of messages to traffic analysis of data and, of course, code-breaking – would be learnt in these years (there were also concerns over privacy). Secondly, the reference point of the pre-electronic age also allows us to see much more clearly what the computer has and has not changed about spying. Britain's First World War system parallels the modern intelligence system that the

UK and US built a century later based on access to the cables that convey worldwide data traffic. But those data pipes carry something far richer than the telegraph clicks of the past, and so modern computer-based 'digital intelligence', with its updated form of packet inspection, offers something qualitatively as well as quantitatively different from what was undertaken under fluorescent lights in the Great War.

Since at least the time of Julius Caesar, those who carry information have used secret codes to shield the content of their messages from prying eyes. What we think of as the computer – a machine – was built at Bletchley Park, conceived out of the urgent need to perform calculations to break German codes on a scale that no man or woman could manage. Alan Turing laid the intellectual foundations for such a machine, while Tommy Flowers built it. The machine – Colossus – performed statistical calculations using algorithms to help deliver victory. That achievement was kept secret, but a deception designed to mislead opponents also deceived Britain itself. The culture of secrecy that surrounded first Bletchley and then its successor GCHQ came at a price. Bletchley also witnessed the first, tentative steps in forming the closest intelligence alliance seen in history, one that persists to this day. It was the US, thanks to the intimacy there between spies and private companies, rather than Britain that would forge ahead in computing. During the Cold War, the desperate effort both to hunt Soviet spies and to provide early warning of nuclear Armageddon took computers beyond code-breaking and into massive data processing and traffic analysis. In the subsequent decades computers would begin to move out of the secret world and into public view.

Anyone listening to experts talking woefully today about 'cyber security' and the manifold vulnerabilities of our systems might think this discussion was uniquely modern. In fact, it is a good half-century old. Much less has changed than people realise from the days when a young captain in the US Air Force worried about computers controlling nuclear missiles and began to wonder how others might subvert or get inside systems. Fears of 'Trojan horses' and 'trapdoors' in both computing hardware and software are far from new. As computers spread, people began to understand more about

the amount of sensitive data they held and their potential vulnerability. The use of secret codes to protect information had been one
of the most closely guarded secrets of spies and states but in the
1970s a pair of Californian academics would face down the power
of the NSA as they took the secret of encryption into the public
domain.

And as computers began to talk to each other, a third option,
beyond those of code-breaking and data processing to support espionage, opened up: using computers to perform the central act of
stealing information. The inherent weaknesses of computers themselves and of a system – the internet – built to share and not secure
provided a means to carry out that age-old task on a scale and from
a distance that would transform the enterprise of espionage. The
KGB would be the first to understand the potential in the late 1980s
as a bunch of drugged-up hackers from Germany rooted around
American military systems in search of secrets for their Soviet pay-
masters. Spies and hackers had met and computer espionage had
arrived. The internet, with its offer of anonymity, became a perfect
playground both for spies seeking to work undercover and also their
targets.

People have spied from land, from the sea, in the air and from
space. And so cyberspace inevitably became a new domain for what
is often called the second oldest profession (following hot on the
heels of the first). Western countries like Britain and America and
even their Cold War rival Russia have their own preconceived ideas
of what espionage involves from watching James Bond films and
reading John le Carré. Espionage involves professional spies doing
dangerous and duplicitous things in shady places. But this is somewhat misleading and misses the important role of computers and
data. Filing cabinets have always been as important to spying as guns
and gadgets; and computers and their connectivity have opened up a
new world of digital intelligence, which in turn has transformed the
traditional role of human espionage, sometimes enabling it, sometimes supplanting it. Today John le Carré's George Smiley would be
sitting at a computer terminal, not walking the streets. Our understanding of espionage can be too rooted both in our own narrow
cultural experience and in out-of-date notions. The internet and

computers have changed spying in ways that the popular imagin-
ation has yet to appreciate fully, in the use of data to search for the
signal amid the noise, to find meaning within apparent randomness
and things – and people – that want to remain hidden. Cyber espionage facilitates spying on a scale previously unimagined and at a
distance, changing the calculus of risk that had previously inhibited
its use.

Another problem that stems from our misconception of spying is
that we struggle to place new sources of cyber espionage into a narrative we know and understand, when there is no reason why they
should conform. China, for instance, pioneered the use of computer
espionage to target Western companies for economic gain and to
seek out dissent. Corporate espionage and intellectual property theft
are not new, but computers have raised the stakes, drawing companies onto the front lines as they have their own secrets exposed, as
Sony realised when it was taken apart by hackers said to be from
North Korea. Meanwhile the hills of Dharamsala in northern India,
home to the Tibetan government in exile, have been at the forefront
of one of the earliest, most sophisticated campaigns by a state that
has entangled both the British and American governments and one
of the world's most powerful technology companies, Google. The
fears over China's rise are also evident in the controversy surrounding Huawei, a telecoms giant. As computers and communications
merge, Washington fears that China will be able to do what Britain
did in the First World War – hit a kill switch and spy through control
of the infrastructure.

Military intelligence once meant sending people undercover to
make maps of far-off lands or, in the run-up to the First World War,
to creep round German dockyards seeking out details of the latest
battleship under construction. Later in the twentieth century it
meant using satellites and intercepts to do much the same thing,
but in the twenty-first century it involves mapping out the computer networks that run another country's power supplies ready for
war, or stealing the latest designs of a stealth fighter so that vulnerabilities can be identified and perhaps even implanted. Computer
espionage overlaps with the world of reconnaissance but also has
moved into that of covert action. The work of intelligence agencies

often extends beyond just finding out secrets into the realm of acting covertly or clandestinely when the state wishes its hand to be hidden – for instance, in sabotaging another country's nuclear programme. Computers and the internet have made this a particularly tempting – and dangerous – possibility for many players.

The phrase 'secret history' is bandied around freely these days. But this book, I hope, merits such a subtitle. Much of the story has remained classified for decades and only now can be pieced together to reveal a picture that had previously been hidden. What emerges is a corrective to the deterministic view that technology was always going to develop in a particular way with particular consequences. National security has been a driver – often unseen – of technological innovation. Companies, governments, individuals, but especially spies, have shaped the world of computing and networking. This process began with Bletchley Park but carried on into the Cold War and then the era of terrorism that hit America on 9/11 and Britain after 7/7. The traffic not just of data but of ideas between the secret world and the commercial world is a central part of this story and it flows both ways. The search for patterns in vast amounts of data was first undertaken by spies in the Cold War to look for signs of abnormal Soviet activity. In turn, as the kind of computing power required became accessible to the private sector in the 1990s, companies developed database marketing to track their customers. The spies after 9/11 and 7/7 then borrowed technology and techniques from the private sector to sift through a vast sea of data to look for unknowns, to find suspected terrorists, turning their focus domestically as well as abroad and making use of what they called the 'home-field' advantage of the commercial world.

Computers have become all-pervasive. The giant Colossus has been replaced by a computer infinitely more powerful that fits in our pockets. Black-box sentries stand guard at the border points at which the internet reaches our shores sifting vast torrents of data. Spies initially struggled to deal with this but eventually worked out how to master the internet rather than be defeated by it, creating what has been called a 'second golden age' of signals intelligence, the first being the days of Bletchley Park. But the exposure of some of these secret programmes by Edward Snowden raised complicated issues about what privacy means in the modern world. Campaigners fear that the power of computers and surveillance will crush dissenting voices, leading to self-censorship that will reinforce existing power structures in many states. The spies believe their methods are the only hope of finding those individuals and groups who wish us harm and who have themselves moved into the online world. The hunger for data is becoming more intense, with more and more people and places seeking to acquire it and analyse it, challenging the very notion of what spying is and who spies are. Espionage over computer networks (hacking) is becoming commoditised and commercialised, and is no longer the domain of just a few states as it was a quarter of a century ago when it began; rather, it is something that affects us all and can be done to us and even by us. We can all retrieve, analyse and correlate information in a way that a spy of a few decades ago could never have dreamt of. So have we all become spies now?

Writing about intelligence is challenging. Some elements of this story have been glimpsed elsewhere – Chinese espionage, the NSA's and GCHQ's activities, the rise of the internet – but this book seeks to explore avenues and stories previously unknown and bring together the disparate strands in a way that reveals the connections between the algorithms of Bletchley and those of Google, the encryption of Enigma machines and that of modern smartphones, and relates them to the spying of Britain, America, Russia, China and others. It is based on first-hand reporting in Britain, the US and China, alongside declassified documents and more than a hundred interviews with a range of people intimately involved, from spy chiefs to hackers.

This book does not seek to cover every aspect of computer security – the issues of cyber crime and cyber war (if such a thing exists) are not explored, other than where they overlap with the central theme of spying. Nor does the book aspire to cover the story of computers and spies in every country in every detail: individual accounts have been selected to convey the broader narrative. Historical parallels are deployed to show where computers have changed espionage and where they have not. It does not approach the subject from a technical perspective (the last computer code I wrote was in the 1980s on a much-loved Commodore Vic20). My aspiration is rather to ensure that the history of spying and the history of technology are portrayed in conjunction with each other and also with events in the outside world.

Spying has always been controversial, raising complicated ethical questions. This is not a book that sets out to tell people what to think. It is a work of history that aims to explain how we got where we are so that people can be informed enough to make up their own minds. We are facing a future in which everything is connected to the internet, in which the physical and virtual increasingly merge. This is a future in which we will leave a rich digital seam that can, for good or for ill, be mined, not just by intelligence agencies but by many others as well. The history told in this book and the issues it raises are not just for technologists or intelligence agencies; they are for everyone.

CHAPTER ONE

BIRTH

By the Second World War, Tommy Flowers, the boy who had stood in his East End garden and watched a Zeppelin brought down in a flaming mass during the First World War, had graduated from Meccano sets and toy engines to building his own machines. In his laboratory at Dollis Hill in late 1943 he was going to test the limits of his new creation. He switched it on and a whirring gathered pace with a click as the telegraph tape turned a cycle every few seconds. The delay between the clicks shortened as the pace quickened. Twenty miles an hour. Thirty. Flowers kept pushing. The tape was now a blur of white. Forty. Fifty. The pulley wheel was spinning the tape around something known as the bedstead because it looked like an upright bed frame. Finally the tension was too much. The paper tape, travelling through the machine at 10,000 characters every second, suddenly snapped in several places. Scraps of paper exploded into the air, shreds falling all around the laboratory like snowflakes amid the noise. 'It was really just pandemonium,' Flowers recalled. Sixty miles an hour, he now knew, was the absolute limit, so a safe speed would be half that. The paper pandemonium meant his machine was nearly ready. Flowers was building something that would change the war and the world. The ambition was reflected in its name – Colossus.

The suburb of Dollis Hill was home to a large, bureaucratic-looking brick building that housed the Post Office research laboratory. The words 'Research is the Door to Tomorrow' were inscribed in stone above the entrance. When the air-raid sirens went off, one of Flowers' assistants, the mildly eccentric Doc Coombs, would grab his tin hat and race up to the roof and shout 'Bandits at 12 o'clock!,' fearing incendiary bombs might destroy their work.

Coombs was one of the few members of the team who knew what
they were actually working on. Even when the King and Queen had
visited they were not shown the creature that Flowers and his team
were toiling on day and night. Most of the engineers were simply
handed diagrams to wire up without any explanation. A few wondered what kind of machine would need so many electronic valves.
Once a door was left slightly ajar and one of the workers looked
in. 'I saw a framework holding paper tape which was being driven
at high speed. There was some sort of control panel with flashing
lights. I recognised the tape as that used for sending teleprinter messages and supposed we must be building special communications
equipment. Naturally I did not tell my fellow technicians what I had
discovered,' he later said. 'One day, when I was helping to load some
of the equipment we had made onto a vehicle, I asked the driver
where it was going. He told me his job was simply to go to a ren-
dezvous point, where someone would be waiting and they would
exchange vehicles. Beyond that, he said, he knew nothing.'[1] The destination was Bletchley Park.

Flowers had been born in London's East End in 1905, a boy who
was good with numbers but struggled with words.[2] His gift for engineering claimed him top place in the Post Office's school leavers'
exam. From 1930 he had worked at its elite research centre at Dollis
Hill, looking at how to update the telephone switching process. He
had realised that valves containing electrons could switch a phone
call faster than existing magnetic relays. This was called electronic
switching. Flowers only just escaped from Germany, where he was
attending a conference, as war began. As he travelled back through
Holland and Belgium by train that night, he could see the railway stations lit up as armies mobilised for war. He eventually made his way
to Bletchley Park. His first morning was spent signing the Official
Secrets Act and receiving a stern warning about secrecy. That same
afternoon, Flowers had a two-hour meeting with the man responsible for suggesting he come to Bletchley – a pivotal figure in the
birth of computing. The mythology that has sprung up about Alan
Turing since the war has often focused on his quirkiness and pecu-
liarities, creating something of a caricature of an eccentric genius
cycling round wearing a gas mask to stop hay fever or chaining his

favourite tea mug to the radiator. But Flowers' first impression of the younger man with his straight dark hair was that he seemed quite normal except for a pronounced stammer. 'He explained the technology of code-breaking,' Flowers said of that first meeting. 'He was concerned with the Enigma.'

These two central figures in the advent of the computer age, neither of whose contributions were appreciated during their lifetimes, could not have been more different. One had followed the path of public school to becoming a fellow of King's College, Cambridge, the other was a working-class East End boy who had been to night school. But despite their different paths, they had been brought together because they had proved to be brilliant innovators in the years leading up to the war, reaching the elite institutions within their own respective fields. Those two areas of expertise – maths and engineering – would then fuse in the white heat of war to forge something new. Bletchley's success was built on the way it threw such different people together.

Turing's wartime work focused on breaking Enigma, but his wider effort before and after the war laid the foundations for modern computing. As a schoolboy aged seventeen, Alan Turing had first encountered the world of codes and ciphers thanks to a maths book he had chosen as a school prize. The book remarked on the 'romance' and 'challenge' in discovering a secret key to a message and Turing was one of those captivated. The prize also had a deeply personal meaning to Turing. It had been endowed in the name of a boy from the year above with whom Turing had fallen in love but who then died. Tragedy drove Turing deeper into science as if in tribute, taking him to Cambridge and advanced mathematical thinking. By 1935, the twenty-three-year-old Alan Turing was wrestling with what seemed an abstract question. Was all mathematics decidable? In other words, could its methods be applied to any assertion to prove whether it was true or not? Turing's mentor at Cambridge, Max Newman, had posed the question in a different way: was there a *mechanical* process which could be applied to a mathematical statement to see if it could be proved? After a long run out of town by the river to the village of Grantchester, Turing lay in a meadow in the early summer of 1935 and pondered what such a mechanical process might be. And, as his biographer writes, 'Alan Turing dreamed of machines.'[3]

The next year, while working on his Ph.D. at Princeton, Turing finished an academic paper that, then obscure, would eventually be seen as a pivotal work of the twentieth century. 'On Computable Numbers' is thirty-six pages long and aimed to answer a theoretical question. Much of it is filled with dense mathematical symbols and equations. And yet within it are ideas whose clarity and importance stand out even to the modern, lay reader. The term 'computer' was not new: previously it had applied to people. They might be performing some calculation, such as Kathleen Lewis working on the correct trajectory to launch an artillery shell to hit a moving target. Or carrying out a repetitive action, a bit like the hundreds of poor examiners who sat opening and reading messages in the First World War. They had to follow strict rules, with the idea of making the process of deciding whether a letter or telegram could be passed to go on its way or stopped for further reading as 'mechanical' or 'automated' as possible. For instance, when scanning a telegram, was one of the names or addresses on a blacklist present or not? A simple yes/no question was needed in order to maximise efficiency when dealing with such a large volume. These kinds of people-computers, Turing said, would have a set of instructions – what he called a 'state of mind'; they would then apply this to the symbols – or the data – placed before them. 'The behaviour of the computer at any moment is determined by the symbols which he is observing, and his "state of mind" at that moment,' Turing wrote. But if this process of performing instructions was broken down into the simplest possible components, could a machine undertake it? Turing imagined a machine that scanned two paper tapes, one feeding in instructions and another feeding in data on which the instructions would compute. Even the most complex calculation, he thought, could be reduced to its simplest form – an elementary operation in which the state of a symbol was either altered or stayed the same. Before reverting back to pages of equations, Turing writes a simple sentence: 'We may now construct a machine to do the work of this computer.'

Until Turing's insight, machines were designed to fulfil a particular function like, say, an abacus or a slide rule, which created a physical analogue of something you were trying to measure. Turing had asked a question: could the instructions be entered into the machine in the same way as the data? He imagined this as happening on tape (the next step would be loading and then storing them in the computer's own memory), which would allow the data to be manipulated by a mathematical set of step-by-step instructions – known as an algorithm. This could provide unlimited flexibility. 'It is possible to invent a *single machine* which can be used to compute *any* computable sequence,' Turing later explained. It marked a profound change in thinking about machines. Turing had realised that instructions, or what we now call software, allow a machine to be programmed to do different things by breaking tasks down into simple, binary, questions. Turing called his concept an 'automatic machine' or a 'universal machine' because of its flexibility. Later people would call it a Turing machine. But in time, the creations that followed would become so complex that they would make the original notion of a human performing the functions utterly implausible. And eventually the machines would adopt the names of the humans who originally employed the machines. These things, not people, would come to be known as computers.[4]

Turing's idea was academic abstraction in 1936. Charles Babbage and Ada Lovelace had conceived of mechanical computing machines a century earlier, and others in America were nearing the same conclusions in the 1930s. But Turing's concept, involving symbols, logic and instructions, also coincided with the arrival of electronics and war would turn his ideas into something tangible. The day after he sat in his Cambridge rooms and heard Neville Chamberlain on the radio announcing war had been declared with Germany, Turing reported for duty at Bletchley Park. He had been recruited to use his remarkable mind on the challenge that was stumping British code-breakers, the same challenge that Turing had told Flowers about at their meeting. It was Enigma – what many thought was an unbreakable code.

The Bletchley Park estate is built around an unlovely country house, an hour or so out of London. It had been purchased by the chief of

MI6 in 1938 as a fall-back location for British code-breakers when war came. After the First World War, the Government Code and Cypher School (GC&CS) had been established under the control of the chief of MI6. Its tasks were the construction, destruction and instruction of codes. The code-breakers were based for a while in the MI6 headquarters at the Broadway Buildings by St James's Park. Already there was a cultural gulf between the two different types of spy. The Secret Service liked military types with a flair for mischief-making and macho adventure. One senior MI6 officer once said he would never knowingly take on a university man. From 1925 the code-breakers of GC&CS began to recruit just such cerebral types. The main target was no longer Germany, defanged by the peace treaty at the end of the war, but the Soviet Union. Russia and Britain had already clashed in the Great Game of imperial competition but the spread of Communist ideology sparked an even fiercer intelligence rivalry.

On 12 May 1927, police launched a dramatic raid on the offices of the All-Russian Co-operative Society (ARCOS), a Soviet trade agency which MI6 and MI5 had identified as a nest of spies. In a chaotic operation, police found a man hurriedly burning documents as well as a secret underground photographic room. But they did not find the definitive proof of espionage they had been looking for. And so the government decided to rely on intelligence to make its case.[5] In Parliament, Prime Minister Stanley Baldwin read out decoded Russian telegrams to justify the raid. The Russians would have had to be truly stupid not to realise that their codes had been broken. So they moved to a new method for their most sensitive communications. This was known as the 'one-time pad' system. It involved combining each character in a message with another from a randomly generated set on a pad held by both sender and recipient. If this method is used correctly – meaning the pads are kept secure, the characters on it are truly random and also that it is only ever used for one message – then this form of code is considered unbreakable. Once the Soviets adopted this system their messages were gibberish, to the frustration of British code-breakers. For much of the inter-war period Britain could read the diplomatic traffic of every country (including the US) except for two – Russia and Germany.

Germany had learnt of British success in the First World War thanks to leaks and public comments and so had moved towards the use of machines to encipher codes. Already by 1925, British code-breakers were looking at patents deposited in London for a new German device called Enigma. The following year the deputy head of the Government Code and Cypher School walked into the manufacturing company in Berlin and simply bought one. He could do this because the primary market for Enigma was banks and other companies who wanted to communicate securely. As they tinkered with the machine, British experts marvelled at the German engineering. When you pressed down a letter on what looked like a typewriter keyboard, an electric current was sent through rotors to light up a letter on a display. The rotors moved each time, meaning that even if the same letter was pressed again, a different path would lead to a different output letter. A message could be scrambled by setting the rotors in a particular way. The crucial point of a code made by Enigma – which holds true for all machine-based codes that have followed – is that it is designed to look random (like the product of a one-time pad). But it is not. It is produced by a machine set up in a certain way to generate complexity, but in a way that another machine can replicate in order to decode it. The advantage of this is that, unlike a Soviet one-time pad, there is no need to create a unique, random piece of gibberish and physically exchange it between sender and recipient for every single message – impractical for extensive military communications. Instead, with a machine-based cipher all that is required is a similar machine set up in the same way and with the same starting position or key applied to the message.

In 1937 Dilly Knox, a classicist and veteran of Room 40, managed to break the code of an Enigma machine used in the Spanish Civil War.[6] But the more advanced military version stumped the British because modifications massively increased the number of routes an electrical current could take. The head of GC&CS thought it was 'a waste of time and public money' even to try to crack the latest Enigma because it was so advanced.[7] In other words, the spies would have to simply give up in the face of encryption. Fortunately, they did not.

On the eve of war, a group of Polish, French and British spies met in a forest just outside Warsaw. The French had benefited from a German who traded Enigma documents, manuals and key settings in return for money and women. They had shared this with the Poles, whose brilliant mathematicians made major breakthroughs. The Poles revealed these to the British, including how they had built a 'bomba' (perhaps so named because of its ticking sound) to work through the various possibilities. But the Poles had lost what hold they had on Enigma when the Germans improved the machines. The British realised that classicists, linguists and crossword puzzle experts were not going to be enough. The Poles had shown that maths and machines were the future. Two mathematicians were recruited to help with Enigma – Alan Turing and Gordon Welchman.

Enigma, like most security devices, was sold to its customers on the vast number of its theoretical permutations. And this was mind-numbingly big. Some models boasted 158 million million million. Others promised more like 100,000 billion billion.[8] This theoretical level of security did have one advantage: it lulled the Germans into a false sense of security. What is known as a brute-force attack – simply going through every possible combination – was not realistic. But while the code may look random, it is in fact deterministic. Trying to understand how a machine was wired was the first task. After that you had to find an individual key being used at any time. Mathematics could then bring down the number of possible combinations by finding patterns. The art of modern cryptanalysis, or code-breaking – developed at Bletchley and still in play today – was to find the pattern within the seemingly random, the sliver of order within the chaos.

The human factor is the key to code-breaking. However good a security device might be in theory, it is only as good in practice as the people who use it. People, especially those in a hurry, take short cuts and make mistakes. If an operator reused a key setting or used a key that was familiar (say, his girlfriend's name) rather than truly random, that might offer a way in – the old-school equivalent of someone using the same password for two different accounts. Bletchley was all about trying to probe for different weaknesses using a mix of techniques. This required collecting and studying huge amounts of data to try to discern these possible patterns or mistakes amid the apparent endless randomness, to understand what looked normal and what constituted something abnormal. This had first been understood in the First World War but was now applied on a much larger scale. Punch-card machines – the precursors to computer databases – were used to store information about messages so that it could easily be retrieved and processed. The value that could be unlocked from data that was properly organised was becoming clear. Building up the data and searching for patterns was laborious, often dull work which required thousands of man – and woman – hours by many people. Bletchley functioned on an industrial scale, employing work patterns and technology in a way that was truly modern. Turing used to refer jocularly to people forced to do these mechanical operations as slaves, and it is worth remembering that 10,000 people were involved at Bletchley, not just a handful of geniuses. For all the impression today of a glamorous country house being home to a wonderful social life punctuated by the odd moment of inspiration, the reality was arduous, repetitive and often deeply frustrating work being done under enormous pressure. Yet out of this came a kind of magic.[9]

'On a snowy morning of 1940, in a small bleak wooden hut with nothing but a table and three chairs, the first bundle of Enigma decodes appeared,' one of the four men who worked in Hut 3 at the time recalled. It was January 1940 and invasion and the fear of defeat loomed. The break felt like a miracle, even though all that was decoded were dull, disjointed, hard-to-understand scraps about the weather. 'Very small beer, and full of foreign bodies,' was the verdict; and yet it was a sign of what was possible. Soon the war began to heat up.[10] The real challenge was breaking the complex Naval Enigma.

German U-boats were savaging shipping across the Atlantic, sinking ships, disrupting desperately needed supplies to Britain and killing men. The war was being lost on the seas. Could intelligence help? That meant doing the seemingly impossible by breaking Enigma. Turing and others like fellow mathematician Gordon Welchman had learnt from the Polish experience that Britain would need to take maths and turn it into machines. The first task was to find something one assumed was in the original text – like a standardised reference to the weather. You would then use your knowledge of the way the system worked – for instance, the fact that no letter could be encrypted as itself – along with the latest techniques regarding statistics and probability to discard a number of key settings and bring down the range of possibilities from the cosmic to the merely astronomic.[11] The remaining list of possible settings could still be vast, but now you could give it to a machine – a bombe. Each bombe weighed about a ton and was six and a half feet high. Young female Wrens would wire the machines according to instructions, working under fluorescent lighting to the click of the equipment and the smell of hot oil.[12] The bombe was not a computer. It did not carry out calculations. It was electro-mechanical, like the Enigma machine itself, and passed a current through rotors that mimicked an Enigma machine, looking for a setting in which the electrical circuit would be completed. When the machine suddenly stopped clicking through settings, it meant the impossible had become possible.

Countless sailors would owe their lives to the breaking of Naval Enigma, and supplies vital to sustaining the war effort would make it to Britain. But the code-breakers were always aware that the advantage could easily be lost because of some simple change of procedure by the other side. The word they used for this (and continue to use) is 'fragile'. The nature of code-breaking means that an air of desperate insecurity and fear haunts even the moments of greatest triumph for those involved. In 1942 the Germans did tighten up Naval Enigma, leading to Bletchley 'going dark' and convoy losses rising. It took two British sailors sacrificing their lives by passing a code book to a teenage canteen assistant from a sinking German U-boat, and also the help of the new bombes, to get back into Naval Enigma again. This paved the way for victory in the Battle of the Atlantic. Germany's Admiral Dönitz noticed the problems befalling his submarines in the Atlantic. Could Enigma be broken? he asked German High Command. They said it could not. He came up with all sorts of other possibilities, ranging from traitors to aerial reconnaissance. There is no clearer reason why Britain constructed such a ring of secrecy surrounding Bletchley. A word out of turn could have led Germany to realise Enigma was breakable and move to a new system. Churchill called the staff at Bletchley the geese that laid the golden eggs but never cackled. He personally struck names off the list of those with access to the intelligence to keep it that way. No actions were taken without ensuring there was a plausible reason other than the breaking of codes in order to avoid stimulating German curiosity. But while Enigma is the most famous code-breaking feat of Bletchley, its importance for the future of electronic espionage is secondary to another system.

In 1940, British police on the South Coast began hearing something different from the normal Morse code that crossed the airwaves as they sat with their headphones clamped to their ears. Their job was to monitor communications – to listen out for enemy spies transmitting messages back to Germany – but now they could hear a noise which sounded as if it was produced by some kind of automatic machine rather than an operator tapping out Morse code. To them the sound might have been entirely new, but to anyone who has lifted up a computer modem and held it to their ear and heard the sound of electronic pulses – ones and zeros – it might be more familiar.

The mysterious new traffic winged its way back to Bletchley Park, where it was met with bemusement. It was nothing like Enigma. The mystery messages were codenamed Fish and the machine that made them christened Tunny. Tunny automated the process of sending a message. Someone simply sat at a teleprinter and typed. The machine encoded it and sent it along a cable to another teleprinter that would decode and then print out the message in clear text. All that was needed was the right settings. A teleprinter would be talked of as 'online' – making the process similar in feel (though not beneath the hood) to sending an email now.

Enigma scrambled one letter into another using wiring. Tunny transformed a message so it could be represented on a piece of paper as 'bits' across five columns – the bit stream. Each of the five columns would have either a hole or not a hole (a mark present or absent, or in modern terms a one or a zero). The letter E would be a mark in the first column and then there would be no marks on the remaining four columns – writing this in binary as 1-0-0-0-0. Tunny would then use twelve wheels (compared to only three or four in Enigma) to scramble this up by adding another letter represented by ones and zeros and then transmit the resultant 'bit stream' along the teleprinter. All the receiver had to do was add the same code letters, which had the effect of revealing the original letter. This was reckoned to be a million times more secure than Enigma. To go through every combination by 'brute force' would take millions of years. Tunny, known to the Germans as 'The Secret Writer', was a tougher nut to crack for a good reason: it carried the communications of the German High Command – including Hitler himself.

Breaking codes was a team effort. Bletchley brought together engineers, theoreticians, mathematicians and classicists. The sum would be greater than even the brilliant parts. Each player would display their skill and then pass the ball on to the next to see where they could take it. First up was John Tiltman. Tiltman's remarkable career began in the British Army in the First World War and ended in the NSA in 1980s America. He learnt Russian after a stint fighting the Soviets in Siberia at the end of the First World War and was then recruited to work at the Government Code and Cypher School. In India he studied messages about Russian intentions in Afghanistan, before working on their codes used in Europe. Because of his experience, Tiltman had the title of Chief Cryptographer at Bletchley and the unenviable task of trying to get to grips with codes no one had seen before. He was no mathematician, but worked on intuition and experience. He found his best thinking took place just below full consciousness, when he was so immersed in a code that his brain would operate on autopilot as he stood bolt upright at his desk. He also worked on Britain's ciphers to protect its own communications (which were desperately weak for much of the war, especially in the Royal Navy, and were easily broken by the Germans). This union of what is known as offence – breaking codes – with defence – building your own – was potentially a great strength. You could make sure the tricks you deployed to break someone else's code could not be used against yours. Tiltman knew that the more advanced and complex a system was, the more likely it was that people would not use it properly. 'The livelihood of a cryptanalyst depends almost entirely on the over-ingenuity of the designers of foreign ciphers,' he would comment.[13] And it was a mistake by a German operator that offered the first way into Tunny when a long message between Vienna and Athens in the summer of 1941 was repeated using the same setting but with some tiny changes (known as a depth). Working by hand, Tiltman was able to decode much of the message. But was there a way of moving beyond relying on mistakes?

Tiltman passed on the fruits of his detective work to a young mathematician, Bill Tutte, still in his early twenties. Tutte, who had first trained in chemistry, used to stare at the wall for months on end, occasionally twiddling his pencil. But in his head he was performing a stunning feat of individual genius. He managed to conceptualise the structure of the Tunny machine and the mathematics that lay behind it without ever having set eyes on the machine (nor had anyone else at Bletchley). Unlike with Enigma, there were no captured code books to work on, no machines bought before the war. Understanding the mathematical properties opened the way to probe for a weakness. Tutte realised there was a pattern that could be discerned. If you tried cycling through all the possible settings and combining the adjacent marks in a certain way, in most cases you would get a random distribution of marks and spaces. But the right setting would reveal a non-random distribution – a statistical bulge. The problem was that this required working through an enormous amount of calculations. By hand it was incredibly laborious and nigh on impossible. The answer had to be a machine.

The ball was passed to Max Newman – who had first brought Turing to Bletchley and introduced him to Flowers. Newman was at the leading edge of the application of technology to mathematical problems. Turing had described in the 1930s how you could break down the solution to a mathematical problem through the mechanical steps of a machine. Newman understood that what was needed now was to take Tutte's maths and turn it into hardware. The first go was a contraption known as Heath Robinson which was plagued with difficulties, frequently making a mess on the floor. Those who worked on it developed a form of sniff- or sound-based test to discern what the problem might be during its frequent malfunctions.

Tommy Flowers had insisted that electronic valves could switch much faster than the Heath. But the bigwigs at Bletchley were initially reluctant. The risks were high. There is a constant tension in technological development – especially during wartime – between building something innovative and doing something guaranteed to work and be ready quickly. 'They said in a year the war could be over and lost,' Flowers recalled.[14] But in a note in the National Archives to the head of Bletchley in March 1943, Newman suggested it might be worth seeing what Flowers could do. Flowers convinced the Post Office to let him try.[15]

Flowers brought together a team of fifty at Dollis Hill to prove the doubters wrong. Like Turing, he was building on work undertaken before the war, in his case on phone exchanges. Manual switching had once involved an operator physically connecting two phone lines by plugging a cable on a switchboard to create a 'circuit' for the conversation. The volume of calls led to the use of a dial to make a sound. A telephone exchange could automatically recognise the dialling noises as a set of instructions to switch the call onto a specified path. This was done through a kind of logic – *if* this sound was heard *then* it meant open or close this pathway to direct the call one way or another through the possible combinations to arrive at the correct destination. The possible pathways for a phone call could be expressed through wiring diagrams – but you could also express that wiring in a shorthand form of algebra following basic logic.[16] This model of inputting information and processing it according to logical instructions could also be applied to mathematical problems, and Flowers had understood that valves containing electrons could switch on and off in a fraction of a millisecond to create electronic switching.

In early 1944, less than a year after starting work, Flowers' creation was ready. His small pocket diary has a simple note for 5 February: 'To Bletchley Park with team,' it reads. 'Colossus did its first job. Car broke down on the way home. Home 1 a.m.' In his understated way, in five lines in his diary, Flowers had recorded the start of the computer age (even though Flowers himself would later reflect that he had never heard the word applied to his creation).[17] The impression it made on others was overwhelming. 'I remember being introduced to Colossus. With other members of the Research Section, I was taken to a large room, where a large box-shaped object, sheathed in sheet metal, stood upon a wet floor,' Bill Tutte, who had played such a key role in the maths behind Colossus' work, later said.[18] '"That," we were told, "is Colossus."' One person gazed at the wet floor and remarked that it had not been house-trained yet. It was time to put it through its paces by testing it on a code that had already been broken. 'Joy knew no bounds when Colossus gave the right answer in a fraction of the time of the Heath,' they recalled. They set it to repeat the same calculation again and again. Every time it came up with the same answer. Alan Turing had conceived the idea of a universal computer in 1936. Newman saw how such a computer could be used to test mathematical statements. Tommy Flowers built it.

Colossus did not 'break codes'. It was not as simple as putting in some coded text at one end and pressing a button so that the answer would be spat out at the other. Data would enter a machine on paper tape at 5,000 characters per second. A scanner would either detect the light passing through the paper (making it a 'one' in binary terms) or it would see no light, in which case it would be a 'dot' or a zero. The one or zero would then pass into the machine for calculation. The wiring of the valves created 'logic gates' which related to the mathematical function or algorithm you were trying to test through a series of propositions ('if this is true, then do that'). For instance, if a 'one' was followed by another 'one' then the machine was to count it. The machine looked for statistical significances and sent the output to a printer. Modern computing is based on binary – ones and zeros. Colossus was where this effectively began. Since we have ten fingers, the standard way people count has been in tens, using a decimal system. A binary system only goes up to two and so uses only two digits to represent numbers – ones and zeros. So two is represented as a 10, then three as 11 and four as 100. This makes writing out numbers much more laborious for humans. But it has a key advantage for machines. The two possible options can be represented in the form of simple signals – one being the mark for a signal being there, or 'on', and zero for the absence of a signal, or 'off'. This creates whole new possibilities for using digital machines – which can detect whether an electrical signal is present or not – to calculate and perform functions using on-off logic.

What made Colossus so novel was not just its size, its speed, its digital or binary nature and the electronic aspect of its components, but also the fact that it could be programmed to undertake different tasks by rewiring the back of the machine. This gave it a degree of flexibility, so that it could be reconfigured to undertake different statistical tests. Remarkably, some of the statistical attacks deployed by Colossus are still secret. It was not quite a universal fully programmable Turing machine. Those building it understood that people's lives depended on the speed with which they could construct something reliable and so they did not try to create a machine that could do anything. Even though it lacked its own memory to store a programme of instructions, Colossus can still claim to be the first digital computer, although others in America and even Germany were working on similar lines. It may have looked – and sounded – more like something from the industrial age, a kind of giant beast, and yet it was the forerunner of the computers that pervade our lives. For the select few who witnessed its work – in a darkened room, to stop light interfering with the reading of the tape – this was a glimpse of the future.

To the wonder of code-breakers, they would soon be reading messages between Berlin and Germany's top generals, making them privy to decision-making and planning at the heart of the Nazi war machine. In some cases they would be reading messages from Adolf Hitler himself.

In February 1944 a man in uniform came to see Flowers with orders from the War Cabinet. Could Flowers build twelve more machines? Could they start arriving by June? The first, improved, machine arrived on 31 May, just in time for D-Day and the invasion of Europe the following week. By the end of the war, ten were up and running. Block H of Bletchley Park was the world's first purpose-built computer centre, with a pack of machines clattering away day and night under the urgency of war, like factory workers on a production line. One American who visited in late 1944 was astonished by what he says was 'British mathematical genius, superb engineering ability, and solid common sense . . . The result is an outstanding contribution to cryptanalytic science.'[19]

Colossus played a crucial role during D-Day. It allowed Britain and the US to map out German defences and their weak points. Britain had also turned German agents in the Double Cross scheme and used them to feed back false information about where the Allies would land. The code-breakers were able to check that the deception had been bought into. Flowers would later take pleasure in describing a story he had heard about a crucial meeting between General Eisenhower and his staff on 5 June, during which a courier entered and handed Eisenhower a note with details of a message broken by Colossus. This supposedly confirmed that Hitler was not going to move troops to Normandy – which meant that a British deception plan had worked and convinced the Germans the real attack was coming in Calais. Handing back the decrypt, Eisenhower announced to his staff, 'We go tomorrow.'[20] The work at Bletchley allowed the Allies to get inside Nazi Germany's controlling mind, to know what it thought, believed, feared, trusted and what it intended to do.

This work out of Bletchley was, one intelligence official later said, 'perhaps the most successful large-scale intelligence operation in history'.[21] It was revolutionary in pushing out intelligence from a central location to commanders in the battlefield as they drove back the Germans. 'As Allied troops moved across France, they moved in sync with the goldmine of intelligence which detailed most of the important German military movements,' an official US report noted. 'Their intelligence officers must have looked like geniuses – they were able to predict German moves before they happened and could advise commanders how to react.'

As the war ended, Flowers and Turing went to Germany together, seeing a Tunny machine for the first time. While they were there news came of the atom bomb being dropped on Hiroshima. What next? On VE Day, Turing had been for a walk with colleagues. 'Well, the war is over, now you can tell all,' one of them said to him. 'Don't be bloody silly,' Turing replied.[22] He understood that the secret of what he had done would be kept that way. The value of electronic code-breaking was clear. If the secret escaped the confines of Bletchley, then new enemies would improve their codes, leaving the spies in the dark.

For the British pioneers of computing this secrecy came at a price. Until near the end of his life, Tommy Flowers could tell no one about what he had done. In a move which must have been heartbreaking for the engineer, he gathered together all the wiring diagrams he had worked so hard on day and night at Dollis Hill. 'I was instructed to destroy all the records, which I did. I took all the drawings and the plans and all the information about Colossus on paper and put it in the boiler fire. And saw it burn.'[23] Flowers was given £1,000 and went back to Dollis Hill, where he struggled to find his place. At first he had thought the secrecy around his creation might give him some advantage, but it proved the opposite. More conservative colleagues were annoyed by what they thought of as his 'pretentiousness' at pushing certain 'fancy' ideas. Flowers, of course, could not say why he knew they would work.[24] He had to endure watching as a US machine called ENIAC (Electronic Numerical Integrator and Computer) was unveiled and hailed as the first computer. When Flowers was interviewed in the last years of his life, by which time the secret was finally out, the bitterness over the lack of recognition was evident in his voice. 'It was a complete shambles,' he said, frustrated, believing both he and British industry had paid a price.

That Flowers' genius went unacknowledged was not only a personal tragedy but also a national one – the opportunity was lost in Britain to build on the skills that had been developed at Bletchley and maintain the momentum required to keep innovating and build a leading industry around computers. The needs of war had accelerated technology in many areas – radar, medicine, atomic power. But in computing – because intelligence agencies were the progenitor – the achievement was never capitalised on in the same way. The moment had been right for computers to be built – the technology and the ideas were ready – and even without the war and code-breaking they would have emerged perhaps a year or two later. But the fact that the secret art of cryptanalysis was behind their initial development determined the initial path of computing in the early years, especially in Britain. There was deception about what had been done – it was designed to mislead opponents but also deceived Britain itself. The culture of secrecy which surrounded first Bletchley and then its successor GCHQ might have been valuable for their work, but it came at a price.

For Turing, like Flowers, there would not be recognition of his achievements in his lifetime. And for the mathematician, life after Bletchley was darker. Although he had helped build the bombes and laid the intellectual foundations for computing, he had moved on during the war while Colossus was being built to work on speech encryption and converting the human voice into ones and zeros. After the war, he returned to his dream of a multi-purpose machine – an 'electronic brain' – but struggled to work with others to produce it. He progressed to questions about artificial intelligence, asking in a 1951 BBC programme, 'Can digital computers think?,' before his mind wandered off into fresh intellectual pastures like mathematical biology. He remained fascinated with trying to understand what constituted human intelligence and how far machines could mirror its subtleties. He would still book in to do his work overnight on an early computer, a sight that an engineer likened to 'playing the organ' as Turing sat and manned the controls, a hooter sounding when new parameters were required.[25]

Turing's house in Manchester was burgled in January 1952. The burglar turned out to be an acquaintance of a man with whom Turing had engaged in a relationship. Homosexuality was illegal and he was charged with 'gross indecency'. Faced with the choice between a prison sentence and hormonal treatment, Turing chose the latter. MI5 was clamping down on homosexuality. It was seen as a security threat. Not just because people could be blackmailed over their illegal acts but also because it was coming to be seen as somehow subversive. Two Foreign Office men, Burgess and Maclean, had just fled to Moscow amid talk of deviant behaviour. Staff in sensitive positions were now being vetted. Bletchley's tolerance of diversity was no more and Turing was stripped of his security clearance. This had been vital in allowing him to continue work as a consultant to GCHQ. His conviction meant he would also be denied entry to the United States. On 8 June 1954, Alan Turing's body was found. By his bed was an apple laced with cyanide.

At the end of the war, the Colossus machines were dismantled. The valves that ran hot were allowed to cool and then removed. The wiring was cut. Technicians returned to the Post Office. 'All that was left were the deep holes in the floor where the machines had stood,' recalled one woman involved in their destruction. 'A sad job. Then we were made to sign the Official Secrets Act again.'[26] In a bit of improvised recycling, the parts of the world's first computer were stripped and used for telephone exchanges. From the telephone they came and to telephones they returned, as if Colossus never was. There was talk that the navy would drop some of the other parts in the sea. But then something happened. 'There was a period when the destruction of all the equipment was stopped as it turned out that the Russians would carry on where the Germans had left off,' recalls John Cane, an engineer who worked on building and then dismantling the machines.[27] At least three of the early Colossi – numbers 10, 11 and 12 – were instead sent to Eastcote in Middlesex and then from there to Oakleigh Farm in Gloucestershire, where they would be modified and continue to run for a decade and a half, working for an institution that grew out of Bletchley and, like it, operated in the dark.[28] The people who wired up the boards at the back – programmers, as they became known – worked out ways to adapt them to new code machines.[29] The efforts to update them were mixed – 'some more successful than others,' a declassified GCHQ document reveals.[30] Only in 1959 were the last of the giants destroyed. By then, the offspring of Colossus were already surpassing their parent in size and speed.

The experience in Bletchley's makeshift huts set the pattern for Cold War signals intelligence and beyond – the notion that the power of computers allied with the human mind could provide a unique insight into your enemy. By breaking codes and processing vast amounts of data, Bletchley had created a 'golden age' that spies continue to hark back to and try to recreate. But they also believed that their place in this paradise was precarious – the fragile nature of their advantage meant a careless word could easily see it stripped away. This sense of insecurity meant that the future for computers and spies was one in which a desire for more power was matched by a demand for utter secrecy. The war was over but Colossus – like the code-breakers in Britain and America – had proved its worth, and so both the machines and the institutions hummed away in the shadows as the Cold War began. War had given birth to the first computer to serve the needs of code-breaking and espionage. That was just the beginning.

CHAPTER TWO

MARRIAGE

It was late when the small band of Americans arrived at Bletchley Park. As they walked up to the front door of the old mansion house, the dark of a January winter was underscored by the blackout curtains that eliminated every last chink of light from the windows. The journey of the four men – known as the Sinkov mission, after their leader Abraham Sinkov – had been stormy, dangerous and utterly secret. To reach their destination they had braved bullets and bureaucratic mistrust, but their arrival marked the opening courtship of what would become the longest-standing, most powerful intelligence alliance in history. If there is one area in which the so-called 'special relationship' has always been most intimate, it is in communications intelligence and code-breaking.

America was not yet in the war when the group arrived in early 1941. Britain had stood alone and vulnerable, fearing invasion and defeat. It needed help, and so the previous summer the first tentative discussions had taken place with the US. As plans for the visit progressed, Churchill and many senior intelligence figures had real reservations about opening up to a country that had not yet committed to the fight.[1] The British were interested in the Americans' progress against Japanese codes, but there was deep nervousness about revealing too much about their own progress against the Germans for fear of the secret escaping. In November, the chief of MI6 told Churchill it might be awkward only to talk about Japan and not open up about Germany. It would be obvious 'we have something to hide', he explained. Churchill agreed they should show a little leg. However, one British official added a rather telling note to a memo: 'What will they think if they find we have been reading their own stuff?'[2]

A decade earlier, in 1930, Abraham Sinkov, the leader of the expedition to Bletchley, had been given a memorable induction into the secret world of American code-breaking by William Friedman, his new boss. Friedman, the flamboyant son of a Hungarian-Russian-Jewish immigrant, enjoyed his work so much that dinner guests would find the menu in code.[3] In 1930 he had just taken charge of a tiny unit called the Signal Intelligence Service within the US Army. His first step had been to double its size by bringing in three young recruits – Abraham Sinkov and Solomon Kullback, close friends and classmates from Brooklyn, and Frank Rowlett from Virginia – all mathematicians who would play a key role in the coming decades.

On a humid June day in 1930 Friedman, in a natty blue suit, asked the three new recruits to follow him, making it clear they were going to be let into something truly secret. Decades later they would be able to recall that moment in forensic detail, memories captured in now declassified files. Friedman took them down the stairs to the second floor of the Munitions building. Friedman swung left into a deserted corridor and stopped outside Room 2742. From his inside coat pocket he took out a small card and began to work a combination lock on the front of a steel door. The bolt swung open. Behind it was another steel door. This time, Friedman extracted from his coat pocket a key to unlock the inner door. Inside the room was pitch-black. There were no windows and foul air spilt out. He then produced a small box of matches and lit one so he could find a pull-cord for the ceiling light. It revealed a room twenty-five-feet square jammed with filing cabinets packed so close the drawers had barely enough room to open. It was the dustiest room the men had ever seen. Friedman turned to them and said in solemn and imposing manner: 'Welcome, gentlemen, to the secret archives of the American Black Chamber.'[4] Today, deep inside the heart of the NSA, sits its most secret room where the records of the codes it has broken lie. It is still called the Black Chamber.

'King Solomon's mines could have offered no greater treasures for us,' Rowlett thought as the new recruits were shown an archive of solved and unsolved codes dating back years. 'We lost all track of time.' Friedman explained that the room contained all the working files of a secret unit which had operated in New York until it was closed a few months earlier and which had succeeded in breaking many diplomatic codes. He explained that the original Black Chamber had been the creation of Herbert O. Yardley, a colourful poker-playing character, who had started working as a code clerk for the State Department just before the First World War. He said he cracked a message from the President in two hours on his night shift. This led to a role with America's military intelligence branch reading all the traffic he could get hold of. He would later say he learnt from British colleagues of their 'long and dark history backed by a ruthless and intelligent espionage,' and believed America needed to match that capacity if it wanted to be a great power. After the war, Yardley convinced superiors at the War Department and State Department to fund an organisation operating under a commercial front (the 'Code Compiling Company' in New York). This spied on the diplomatic traffic going in and out of Washington of around three dozen countries, thanks to messages provided by cable companies.[5]

'Gentlemen do not read others' mail' was Secretary of State Henry Stimson's sniffy verdict in 1929 when he learnt about the work. That had led to the closure of the Black Chamber. But the military simply snaffled all the files and created its own new team under Friedman to protect its own codes and prepare to attack those of an enemy in time of war. Closing the door behind Sinkov, Rowlett and Kullback, Friedman explained it was now his new team's job to catalogue the dust-laden files and understand what they contained. Secrecy was vital – not least from the State Department.

The team of code-breakers began work. Resources were limited with the Great Depression under way, meaning they had to pay for their own pencils and paper from the dime store or write on the back of old weather reports. They would sit in the Munitions building in Washington with no air conditioning during the hot summers. The fans needed to keep them cool meant the team had to hold down the pieces of paper they were working on to stop them blowing away. The first traffic they focused on was known as 'Rum-Runner'. Prohibition, which banned the sale and transportation of alcohol, had led criminal gangs to employ retired navy men to help smuggle liquor by boat from Canada, Mexico or the Caribbean. They were organising by radio, using codes. Friedman's wife worked in the Coast Guard and passed on messages for his team to crack. The team also had what they called a 'nut' file full of people who had written to the government saying they had invented the perfect cryptographic system for keeping secrets and they were willing to sell it for $1 million, but they regretted that if their own government was not interested they might have to go to another country. The team asked for samples and normally solved the codes within minutes.[6]

At first Japanese messages were intercepted by an industrious army colonel in San Francisco. He rigged up an alarm clock with a clothes-pin to act as a time switch to tape-record the radio traffic sent to Tokyo at a regular time. The tapes would then be airmailed back to the code-breakers to work on. 'What we were doing was theoretically illegal,' Solomon Kullback reckoned, and the team discussed whether they would ever end up on trial. Friedman told them it was in the national interest and if they had any qualms to get out. Eventually the group received an opinion from the Attorney General legitimising the act after the event, which was kept in the office desk. Until war led to a sounder footing, this seems to have been the rather thin legal foundation for asking telegraph companies to photograph traffic.[7] Everything was a little informal. Rowlett himself never had any kind of security check before the war.

But the secrecy of the Black Chamber did not last. Whereas in Britain loose-lipped politicians revealed the code-breakers' work after the ARCOS raid, in America it was one of their own. The demise of the Black Chamber had left Yardley out of a job. Broke, angry at his treatment and believing both he and the art of code-breaking deserved recognition, Yardley decided to tell his story. His book *The American Black Chamber* appeared in June 1931, with excerpts running in the papers. The book included a picture of a reconstructed 10,000-word British Foreign Office code book which the Black Chamber had worked on and made it clear that London's diplomatic traffic was not immune from American interest, just as Britain had been targeting the US.[8] It also revealed that America had exploited the discovery of Japan's fall-back position in negotiations about disarmament in the First World War. Japan realised it was vulnerable and began to change all its codes (Rowlett later thought this actually helped, as it made the team up their game). Even countries not directly mentioned in the book began to tighten their systems. There was panic on a scale not matched until another insider revealed secrets three-quarters of a century later. And in Yardley's case, this was a man who knew everything. Congress passed the first legislation criminalising the revelation of secret code information. The military tried to track Yardley down, fearing he had taken classified material with him. When three officers turned up on his doorstep he denied possessing anything and began writing spy thrillers (*The Blonde Countess*) before going to China, where he was paid $10,000 a year to help target Japanese communications.

The team in Washington eventually began breaking back into Japanese systems. But in March 1939 an unreadable message was intercepted between Warsaw and Tokyo. The new system was codenamed Purple. It was Japan's highest-level diplomatic cipher machine and it stumped American code-breakers for a year and a half until it was broken in September 1940. Friedman had suffered under the pressure. He had a breakdown and was briefly hospitalised just as the trip to Britain was organised. So it was Sinkov who left for Bletchley carrying an incredibly precious cargo – their reconstruction of the Purple machine along with a hoard of related documents.

Together with Sinkov and an army colleague, Leo Rosen, were two men from the navy – Robert Weeks and Prescott Currier. The problem for America's code-breakers was that there was not one but two code-breaking teams and they competed fiercely and destructively. Bureaucratic turf wars have been fought back and forth across the US intelligence community for decades. Before the war, bitter rivalry between the army and the navy meant they each duplicated the other's work and jealously guarded what they produced. In a sign of just how absurd the fights could be, early in the war US naval analysts worked on Japanese diplomatic codes on odd days of the month, the army on even days.[9]

The four men left Annapolis on a cold January day. They spent hours on a small boat alongside the British battleship *King George V* waiting for the new British Ambassador to Washington to disembark so they could begin their mission.[10] The ship headed for the Orkney

Islands, just off the north coast of Scotland. By 7 February it arrived
at Scapa Flow. The plan was for two flying boats to take them down,
but the weight of the thirty crates they had brought – adding up to
nearly two tons – meant they simply could not fit everything through
the hatch of the flying boats. So they were left stranded. 'It was raining,' recalled Prescott Currier. 'It was kind of unpleasant.' *Neptune*,
a battered, barely afloat cruiser, was diverted up. Captained by the
colourful, red-cape-wearing and poetry-reading Rory O'Connor, it
got halfway down the coast before it passed a convoy tailed by a
German naval reconnaissance plane. 'We all knew what this meant,'
Currier later recalled. Two German dive-bombers appeared. 'I was
trying to eat some soup but my mouth was so dry I couldn't swal-
low it,' Currier said. 'I've never been so scared in my life. We heard
a bomb landing on one side and one on the other side and the ship
would bounce out of the water and back down again. Then I heard
something that sounded like someone dragging chains along the
deck.' The chain sound came from German guns strafing the deck,
half a dozen times in all. 'God, I thought, all that gear is stacked up
on deck unprotected. Absolutely unprotected. And I was so scared
I couldn't go up and do anything about it anyway. Well, that's the
Purple analogue gone,' thought Currier. After the attack he and the
three other Americans went up on deck. It was strewn with spent
bullets. To their relief, the Germans had been using copper-jacketed
explosive bullets which never penetrated the crates. It was a lucky
escape for the Purple machine (although a year later the *Neptune*
would strike a mine off the Libyan coast, killing all but one of the
767 men on board). At Sheerness the Americans were met by a delegation from Bletchley. 'And there was Brigadier Tiltman standing in
his full regimental [uniform] with his legs spread apart and his hands
behind his back.'

Waiting at Bletchley behind the blackout curtains for his guests'
arrival, the Director, Alastair Denniston, had given strict instructions to his assistant. 'There are going to be four Americans who
are coming to see me at 12 o'clock tonight. I require you to come in
with the sherry. You are not to tell anybody who they are or what
they will be doing.'[11] When they entered his study at the front of
the house, the Americans were met by the senior staff of Bletchley,

stood in a semicircle, who were introduced in turn. Denniston's assistant, who had never seen an American before, 'except in the films', poured out the sherry (whisky was hard to come by). The British had been ordered to help but were nervous.

The American gift of the Purple machine was an ice-breaker though, 'a tremendous gesture,' recalled Tiltman. 'Somebody had to make the first step and the Americans made it.' Tiltman was a pivotal figure in the Anglo-American relationship, seen by the Americans as the embodiment of British eccentric brilliance but without falling into their other caricature of officious condescension.[12] He had worked on Japanese codes and shared what he knew. But what of Enigma? Tiltman's superiors feared a leak could endanger the nation. Many senior figures in Britain were not even allowed to know the truth – deceived instead by the idea that the rich intelligence was coming from a German spy codenamed Boniface. Tiltman thought full exchange meant just that and went to the head of MI6, who agreed that the Americans could be told if they limited the knowledge to an agreed list of people.[13] And so, right at the end of their visit, the Americans were let into the great secret. They were shown the bombes and told most (but not all) of the details of the break into Naval Enigma. It is easy to forget quite how remarkable this was. A year before America actually joined the war, the two countries were sharing their most precious secrets. The visit was a success – professionally and personally. The Americans were taught the game of rounders, which they thought of as baseball played with a broomstick. Sinkov could recall being put up in a fancy country house where the butler looked down on him for failing to use a butter knife correctly. There were reminders that this was a country at war. They visited a nightclub one night to see the band playing American music. The next night a bomb hit the club, killing all the musicians.[14]

The team returned after an exhausting, rough crossing and arrived bedraggled back in America. Their destitute state may have been one of the sources of gossip in Washington that they had come back empty-handed – after all, they had given the British a Japanese machine but had not brought back a German one. Sinkov's notes in America's National Archives make it clear he did not put down on paper everything he knew when he returned, and the fact that the team had to keep secret what they had witnessed may also have fuelled the speculation of those not in the know that America had somehow got the raw end of the deal and was being given the 'run-around'.[15] There was tension in the early days. By November 1941, the US Navy – less keen on working with Britain – was making a fuss. The US Navy and Army would both end up developing independent relations with Britain. 'British officials regarded negotiations with the Americans as a little like dealing with the former colonies after the American Revolution – disorganized and frustrating at times, but they could still play one off against another to achieve objectives,' an official NSA history notes.[16] The American army-navy rivalry was believed by some to have contributed to the disaster of December 1941, when the Japanese launched a devastating surprise attack on Pearl Harbor which the code-breakers had failed to predict. It would draw the US into the war, force the services to work more closely together and also lead to a long-standing desire for communications intelligence to provide warning of 'strategic surprise'. A division of labour was also agreed, with the US leading on Japanese systems and Britain supporting, while the roles would be reversed when it came to Germany.

America sent a stream of about 100 experts to Bletchley who were integrated into the work of each hut rather than operating as a distinct team. Already some of the cultural differences between the US and the UK were evident, which would persist to the present. The Americans noted how at their new headquarters, Arlington Hall, everyone would wear their military uniform if they had one. The atmosphere was formal. That was not the case at Bletchley. To make the point, one American liked telling audiences the story of two wounded personnel assigned to Bletchley to help out who, while convalescing, looked at the way people dressed and acted and believed they had been sent to a mental institution. The Americans also seemed an exotic bunch to those at Bletchley (including a number of the young women, some of whom found their manner infuriating; others thought it more appealing).

America offered industrial as well as intellectual muscle. Alan Turing visited the US in November 1942. After some initial problems getting through immigration at Ellis Island in New York, he saw the way in which the US was adapting his designs. He was not entirely convinced by their understanding of all the concepts and expressed it in the rather superior British attitude often found at the time. 'I am persuaded that one cannot very well trust these people where a matter of judgement in cryptography is concerned,' he wrote in his official report before adding, 'I think we can make quite a lot of use of their machinery.'[17]

Britain had been struggling to produce enough bombes; America's industrial war machine would not. The US had decided, when Britain had not been able to read German Naval Enigma in 1942, that it would develop its own machines, especially because it sensed Britain was not sharing everything and it too was suffering losses in the Atlantic. The US took what Turing had designed at Bletchley and industrialised it on a scale that mirrored the imbalance in resources between the two Allies. The National Cash Register Company (NCR) in Dayton, Ohio, built bombes at a pace that both sides knew foretold a shifting balance of power in the future. By the second half of 1943, the US produced seventy-five bombes – more than Britain managed throughout the whole war. These would process requests sent from Britain in an hour.[18] The Americans were happy to take on some of this work as they understood that it gave them the chance to master a process that Britain had developed. They wanted to be in a position to be independent.[19] The work at NCR and with companies like Kodak and IBM displayed the ability of the growing American private sector to re-engineer products to help the war effort. America had been at the forefront of experimenting with machines in the 1930s thanks to pioneers like Vannevar Bush, including support in code-breaking, but the focus on building bombes diverted manpower and expertise from the dream of building a proto-computer like Colossus.[20]

The Americans who made the pilgrimage to Bletchley during the war were particularly taken with the work on traffic analysis which logged data about German communications. An elaborate cross-indexed system of five-by-eight-inch cards was used, including punch-card machines. America had been using these machines from the 1930s, but Bletchley was more systematic in its application. This was used to build up knowledge about the communications pattern of existing known targets. But it had other uses. If you had a systematic understanding of what you already knew, it was also easier to spot something that was different. This might be, for example, a new type of signal or communications device – a target which, once discovered, might need monitoring. And as had been learnt in the First World War with the Zeppelins, even if you could not break the code of a system, you could glean useful intelligence. Based purely on the externals of the traffic in the Second World War, Britain was able to collate an order of battle for the Germans and scramble planes to intercept German bombers when a certain pattern of signals was detected. The latter process was performed by people called 'computors' who studied German air force radio call signs. They were able to report in near real-time the take-off and course of German fighters (and also tell the difference between fighters and bombers, which early radar struggled with). Their accounts in the US National Archives show how the Americans reported home that Britain might even have underestimated the value of this systematic use of data. The US became determined not just to replicate but to improve on the system, something British visitors would acknowledge by the end of the war. To sort this sea of data, a liaison unit from IBM was installed inside Arlington Hall to work out how best to apply the technology. This understanding of the hidden secrets that could be unlocked from raw data was a sign of things to come.[21]

Arthur Levenson, whose talent for maths was in part expressed through an amazing ability with baseball statistics, was one of those visitors who would build on his Bletchley experience to become a leading figure in American code-breaking. He always remembered the freezing outdoor toilets, as well as working with one colleague with a refined Cambridge accent and another from Glasgow. While he was made to feel very welcome, he never had a clue as to what either was saying. Levenson worked in Hut 6 developing the menus to be programmed into bombes breaking Enigma.[22] There was one decrypt, he would always remember, in which Rommel asked why every supply ship was being sunk by a British destroyer. 'That can't be coincidence. That damn machine,' Rommel had said. The team held their breath but the Germans shied away from facing up to the possibility that Enigma was broken and having to upgrade their entire communications system in the middle of a war. Another message Levenson decoded was a 70,000-character account by Rommel of his inspection of defences along Normandy before D-Day which mapped out the entire Western defences. This located a German Panzer division just close to where they were going to drop the 101st Airborne. 'They would have been torn to pieces. So they moved it, based on intelligence . . . That – I'm sure – saved lots of young men in the 101st Airborne,' he explained in a now declassified account.

With the invasion a success, Levenson was in Paris on VE Day heading towards Salzburg as part of a special team belonging to TICOM (Target Intelligence Committee). These teams, which drew on Bletchley veterans, fanned out across Europe hunting the secrets of Nazi Germany in a race with the Soviets. The full story of their work remains secret, but both the US and UK knew that getting hold of Nazi people and equipment would help them discover which of their own codes had been broken and also find out what progress had been made against Soviet codes, which they could then take advantage of.

Levenson went to Austria looking for an original Tunny machine, something which no one at Bletchley had actually seen. 'Occasionally you'd get a guy and then we told him, if he doesn't want to answer questions we're turning him over to the Russians. And they talked,' he remembered. His team drove an entire German communications train of six or seven vehicles packed with encryption equipment to Britain. Also retrieved were the Germans who demonstrated how it was used (some Germans would end up working with the NSA). Howard Campaigne, another Bletchley veteran, found the revelations from interviewing the captives were instructive. 'We found that the Germans were well aware of the way the Enigma could be broken, but they had concluded that it would take a whole building full of equipment to do it. And that's what we had. A building full of equipment. Which they hadn't pictured as really feasible.'[23]

That's what Britain had constructed – a building full of equipment at Bletchley Park. Howard Campaigne would go on to lead America's quest to master machines in the years to come. But the more immediate question at the end of the war was: what to do with the rooms of equipment and the huge spy agencies that ran them?[24] The seamless transition from Bletchley's huts into GCHQ and the NSA was not as smooth as sometimes portrayed. There was fear, especially in the US, that tightened budgets would mean that all but the most cursory code-breaking capability would have to be abandoned, as happened after previous wars. But an important difference was that the most senior military and political leaders knew what the code-breakers had achieved.[25] The goodwill would buy time in the difficult years ahead.

Britain knew that breaking Tunny was an immense achievement that might offer hope against future systems. The US meanwhile had never managed its own independent attacks on Tunny during the war. In August 1945, according to declassified US documents, Britain made a remarkable proposal that has never been talked about since. It offered to give one of its Colossus machines to the US and 'hinted' more would follow. To Britain's shock, the offer was rejected. 'They found it difficult to understand why and so did some Americans,' a recently declassified NSA report notes.[26] The reason was that the US was determined to build its own version. It did not want to be tethered to Britain but to be independent. Refusing Colossus was part of a broader strategy to be friends with Britain but also to ensure that America would lead and not follow. This was evident in wider relations between the two countries, including across the intelligence field. There were even some in the US signals intelligence world who believed the best way for them to stay afloat was to break the alliance with the UK. A number of 'influential men in the intelligence community' suggested ties with Britain be cut. The theory was that if policymakers knew they had to depend on the US code-breakers alone for the type of intelligence the alliance had provided, then there would be less chance of their budgets being slashed. But some early British successes against Russian cipher machines and their ability to intercept and process non-Morse transmissions like Tunny proved critical in persuading the Americans that continued co-operation was worthwhile.[27]

An Iron Curtain was rising across the centre of Europe, Stalin's Soviet Union erecting barbed wire and border patrols to keep citizens in and spies out almost as soon as Nazi Germany was defeated.

During the war, the US and Britain had both been – independently and secretly from the other – targeting the communications of their Soviet ally. Now, Stalin's Soviet Union was steadily moving to take control of Eastern Europe, crushing democratic parties and dissent. It would come to be seen as the Cold War, but there was every chance at the time it might turn hot. The two Western Allies realised co-operation was going to be vital. The British, who had almost immediately turned their radio receiving stations towards Russian traffic when the war ended and had more history with Russia, were initially ahead. 'The British seemed to be reading almost everything; the Americans virtually nothing,' a declassified American study noted. 'The British provided much of the cryptanalytic expertise, the Americans most of the processing capability,' an NSA history recalls. Work against the Soviets was given the cover term 'Bourbon'.[28]

Secret it may be, but code-breaking is also a team activity – as the work against Tunny showed. Two heads – and two countries – were always better than one. And so what might have been a brief affair solidified into a marriage, albeit one based on a formal written agreement rather than pure emotion. Recently released documents have for the first time revealed the nature of that marriage contract: a May 1943 agreement was expanded and eventually christened UKUSA. Those who negotiated the marriage had often worked closely together at Bletchley Park and would go on to rise to senior positions on both sides of the Atlantic, providing a deeply personal bond which extended beyond the formal provisions of the treaties.[29]

The two partners promised honesty and openness with each other. This meant they agreed to divide up the world in terms of what they intercepted and to unrestricted sharing of all raw traffic and technical work in order to eliminate duplication – with exemptions only when specifically requested by one party and agreed by the other.[30] The aim was marriage, but with a bit of space to pursue their own interests so they could still respond to their own national requirements.

They promised fidelity – a clause said there would be no deals or sharing with any third parties unless one told the other. The British Dominions of Canada, Australia and New Zealand were eventually included, but only after Britain allayed concerns over leaky security and penetration by the Soviets. They also promised not to exploit the relationship for money. A clause outlines a blanket prohibition against passing material to any ministry, agency or individual who might use it for commercial competition or economic gain or advantage.[31]

They promised secrecy. The fear was that if other countries understood the vulnerabilities that were being exploited, they might move to more secure systems. 'The value of Communication Intelligence in war and peace cannot be over-estimated; conservation of the source is of supreme importance . . . The time limit for the safeguarding of Communication Intelligence never expires.'[32] It was agreed that no one who knew the secrets should be put in a position where they could be captured and subject to interrogation.[33]

They promised commitment – for better or for worse. And they made preparations for the worst. Not divorce but war, with detailed plans for how they could cope (GCHQ would grow to 9,000, with two-thirds from Britain and the rest from America).[34] The British operation had moved to Cheltenham, a location chosen to be away from London since it would be a prime target for bombs. Some thought that even this was not far enough away and that the whole operation should be moved to Canada. But in 1947 a private visit to Cheltenham led to word of some old Ministry of Pensions buildings which had become vacant. Someone had a look and reported back that there were good communication lines from when the Americans had used the area as a wartime base and the town seemed a nice place to live. And so GCHQ, as it became known, found itself in the West Country.

Meanwhile in the US, the army and the navy had eventually begun to co-operate – motivated partly by a fear of budget cuts but also by the need to agree joint collaboration with the UK. The disaster of Pearl Harbor had increased pressure for a more centralised system, a push driven further after failings in the Korean War. This led in 1952 to the creation of the NSA. A military officer would be in charge but with a civilian deputy. Its home at Fort Meade was the third largest government building after the Pentagon and State Department. A pneumatic tube system could carry papers at twenty-five feet per second and handle 800 message tubes per hour.

Joined in matrimony, British and American code-breakers expected to be inside all Eastern Bloc intelligence and diplomatic communications, and even inside the top Soviet machine that carried the highest level of communications as the Cold War began. Engineers were building analogues of the Russian ciphers as they had done for Enigma and Purple and building the equivalent of bombes to focus on solutions to specific machines.[35] Everything looked set for a repeat of wartime success against Germany. Soon, they thought, they would be in Moscow's mind, as they had been in that of Hitler's Berlin. They were wrong. And it would be that failure that would shape the history of computers and of spies.

CHAPTER THREE

INTO THE COLD

Frank Rowlett wore civilian clothes as he made his way to an isolated, lakeside summer cabin about ninety miles from the Canadian capital of Ottawa. It was autumn 1945 and the American code-breaker was going to meet a Russian spy. Rowlett was not there to betray his country but to talk to a member of Russian military intelligence who had made that choice. It was a defection that helped instigate the spy fever that gripped America and Britain and that would be a defining feature of the Cold War. Igor Gouzenko was under armed guard in the cabin. He worked in the Soviet Embassy in Canada and had only just escaped the clutches of the KGB, who had been searching for him. What no one knew at the time was that the KGB had been hot on his tail because they had been tipped off about his intention to defect by their master spy in London, Kim Philby. Gouzenko's inside knowledge would help, in time, expose Philby's Cambridge spy ring and others offering secrets to Stalin. This was in part thanks to the first use of computer technology to catch spies.

Gouzenko revealed that the Soviet Union was running an espionage campaign against the West on a scale no one had grasped. Treachery – motivated by money or sympathy for Communism – had reached deep into America and Britain – even into the most secret parts of the state, like the Manhattan Project building the atomic bomb. Gouzenko was particularly valuable because he was one of those important people in Cold War espionage – a code clerk, in his case for Russian military intelligence, the GRU. That meant his job was to scramble the messages sent from a Soviet embassy back to Moscow so that they could not be read if they were intercepted en route. During his time in the cabin with Frank Rowlett, Gouzenko was able to explain precisely how Soviet spies enciphered their messages. This included the way in which they used their ultra-secure one-time pads to add a random letter to a message to scramble it. Back in Washington, a small team of code-breakers hungrily devoured those details as they sought a way into a stream of Soviet diplomatic traffic that the US had been intercepting but not able to decipher.[1]

The first attempts to break into Soviet codes during the war were assigned to a team numbering just two. The pair (one a high school teacher, Gene Grabeel) worked in the corner of a room also occupied by the British liaison to Arlington Hall. He was not supposed to know what they were doing, which meant they had to communicate in whispers. By 1943 the team had grown and made a breakthrough using data analysis. They had run 10,000 messages through a punch-card machine made by IBM. This was not a computer but a processor which looked for what were known as 'depths' in messages – when a key had been used twice. They were amazed to find that this appeared to have happened in the Soviet traffic. But they were not yet sure what that could do with this insight. In late 1945, Gouzenko's explanation of how the one-time-pad system worked gave them further clues (although no final answer). It took a brilliant analyst called Meredith Gardner to make the crucial breakthrough. As the Nazi war machine had pushed into the Soviet Union, the carefully controlled system to protect codes had reached breaking point. The Soviets made the mistake of accidentally duplicating about 35,000 pages of one-time pads. This meant a small proportion of the one-time pads in circulation had become two-time pads and were no longer truly random.[2] Work out which two messages had used the same pads, and you might be able to read them. But that was a huge job. A job for a machine. A machine might be able to work through all the traffic that had been collected in the past to identify the messages that could then be broken by hand. This project was known as Venona. The stakes were high. The KGB traffic out of the US would contain details about Americans and other Westerners – some perhaps in senior positions – spying for the Soviets. The names might be in code (agents being referred to as 'Stanley' or 'Homer') but with careful detective work to correlate details with other information the traitors might still be identified.

Klaus Fuchs, a German-born scientist who worked on both America's and Britain's atom bomb projects, was one of those exposed. Venona revealed him passing on secrets to his handler that allowed the Soviet Union to build its own bomb. Keeping Venona secret was so important that it could not be revealed in court, so in the Fuchs case an MI5 interrogator had to slowly tease a confession out of him. Venona's revelations helped fuel the McCarthy era of witch-hunts in Washington amid fears that the Soviets had reached deep into the establishment. Oliver Kirby at the NSA worked on the traffic and was shocked by the evidence he saw of Americans taking money from the Soviets, much of which has still never been made public. He would personally brief his boss about the discoveries. 'He'd say "Don't you say a damn word about that because if we ever become Big Brother we'll lose every nickel of funding that we might get." So you never talked about that stuff, but it's there.'[3]

There were tantalising glimpses of a mysterious British spy ('Homer') operating out of the embassy in Washington. The investigation was being run jointly with the British. Unfortunately for the spy-hunters, the MI6 station chief in Washington was Kim Philby. He was indoctrinated into the Venona secret (no one realising he was 'Stanley' in the messages) and he tipped off fellow Cambridge spies Donald Maclean ('Homer') and Guy Burgess ('Hicks') who fled to Moscow. That intensified the spy hysteria sweeping Britain and America. In Britain, Philby's defection heralded a dark period for British intelligence as its spy-hunters became lost in a wilderness of mirrors, seeing Soviet agents everywhere around them, a world captured in the fiction of John le Carré.

Venona did not provide real-time decoding of Soviet messages, only the ability to work back painstakingly through messages that dated mainly from the war. But it was considered so valuable that this piecing-together of historical fragments to identify spies continued in secret until 1980. The desperate desire to break the messages led to the construction of computers specifically to work on the problem. The reason why Venona was pursued with such intensity was because this tiny glint of light was almost all the code-breakers had to hold onto.

In the first years of the Cold War, Britain and America had made progress against current Soviet cipher systems as well as the historic Venona traffic. But then, on a day in 1948 known as 'Black Friday', the world went dark.[4] The Soviets changed systems. Nothing could be read. By 1949, the four major Soviet cipher systems that had been broken were no longer being used. It was what Bletchley had always feared with Enigma. The new machines that appeared were much better protected. The cause was almost certainly betrayal. An American called William Weisband, who had worked in signals intelligence during the war, was spying for the Soviets and is believed to have passed on details until he was spotted (codenamed Zveno) through Venona. There would be other betrayals like his during the Cold War but few were so devastating. The costs were soon clear. Intelligence failures followed the blackout – the lack of any warning for the Soviet testing of an atomic bomb; the Berlin Crisis; the smothering of Eastern Europe by Communist regimes; the triumph of the Communists in China and the Korean War. The 'golden age' of reading your opponent's tactical and strategic messages – with Nazi Germany and then, all too briefly, the Soviet Union – was over. There was a view in some quarters that signals intelligence was also over, a view much like the one heard when Germany had introduced its new Naval Enigma machine a decade earlier. But, rather than give up, a huge investment was made over decades to try to recapture that wartime success.

The answer to 'the Russian problem', thought men like Howard Campaigne in the late 1940s, had to be computers. Could they break the Soviet messages as Colossus had with the Germans? The secret world would invest heavily to find out. In the late 1930s there had been those in America who were hoping to build something like Colossus, but institutional inertia and bureaucratic battles slowed them down. Instead they focused in the war on the pressing task of building bombes and other machines to help with Enigma.[5] After the war, America began to forge ahead.

The secrecy that surrounded Colossus meant that the first machine to emerge in America after the war would claim its crown as the first computer. Cryptanalysis was the highly secret thread to the origins of computing, but at the same time there were other more public efforts allied to the military cause. Those included work on radar and firing artillery. The challenging job of calculating – or computing – how to fire an artillery gun so that it would hit a distant, perhaps moving, target while also taking account of speed, bearing and wind, and doing so in a time that was practicable, had long involved machines. From the First World War, analogue machines were being developed to help with this and by the Second World War electronics were being used. At the University of Pennsylvania ENIAC was unveiled in 1946 to carry out ballistics calculations. It did not yet employ binary and had no memory to store its programmes but – with Colossus a secret – it claimed the public title of the first computer and set off a race to build a more advanced model. In the secret code-breaking world there were two paths and two different camps. One side favoured building what were known as 'special-purpose' cryptanalytic machines that did one thing but did it fast. In code-breaking, these were initially wired like bombes to work very quickly on a very specific attack against a specific cipher machine such as the Hagelin, similar to Enigma. Early special-purpose machines were codenamed Alcatraz, O'Malley, Warlock, Hecate and Sled.

But then Howard Campaigne and a colleague attended a hugely influential public conference at the University of Pennsylvania. They returned to the secret world to make the case for universal or 'general-purpose' machines.[6] These, as Turing had imagined, were flexible because they could be programmed with instructions. This became known as the 'von Neumann' model after John von Neumann of Princeton and Pennsylvania, a leading mathematician and thinker on computing. A special-purpose computer worked on one type of calculation. The problem was that if your opponents changed their system then your machine might become irrelevant. A general-purpose machine promised flexibility (even if it might be slower). Today, the NSA and GCHQ deploy both types of machines in their computer halls.

The rapid advances made under pressure of war had meant that a long-dreamt-of vision was in reach – a universal or Turing machine – a high-speed electronic computer that could be programmed to undertake any mathematical or logical process. What was also needed was a way of storing instructions in a memory so that a computer could work through various options and become more than just a giant calculator. The technically challenging notion of giving a computer a memory would move beyond Turing's abstract concept of feeding in instructions on tape and also the laborious task with Colossus and its early successors of having to wire up the machine physically.

In post-war Britain, struggling with austerity and rationing, the influence of the Pennsylvania work led a handful of teams to begin work on a general-purpose machine. Turing himself had an unhappy experience at the National Physical Laboratory (NPL) in Teddington, where his desire to do everything from first principles clashed with more practically minded engineers. Other centres at Cambridge and Manchester (under Max Newman, who had brought Turing to Bletchley) made more progress with help from the people who had worked on radar during the war at the Malvern research centre. There was ample transatlantic cross-fertilisation of ideas, but Britain's progress initially seemed strong with the 'Manchester Baby' and EDSAC (Electronic Delay Storage Automatic Calculator), the first stored-program computers.[7]

GCHQ was determined to build the successors to Colossus – but in secret. Six lectures were given to staff between July and November 1948 that gave an overview of what computers could do and developments under way in the UK and US.[8] A few details of the organisation's early computers have been released into the National Archives. Images of a machine called Colorob reveal a vast steel structure with a veritable spaghetti of wiring spilling out of the back which had to be replugged for every different function. Being a computer programmer in those days was essentially being an engineer. Much about these early machines remains classified though.[9]

In Britain, the desperate need to break Venona and learn the identity of Soviet spies seems to have been a driving force in secret computer development. In 1954 GCHQ took delivery of a special-purpose machine codenamed Oedipus. Like Colossus it worked on statistical probabilities. It looked for the likelihood of a message having been encoded with a particular one-time pad and code book, processing 10,000 attempts in 1.6 seconds.[10] There were hundreds of thousands of Russian diplomatic messages that had been collected, each of which had to be analysed to see if there might be a match. By 1957, Oedipus appears to have had some success against Venona. The work against Venona was undertaken in a large wooden hut which sat off one of the main avenues at the Cheltenham site. A young analyst supervised the work (described as 'a joyous menagerie' by one visitor) from a small office at the front. Behind him were teams of linguists working under 'harsh lamps, toiling for matches' looking at row after row of apparently random numbers to try to tease out clues which would then be sent on to MI5 and MI6.[11]

The effort against Venona remained a priority. The first computers were of little help, but by the early 1960s new computers would offer the chance to look for matches at a much faster rate once the messages had been transferred onto punch-cards in order to ease the burden on the teams beavering away. That job was farmed out to the Atomic Weapons Establishment's computers to begin with. At first there were high hopes. The first message to be decrypted in 1963 by the new computerised system related to an agent codenamed Stanley – Kim Philby. A reference to Mexico, which Philby had worked on during the Second World War, might have provided a useful piece of evidence of his betrayal if it had been extracted earlier. But it was too late. Just months earlier on a rainy January night Kim Philby had fled Beirut on a Russian freighter. His arrival in Moscow was a shattering blow for British intelligence. And after that early promise the intelligence take from Venona in Britain proved to be slim pickings, even as the workers toiled in their huts in Cheltenham.[12]

When the GCHQ code-breaker Hugh Alexander visited the US in October 1961, he could only wonder at the power of the machines the Americans were developing. He understood they might be strong enough to find the non-random patterns hidden in even the most advanced Russian cipher machines that British intelligence was having to find other ways of cracking (for instance by breaking into Russian embassies).[13] A few years earlier, Britain's Joint Intelligence Committee – a place where Latin and Greek pentameters and cricket were the stuff of conversation – found themselves discussing their American cousins' enthusiasm for the use of computers in processing and recording intelligence and whether Britain too should follow suit. The idea was kicked into touch. Britain – and especially its establishment – was still largely old-fashioned in its ways, with its bowler hats and briefcases. Computers might be vital for code-breaking in Britain, but they were niche. Even within GCHQ most of the intelligence team remained paper-based until well into the 1980s, with senior intelligence officials having no need or clue how to actually operate a computer. GCHQ officials who visited their American cousins often came away in awe of the use of computers to do things like track missile launches in real time (an aspect of signals intelligence which was not about communications). Meanwhile, American code-breakers who visited their British colleagues were struck by two things: the role of trade unions and the concept of 'public footpaths', which meant ramblers tried to wander through sensitive sites.[14] Post-war Austerity Britain was broke. Spies in MI5 and MI6 were much more comfortable with pen and paper than these expensive computer gadgets.

America – less rooted in the old ways, with a stronger faith in technology and closer ties to companies – drove ahead. The different approach was most evident in the relationship between spies and the private sector. The reports from Pennsylvania led the navy to build a machine called Atlas. Twenty feet long, it was named after a comic-strip character who was a 'mental giant'.[15] It was constructed primarily by Engineering Research Associated (ERA), a 'captive' company linked to the navy, who delivered a version at the end of 1950 for $1 million (triple the original cost). ERA was then, with navy permission, able to spin off a version for sale to the wider world. Already individuals were moving back and forth between industry and the secret state in America in a way they did not in Britain.[16] What did they use the new computer for? It is telling that the very first operational programme written for Atlas was designed to attack anomalies in Venona messages, another little-known sign of just how important this task was in advancing computing.[17] The old rivalries were still there though, and the army had worked on their own system. This was called Abner, after yet another comic-strip character – a powerfully built country bumpkin.[18] These machines spent as much time not working as working. But when they did perform the results were astounding, and everyone could see the potential, perhaps explaining why they were given the names of comic-book superheroes.[19]

In England commercial computing was initially motivated by providing a nice cup of tea. The Lyons Catering Company led the way in building a machine that could be sold. It may seem an odd fit, but with all its tea shops around the country Lyons was at the leading edge of management techniques and data processing, looking for ways to become more efficient in supplying its shops with what they needed when they needed it and as automatically as possible. The company worked with Maurice Wilkes, who had explored electronic radar at Malvern to develop Leo, which was used from 1953 to replenish tea shops each morning and do the payroll (more National Insurance than national security).[20] A hundred Leo computers would be sold abroad, but the Lyons computer team eventually merged unhappily with English Electric. British governments tried to create a national champion in English Electric and then ICL, but each merger meant effort was expended trying to blend incompatible computer systems and management teams rather than innovation.[21] The story of the British computer industry after the war was one of brilliant engineering but business failure.

Those early British computer companies would become roadkill, crushed under the juggernaut of the American giant IBM once it moved into computing. British declassified files show how in 1960 Britain's Atomic Weapons Establishment was desperate for a new computer for warhead design and was being offered a rental deal by IBM. In Whitehall, top officials all agreed that IBM was using its contracts to eliminate British competition, but they felt cornered.[22] IBM built fast and offered compatibility and upgrades. In 1955 there were thirty computers running in the UK and all were British, but by the end of the 1960s IBM was dominant.

America's success emerged from a unique relationship between spies and companies that neither side liked to talk about. It is ironic that the free-market US ploughed massive amounts of NSA and Pentagon money into the private sector, providing a state subsidy that 'socialist' Britain with its nationalised industries in other fields never matched (the US also had a larger domestic market, which helped). The NSA was so desperate for a breakthrough in the early

Cold War that it paid millions to companies for computers which almost never lived up to the promises and which went wildly over-budget. All the major US computer companies would open up their own special security-cleared departments solely to work with the intelligence agencies. In the shadow of what President Eisenhower called the 'military-industrial complex', a spy-industrial complex was emerging involving the computer industry, centred partly around Washington DC but also northern California, where a raft of high-tech companies were springing up close to Stanford University in a region which would become known as Silicon Valley. IBM grew to be dominant as it realised that government contracts could subsidise its research, which would then spin off into commercial work. There were those in the NSA who felt IBM was too close and was taking the agency for a ride. The most ambitious project was Harvest, in which special-purpose machines were controlled by a general programmable computer. It aimed to be 100 times faster than anything else. After the NSA originally signed up IBM to build Harvest it was told it would cost nearly double the amount planned, take longer and not be able to do quite everything that was originally agreed.[23] The NSA also realised that the near-$10 million bill did not include the software or programming (eventually costing another $4 million). 'As usual the agency has a firm hold on the IBM leash and is being dragged down the street,' an engineer wrote.[24] Harvest was delivered in February 1962 and ended up being used for fourteen years on cryptanalytic projects. Working on Harvest involved sitting in a busy operational area with a series of desks, terminals, printers and tall storage cabinets.

The 1950s were years of growth and optimism in America, of cars, consumerism and suburbs. And behind all of this was technology. In secret, the desperate desire to break stubborn Soviet codes was driving computer development harder and faster. The $25 million project Lightning was authorised by President Eisenhower in the 1950s. 'A jet plane can go one hundred times as fast as a man can run. A computer can go ten thousand times as fast as a man can compute. Lightning will go ten million times as fast,' Howard Campaigne wrote.[25] Working with a range of contractors, the project helped develop techniques like high-speed circuitry that then bled out into the commercial world. One company in California said it could develop a tube that could switch in three nanoseconds. No one could even measure that to know if it was true.[26] Campaigne argued that the NSA played a crucial but hidden role in leading the way in fields like solid-state transistors and even voice recognition. By the end of the 1950s there were two dozen special-purpose computers that could digitise an audio message and scan the text for keywords. If a keyword was not present it could be discarded, thus reducing the burden on human operators.[27]

And yet, for all this technology, the 1950s remained barren years in the core task of reading Soviet messages, with talk of giving up. In early 1957, a possible break called 'Hairline' led to some excitement for the few in on the secret: $20 million was allocated and six new special-purpose machines were built to run through the 638,073,495,557,089,200 mathematical possibilities. But by the early 1960s it was only opening up a small percentage of traffic which carried low-level information. An official review suggested the 'age of heroic cryptanalysis had ended'.[28] The future, many thought, lay in the CIA's new toy, the U-2 spy plane, which could carry out aerial reconnaissance (the CIA was also pushing its own cryptanalytic team).

At the CIA, the spies engaged in recruiting and running agents were learning about the potential of computers to store and retrieve data as the sixties began. Just after Allen Dulles stepped down as Director of the CIA he spoke of his wonder at what the machines were capable of: 'The intelligence service needs a man who speaks Swahili and French, has a degree in chemical engineering, is unmarried and over thirty-five but under five feet eight. You push a button and in less than forty seconds a machine – like those commonly used in personnel work – tells whether such a man is available, and if so, everything else there is on record about him . . . What, before the advent of the machine, might have taken the analyst weeks of search and study among the files, the machines can now accomplish in a matter of minutes.' Declassified files from the time reveal that the agency, though, also had concerns about this. Considerable thought went into ensuring that highly sensitive information about the identities of its undercover agents was kept 'compartmented' on computers so that even others inside the CIA could not access it. And already there were fears of being overwhelmed by the multitude of tiny fragments of intelligence flowing into the CIA from all its different sources – there were already too many snippets for people to be able to catalogue by hand. And, without a working index 'we become a large but unusable library', one 1966 report warned. It was no good just holding information – you had to know you had it. The study argued that what was needed were 'powerful electronic arms which will continually reach out and gather in relevant information, arrange and correlate it, and present it to us. It is also likely that we can look towards machines which are self-improving.' Three decades before Google and its algorithms came along here was the idea of a 'search engine'.[29]

The real secret of American and British code-breaking in the Cold War is that they did not really succeed in breaking Soviet codes. Instead they did something else with the help of computers. They had learnt from Bletchley the power of traffic analysis – studying the externals of a communication even when you could not break into its content. This required collecting and analysing every detail and intricacy of the system you were trying to break – how it was structured, how people used it and what was produced – so that you could find that weakness or mistake which code-breakers and their computers could work on. 'You must know much more about the other man's signals than he knows himself,' Oliver Kirby, an American veteran of Bletchley turned NSA man told his successors. You needed to understand the normal so that you could spot the abnormal, which might be a chink in the armour you could then target. But there was a further point. Even if you could not break the code, the data also had a value in itself if enough was collected and analysed properly. A bit like tracking Zeppelins through call signs in the First World War, you could watch the outward behaviour of Soviet military communications. That required a huge effort. And the data required to feed this beast in turn led to the creation of a vast, global enterprise.

By the early 1960s, over 100,000 Americans were involved in intelligence work – more than half in signals intelligence. Of those, half were at overseas sites, collecting signals in remote stations or on board clandestine ships or from airborne platforms.[30] By 1955 there were more than 2,000 listening stations. These sent thirty-seven tons of intercept material to the NSA each month. Another 30 million words of teletype intercept were also arriving. The traffic analysis section was dealing with 3 billion message groups a year. They might be punching a million IBM cards a month for just one problem. There were also the 250,000 Chinese enciphered messages a year.[31] This was a gargantuan task – one based around mammoth large-scale data processing and computers which could seek to tease out meaning from this torrent.[32]

The analysts in the UK and US studied in great depth the call signs used by each part of the Soviet military to communicate. Through this alone they were able to establish the Soviet order of battle on the other side of the Iron Curtain – which units were where. This information would then be supplied to the land forces' commanders on the ground so that they knew what they would be up against in the event of war. Every month when new call signs were introduced by the Soviets, the GCHQ and NSA analysts would race each other to be first to get it done. In this game, the British always liked to make the most of the advantage the time difference gave to try to win. As well as tactical military help in the event of war, this system also offered the chance of early warning.

At GCHQ, traffic analysis involved looking for tiny changes in the normal pattern of Soviet communications by watching behaviour and the externals of messages (rather than the content which was encrypted). The aim was to offer ministers seventy-two hours' warning that Soviet forces were on the move and that the Third World War might be beginning. Analysts reckoned they could provide up to three months' warning of the earliest preparations. In practice, this involved physically printing out pages and pages of sheets that would be deposited on an analyst's desk in the morning. He or she would then go through it line by line with a pen looking for unusual patterns. This technique remained in use through the 1980s (with personal terminals only arriving that decade).

The parameters of Cold War signals intelligence were soon set. Bletchley had showed Britain and America that code-breaking required a mix of people and machines. Computers did not replace people. Quite the opposite. They often ended up needing more people to collect and then input all the data and then work on the output.[33] Solomon Kullback, who had started in the days of pen and paper with Friedman in 1930, was careful to avoid making computers appear as either the problem or the solution. 'The computer never will replace the human element,' he said, explaining that the machine was 'a very stupid but very fast and accurate clerk who will do what you program it to do. And all these stories about computer error or that the computer fouled up ain't so. All the computer did was what it was told to do, and if whoever told it what to do fouled up the instruction, then the computer did too. No, the computer will never replace the human in terms of judgment.'

Machines were not enough by themselves, but allied to people they were proving immensely powerful, able to work on a vast scale. At the time, some called the era after the Second World War the atomic age, but in reality it was as much the computer age. Both were defining symbols of the end of that last conflict and the arrival of the Cold War. And computers were already breaking out of the confines of the secret state.

CHAPTER FOUR

COMING OF AGE

'Just out of its teens, the computer is beginning to affect the very fabric of society, kindling both wonder and widespread apprehension,' *Time* magazine proclaimed in an April 1965 cover story entitled 'The Cybernated Generation'.[1] The world beyond the confines of spying was beginning to wake up to what this new arrival might mean. The article described how, in air-conditioned rooms, the machines were lined up in rows 'waited upon by crisp, white-shirted men who move softly among them like priests serving in a shrine'. In 1951, the magazine reckoned, there had been fewer than a hundred computers in the US, but fourteen years later there were now 22,500 – four times as many as existed in the rest of the world put together. The machines were performing tasks from navigating planes, sorting out taxes to mixing cakes. The computer industry was reckoned to provide 650,000 jobs on $5 billion of sales. Yes, jobs were being lost by automation, but an IBM economist assured everyone that soon the process would bring about a twenty-hour week and a mass leisure class. In Chicago, a drive-in computer centre was offering to process information for customers while they waited. The New York Stock Exchange had built a talking computer that could provide information over the phone. 'The day is clearly coming when most computers will be able to talk back,' *Time* predicted. American Airlines ran the single most expensive system (at least as far as the public knew). Its genesis came when a young IBM salesman sat next to the president of the airline on a 1953 flight. That chance meeting led within a decade to the $30 million SABRE (Semi-Automated Business Environment) system, in which a computer in New York processed 84,000 requests a day that came over phone lines to tell travel agents what seats were available on flights.

But *Time* also asked a question: 'Is the computer a friend or enemy of man?' The reliance on machines was already apparent. 'If all the computers went on the blink, the country would be practically paralyzed,' it wrote, 'plants would shut down, finances would be thrown into chaos, most telephones would go dead and the skies would be left virtually defenceless against enemy attack.' And so, hot on the heels of the computer age had come the realisation that the seductive power of computers brought with it dependence. All the twenty-first-century talk of 'cyber security' is far from new – it is merely the modern reworking of much older fears over the vulnerability of computers, vulnerabilities that spies would come to exploit. But in the 1960s it was the US Air Force that was at the leading edge of understanding those dangers.

Bletchley was not the only destination for those first visiting American code-breakers in the Second World War. They were also taken to Dover Castle, which overlooked the Channel that separated Britain from occupied Europe. The Americans had already been to the RAF station at Cheadle and seen how Britain used traffic analysis and direction-finding to scramble its fighters in as fast as ten minutes from the time a message was intercepted indicating that the German bombers were on their way over the water. But at Dover, the Americans were shown how another new invention – radar – worked.[2] Prescott Currier could remember two things about his Dover visit. Firstly, the prettiness of the female operators who sat at screens monitoring the radar signals. And secondly, the way one of them had tapped him on the shoulder. 'In about three minutes go outside and look up,' she said. Standing on the famous white cliffs, he saw a squadron of German bombers approaching right on cue. But coming the other way – alerted by radar – were British Spitfires. They came together in the skies above the cliffs and intermingled, making it hard for a moment to tell who was who before the Germans were forced to retreat. That was what radar could do. Brave pilots and superbly engineered planes counted for a lot, but so did information and data flow to make sure the planes and pilots could get where they needed to be. In the Cold War, the importance of this technology, including interconnected computers, would only grow.

The 1950s and early 1960s were years when the Cold War was at one of its most dangerous points, with annihilation only a few moments way. The fear was of Soviet planes dropping nuclear bombs on America. Anyone who thought about it for a moment understood that the drills for people to hide behind a desk or a door when the alarm sounded would be little hope against an atomic blast that could flatten a city in an instant, spilling radiation across the well-groomed gardens of the suburbs. America's own nuclear-armed bombers would constantly circle the Arctic ready to head to the Soviet Union to drop their own bombs in retaliation. But could anything be done to stop the Soviet planes getting through? The only defence was fighter aircraft and missiles. Early warning and radar – as witnessed at Dover Castle – were vital. Electronics had been developed for radar in the war and computers like ENIAC for ballistic artillery calculation, and now the hope was to merge the two by having computers help perform the complex task of finding targets as they approached.

America began building a vast array of radars spread as far afield as possible to alert it to incoming Soviet bombs so it could get its own planes or missiles up as quickly or 'automatic' as possible. These radars were initially in the continental US but spread further afield out to the remote northern tip of Greenland and also to North Yorkshire in Britain, where strange golf-ball-shaped structures appeared on the moors at Fylingdales in the early 1960s. Britain too was developing its own computer-controlled fighter intercept system in the 1950s and a declassified report explains the vital role machines could play in meeting the challenge of a massive attack:

> When an air attack develops at a rapid rate, and the fighter defence has little control capacity or warning in hand, the task of allocating the right fighters to the right targets at the right time is beyond the capability of the unaided human brain. On the other hand, certain rather subtle considerations, based on intelligence reports, identification problems, psychological factors, and on a broad appreciation of the tactical position, ought to enter into the decision, and these considerations cannot readily be built into a simple electronic computer.[3]

Man and machine would need to be melded. A computer, it was hoped, could draw in all the information and help a person make a speedy decision. American Airline's SABRE was an outgrowth of America's vast air defence system called SAGE (Semi-Automatic Ground Environment), built to do just this. First conceived of in 1951, SAGE would take a decade to bring online but then survive all the way through until 1983. It cost $8–$12 billion – not in current money but in 1950s prices – making it more expensive than the Manhattan Project. IBM 'threw everything' at making sure it got the contract, which generated 80 per cent of the company's revenue from computers in the mid 1950s. It was the largest, most expensive and ambitious computer system ever built – 'the first geographically distributed, online, real-time application of digital computers in the world,' IBM would boast.[4] It was also a creature of its time, when nuclear oblivion was only a moment away.

The control centre for SAGE, weighing in at 250 tons, filled up an entire concrete building, half a block square, able to withstand attack. SAGE offered many novelties beyond just scale. Previously computers like Colossus had produced printouts of their results. SAGE offered a visual aid – a display screen ('a giant picture tube,' as it was called at the time) that would show the results of the calculations in a way designed to help the operator digest the information. An operator could even interact with the screen using a light-gun pen. This would all help them understand as quickly as possible what was happening in the skies and offer them their best options as to how to engage different targets based on their location, velocity and position. They could also use the computer's memory to go back and look at how a situation had developed – 'by analysing the past, SAGE can project into the future,' IBM explained. 'Aladdin's lamp couldn't do more.' The system needed to help tell the difference between civilian airliners and Soviet bombers (and so would become the forerunner of modern civilian air traffic control). All this data needed to be collated from a number of sites, so information went across phone lines to the central computer using a device that became known as a modem. This was the first international computer network (although it was a closed and very specialised system) and marks the next stage in the merging of computers and communications.

Strategic Air Command ran the bomber and ballistic missile force and had a sprawling system using computers for command and control, war planning and data processing, with 900 programmers and systems analysts working on thirty-five machines. As the 1960s progressed, nuclear-tipped missiles designed to shoot down incoming Soviet bombers over the ocean were being controlled by computers. One or two people started to wonder if having these missiles stationed within the US, largely outside of human control, involved a degree of risk.[5] What if these systems went wrong or someone interfered with them? Could a nuclear-tipped missile be sent the wrong way and back into the United States?

A young, technically savvy air force captain called Roger Schell, who had worked on SAGE, became one of the first people to worry about computer security. He was the kind of person who annoyed senior colleagues because he actually knew what he was talking about and was willing to point out when they did not. His superiors assured him there was no risk. After all, under the two-man rule, a pair of 'real' people needed to turn their 'real' keys to launch a missile. They told him a study proved this. 'I looked at it and I said it's garbage,' Schell told his colleagues after reading the study. 'You can't say that. No, all that stuff doesn't matter because at the end of the day, the keys that you are using are not in the electrical wires, their input is to a computer and the actual launch command is given by a computer. And the controls that are provided, are provided by a computer. There's no positive control; it doesn't matter what the human did with keys or didn't do, it's a computer that sends the signals.'[6] Schell refused to sign off on the system as secure. It was vulnerable to subversion, he maintained. 'People were not particularly impressed,' he recalls. 'It caused a fair degree of turbulence.'[7] The 1964 film *Dr Strangelove* brutally satirised the dangers of hair-trigger nuclear responses and the idea that both sides in the Cold War were increasingly escalating and automating their response to the first sign of attack by the other. In the wake of the Cuban Missile Crisis, these fears felt all too real.

Roger Schell was getting a reputation for being a troublemaker who asked difficult questions. But some recognised that he knew what he was talking about. He was brought on board a team who went on to produce a ground-breaking but almost forgotten 1972 report into computer security for the US Air Force. It was known as the Anderson Report, after the principal author, James P. Anderson, an engineer and expert who had done work for the NSA and CIA. But because the NSA was so secret it kept its computer systems inside its walls and unconnected to the outside world, meaning it fell to the air force, whose systems were far more open and global, to be the first to worry about security.[8] Anderson's report is the blueprint for every cyber security report today.

Anderson saw that the air force was becoming increasingly reliant on machines that had vulnerabilities which opened them up to subversion. A central issue was how to deal with what was known as the 'multi-user' problem. Computers were not the personal terminals of today but huge, expensive monsters. Their processing power needed to be shared among lots of users – not least to make them economic given the vast costs. This was true both of the military and of universities, where staff had to log how much time they used a computer for so they could be charged 'rental'. In intelligence and military institutions this created a problem. If lots of individuals were using the same computer, how could you keep information secure when some were cleared for Top Secret and others were not? In the earliest days computer security was about physical security – making sure there was a guard or that anyone who wanted to enter the processing facility had a swipe card.[9] But with different clearance levels in an institution, it became inefficient to restrict computers on that basis. And what about the problem that now computers were being hooked up so that they could be accessed remotely from terminals? Computers were being under-utilised in the air force, Anderson found, because of the fear that users would see something they were not supposed to.

This was not just about 'data spillage', but also the risk of a system being open to 'hostile penetration' from what were called 'malicious users'. In 1972, the Defense Intelligence Agency was creating large, multi-access databases to share information. A test was run to check the security of the system. The NSA easily got inside. It even proved possible to gain access from a remote terminal and seize control of the entire system. It was a terrifying insight into what could be done.[10] Computers were an especially tempting target since so much data was in one place if someone could get in. This went beyond the possibility of passive interception of communications like wiretapping (although that was still a worry in case someone put a monitoring device on the hardware during manufacture or maintenance). The big problem was 'active infiltration' or 'subversion' – someone being able to get inside the computer and issue instructions to gather information or do certain things. They could even play with the operating system itself – the instructions for how a computer should work – and thereby subvert and bypass all the security you put in place. The fact that the operating systems were obtained off the shelf and were not developed in house just added to the concerns. These systems had never been designed with security in mind and now were becoming hugely complex. This meant there were flaws – vulnerabilities – in all the systems, which were hard to spot and which a skilled programmer could exploit.

The beauty of computers was that they could be programmed to do anything. The problem was that it was possible to manipulate these instructions in a malicious way to do something untoward. The dense complexity and increasing length of computer code as computers were asked to do more and more would make such malicious code increasingly hard to find. If everything in a computer's system is built on code then it could always be altered by code. Someone could just reprogramme the computer to give themselves top-level access, for instance. Vulnerabilities plus code equalled a problem.

Trapdoors were one of the fears the Anderson Report raised. Programmers sometimes put in secret ways to bypass controls and security checks, to make their lives easier if they needed to change something. But what if these were introduced deliberately in a malicious act? A few keystrokes and an attacker would be in. Since commercial systems were being used, what happened if someone put in a trapdoor during the manufacturing process or when they updated or revised the software? How do you check for that?

The variety of possible attacks was dizzying. The most fascinating aspect of the Anderson Report is the appendix that outlines penetration techniques. Although written from the point of view of defenders, it is effectively a how-to guide for hackers. The appendix goes point by point through the myriad ways in which someone can exploit the systems and uses real examples. The 'Trojan horse', the report notes, is a 'rather interesting attack' in which code with trapdoors is placed into a target system. This is done by presenting the operators of a system with a programme that appears to them to be so useful that they will ignore any concerns and run it on their system even though they may not know what is inside. Like the Greek myth, it was a gift horse that contained a surprise. The Trojan horse, once accepted through the gate, might then be able to record the user ID and password of people using the system or copy all or part of the file to make it accessible to a penetrator. This had the potential to 'bypass' all security controls and is described as the 'quintessence of the malicious threat against contemporary systems'. Here, four decades ago, is the realisation that the most dangerous point in computer security came when the human met the technical – where an attacker might play on someone's desire to take a short cut or utilise some clever programme, but in doing so let the enemy in. Trojan horses remain a favoured method of getting inside a system today.

The report said it could not emphasise enough that 'patching' known weaknesses or vulnerabilities individually was ultimately 'futile'. Creating 'firewalls' to stop data spilling between users, encrypting all stored information and carrying out surveillance and monitoring of activity on machines were all useful but were really just papering over cracks without dealing with the fundamental problems. These are pretty much the same techniques that cybersecurity companies talk about today (often portraying them as some great innovation), and a modern security expert might allow themselves a little chuckle at the failure to make that much progress on this more than four decades on. The size and complexity of systems that were not designed with security in mind were a headache then and a much bigger one now as the scale has grown enormously. The great irony of computer security is that many of the flaws that allow an attacker to get in have been known for decades – many since the Anderson Report – but they have not always been fixed. Why? Because companies desperate to get products to market first in a fiercely competitive and innovative environment have never prioritised security enough to slow down and redesign their systems. It was never worth it. The Anderson Report recognised that a whole new way of conceptualising security was required, rather than just plugging holes. It came up with the concept of a 'reference monitor' or 'security kernel' to verify activity on a system, but the panel could not yet see a way of implementing it and so was left pessimistic about the central danger of 'subversion' in the future.

Among the things Anderson did was look at the relative costs of security. Finding a vulnerability to exploit was reckoned to involve a month of one person's time (costed at $2,000) and less than $1,000 of computer time – a total cost of $3,000. If you were to patch your system to close the exploit but not fundamentally change it then you might raise the attacker's costs to $8,000, but they would still get in. The cost of redesigning and creating a fully secure system which no one could get into was reckoned at a sizeable $10 million per system. Security did not come cheap. This is the other truth that remains today. Proper security costs, but most commercial companies (and even most parts of government, apart from the most secret) are not willing to pay what it takes to be truly secure. The economics are in favour of the attacker and the spy.

Anderson's report ends with a stark warning: 'The situation will become even more acute in the future as potential enemies recognize the attractiveness of Air Force data systems as intelligence targets, and perceive how little is needed to subvert them.' As you read the report you sense that, while doing the research, an awful truth had dawned on James Anderson, Roger Schell and the others involved which they were now desperate to try to communicate urgently to everyone they could. The truth was that computers were not secure. They had not been built for security and it was not easy to engineer or 'retrofit' them to be secure. Security should have been there from the start. But it was not.

When the report was being finished, the team behind it began to debate what to do about what they had found. Should it be published? Should it be classified? The NSA representatives had concerns, perhaps worrying it would alert attackers to what could be done. Those with links to industry wanted the information out there so people could understand the dangers and improve their defences.

So another early truth about computer security was exposed – the tension between keeping secret the vulnerabilities you found so that others could not exploit them (and perhaps your own spies could) and making them public so people could improve their systems. Roger Schell, who was project-managing the report, wanted to publish it openly and printed between 300 and 500 copies and simply sent them out. The next day he got a phone call. 'NSA has determined that this is classified and they are exercising their classification authority,' Schell recalled being told by the official. 'I said, "good luck, it's mailed".' [11]

Schell also participated in what were called 'tiger teams' in the late 1960s and 1970s – groups of people carrying out 'penetration testing' to find the flaws in systems used in the defence and intelligence community. They were never popular with people who thought their systems were watertight. In house at the NSA, they knew all the dangers of Trojan horses and operating system vulnerabilities, but commercial companies were going round claiming their machines were safe. This was becoming a selling point, since companies like Ford and General Motors were starting to think about the need to protect corporate secrets like technical data and intellectual property. 'As a penetration team, we demonstrated that that wasn't true; that we could get to, essentially, things that they didn't think anybody could get to; which didn't endear us, particularly.' Honeywell Systems had taken out full-page adverts boasting of their security. Schell orchestrated a meeting with a senior vice-president of the company and its technical team on one side and Schell's team on the other. Schell got one of the over-confident technical staffers from Honeywell to change his password to something new and then had one of his team tap at a keyboard. A few seconds later Schell's man printed out the new password and handed it to the corporate bigwig to show how easy it was (a trick still used to scare executives). At the same time IBM was showing off that it had a $40 million budget for security. At a conference Schell piped up, 'From what I can see, the IBM budget allocation of the $40 million is roughly $39 million for marketing and $1 million for travel.' That did not go down very well. What was clear was that none of the companies wanted to change the fundamentals of their systems. They just wanted to patch up the holes they could find to stop the ship sinking, not build an expensive new boat. Companies – right up to today – prefer to get their products out first before a competitor and then patch any security later rather than spend time working out how to secure them from first principles and risk losing out in the market. The pace of private-sector innovation behind the rise of computers had many advantages, but also a distinct disadvantage when it came to the priority given to computers.

The Anderson Report had come at the cusp of a new era for computers. Computers were beginning to talk to each other rather than sitting like kings in splendid isolation on their thrones. This led the report to highlight fears of a 'man in the middle' who might be able to interfere with the lines between Pentagon machines, requiring security to be 'end to end'. But at the same time as the Anderson Report was issued, a new type of networking was emerging in the outside world that was going to take on a life of its own and revolutionise computers. It would become known as the internet.

There are plenty of myths about the creation of the internet, including the idea that it was a creature of the Pentagon. The real story is more complex. Part of the problem is the occasional confusion as to what is being talked about – there is a series of concepts ranging from packet-switching to computer networks talking to each other through the World Wide Web, all of which are distinct but which tend to get wrongly lumped together as 'the internet'. Innovation is often a collaborative process, but there is also the problem that 'success has a thousand fathers', with many individuals laying claim to their critical role. In simplest terms, the internet is the ability of computer networks to communicate and connect with each other.

The Cold War fear of nuclear Armageddon did play a role in some of the initial thinking. Paul Baran, a Polish-born engineer who worked on early computers, joined RAND – the US Air Force's outsourced research arm – in 1959. He saw there was a 'glaring weak spot' in the way in which military command and control depended on shortwave radio and a national telephone system run by AT&T.[12] Tests in the Pacific showed that a high-altitude nuclear bomb blast would disrupt shortwave radio for many hours while RAND computer simulations showed that the telephone system within the US would collapse after an attack. In order for the concept of mutually assured destruction to hold, there needed to be the guarantee that retaliation could be launched. This required a command and control communications system that would survive even a surprise attack so the President could issue his order to retaliate. With nuclear apocalypse only narrowly avoided in the 1962 Cuban Missile Crisis, Paul Baran began to look at what kind of communications networks might be more likely to survive. The obvious answer was one that was decentralised, without a single centre that could be knocked out. But then there was a further idea. What about not just a *decentralised* but a *distributed* communications network, meaning that there was no hierarchy at all and therefore no single point of vulnerability? A distributed network was like a web, with all the different parts connected to each other rather than being connected into one or two single central points. This offered the possibility of redundancy, so that even if one or two connections were destroyed by an attack, a message could take a different path.

Baran says he realised that the best way of doing this would be by breaking up radio communications into the digital form of ones and zeros (so that audio quality would not degrade on a long route). This could then be sent by breaking down a message into smaller fragments and by sending the routing information (the 'to' and 'from') with each of these blocks. One colleague suggested the analogy of a postman at each node of the network who would receive a letter with 'to' and 'from' details so he knew where to send it on to or back from if it got stuck. Also attached would be the route the message had already taken which, over time, would help him establish the quickest route to send future messages. Another analogy was the hot potato: each bit of a message was a hot potato tossed from person to person with the aim of getting it out of your hands as soon as possible before you got burnt. If the first person you wanted to toss it to had their hands full already, you would look for the next-nearest person until you found someone who could receive it. If everyone was busy then you would throw it back to the person who gave it to you and ask them to find another route. An algorithm could work out the best available route based on the experience of the system so far. The aim was to move the data round the network fast until it found its destination. Since it was broken into chunks, parts of the message might arrive by different routes and out of sequence, so a serial number in the header would allow the original message to be reconstructed. Baran's concepts proved prophetic, but were not quite adopted in the way he envisaged and were not yet married up to computer networking.

Britain had its own candidate for developing some of the underlying concepts for the internet, although again it missed out on the credit. The National Physical Laboratory in Teddington was home to a scientist called Donald Davies. Colleagues remember Davies as a quiet, self-assured figure whose large head seemed designed to contain an unusually large brain. Every year he would study something completely different from his normal work – like contemporary art – to keep him stimulated. He had begun his career during the war (working with Klaus Fuchs, later found to be a Soviet spy) and then worked on the early ACE (Automatic Computing Engine) computer that Turing had been involved in. To use ACE, a researcher had to book time-slots or give their instructions to computer operators. The next step was to lay cables between the computers to other buildings on the site where users worked. Davies looked for another solution to make this sharing of computing power easier.

Telephone switching had long fascinated Davies – the kind of work Tommy Flowers had undertaken years before – and Davies often visited the Post Office research team at Dollis Hill.[13] But connecting up computers along phone lines was proving a wasteful business. A phone line used a single circuit or connection from origin to destination, and that line would be monopolised even if the data were only flowing a small proportion of the time and at irregular intervals, perhaps only using 1 per cent of the capacity. In 1965 Davies came up with the idea of breaking data into small chunks and then storing it. When a line was free it could be forwarded in an efficient burst with other messages before being reconstituted at the destination based on the 'to' and 'from' headers linked to the message. Davies gave the concept its formal name – packet-switching – and in 1966 wrote a paper called 'Proposal for a Digital Communication Network'. Colleagues say Davies had foreseen packet-switching as the backbone of a very large global communications network, delivering sound, video – even, he said, railway timetables. At least one copy of the paper went over to the US and the Pentagon, where it was read by Larry Roberts. The two men met at a conference in the US in 1967 where Davies talked about his ideas. Davies was sure his paper had an impact. 'When I visited him in the Pentagon on one occasion, it was lying on his desk in tatters,' Davies later recalled. 'It had obviously been very heavily thumbed and turned over.' Roberts, then at MIT, was also looking at ways of connecting up computers to share their workload and wrote his own seminal 1966 paper called 'Toward a Cooperative Network of Time-sharing Computers'.[14]

Davies began experimenting with how to actually send messages. He briefed the Minister for Technology (Tony Benn) and the Postmaster General (John Stonehouse – another of those Communist spies). In January 1970 the Mark I network first operated successfully and it went live in July 1971. By then, though, it had conceded pole position. Larry Roberts' larger, more ambitious US project had forged ahead.[15] The idea for something like the internet was in the air at the time, but the Americans moved faster and built bigger. Roberts' project had been backed by – and named after – the Advanced Research Projects Agency (ARPA), part of the US Department of Defense, and so was called ARPANET. Davies was always clear that ARPANET had made more progress, but could not hide just the touch of regret given that he felt his paper may have had a role and that it had been largely overlooked.

In October 1969, a UCLA computer science professor sent a message from UCLA to the Stanford Research Institute through ARPANET. The first attempt crashed. The second worked. Two different computers had begun a tentative, awkward conversation, the beginnings of the vast cacophony of global noise that the internet has today become. In late 1970 the man in charge of ARPANET suggested to Britain's Davies that they get their two networks to talk to each other. Bizarrely, this was scuppered by politics in London. The Heath government was keen to build links with Europe rather than the US. As a result the NPL – which came under the Department of Technology – was instructed to work on projects with Europe, not America. So instead the task fell to academics at University College,

London (UCL). After funding crises and tangles with British bureaucracy, in September 1973 ARPANET reached out over the Atlantic
and the two sides were able to converse in real time.[16] That was
not what the system had really been built for – sharing computing
power was the original intent – but, as was often the case with technology, people used it for what they enjoyed doing or found useful,
and that increasingly became person-to-person communication. In
1972 Ray Tomlinson started to use the @ symbol to show a message
was from a particular person at a particular computer network. In
1976 the Queen visited Malvern and became the first head of state to
send an email (with the username HME2, standing for Her Majesty
Elizabeth the Second).

So by the early 1970s different systems from different universities
and even different countries began to connect up with each other.
Modems were converting digital data into sounds which could then
be transmitted along a phone line and reassembled. Networks were
talking to each other – a network of networks. The inter-network
traffic was then abbreviated down to 'internet' (the definitive article
at the front tended to come later). A protocol was needed for these
different networks to be able to understand each other. This led to
work by a group in California including Vint Cerf, Bob Kahn, Bob
Braden and Jon Postel creating a common standard (known as TCP/
IP).

At the end of the 1980s a mild-mannered British scientist, Tim
Berners-Lee, was working at the CERN laboratory in Switzerland.
He was interested in how to organise information so that the large
community of researchers could keep track of each other's work
and draw connections between different projects. He developed
the idea of a web in which information held on servers could be
pulled off using a browser which navigated the growing stack of
information. Again this idea proved, to the surprise of purists, to be
useful in realms far beyond pure physics research. It was christened
the World Wide Web. Once it became clearer that it was getting
easier for ordinary people to navigate the internet, the private sector
began to move in on a world previously the domain of academics
and researchers.

The origins of the internet are vital to an understanding of how it shaped spying. Its defining features were the key to its success, but they also engineered a form of insecurity that laid it open to espionage. The network is 'dumb' and it has no central control or security.[17] It is simply a mechanism for sending data packets from one place to another as fast as possible. It does not care who sends them or what they contain. There is no guarantee what route a packet takes, no guarantee that it will arrive in one piece in a timely fashion, or without being copied or interfered with. There is no guarantee that it is from the address – or person – it is supposed to be from. This meant it was incredibly open and adaptable: people would eventually be able to send different types of data (like audio and video) and design their own applications and systems to sit on the internet and send that data. There was no authority restricting or defining how it would be used. This would fuel tremendous innovation but also make it susceptible to all kinds of mischief. These attributes provided speed and flexibility but also allowed people to do things like pretend to be somebody else, divert traffic and implant malicious code. The internet would be used to do things the original creators never dreamt of, and that would include spying.

The openness of the internet's structure is partly because of Paul Baran's concept of survivability, but later because it was built by a small academic community who wanted to share information, a trusted community in which people often knew each other personally. And so it was not designed with security in mind. And because the internet was not the creature of government or commerce, but academic research and co-operation, those values infused the system. There was also an element of the late 1960s California counter-culture to the whole project – a focus on freedom and sharing – which might have been very different if the government had planned the whole thing from the outset or if companies had been in charge. This mindset would go on to infuse the tech sector as it grew (with companies and the state trying to wrestle back influence at various points). As more and more people took to the internet, the openness ensured its spread. Security features like verifying the route a message took or the identity of the person sending it would have slowed the whole project down, added to the cost and been contrary to the co-operative mindset. But the weaknesses could be exploited as the internet organically expanded.

No one predicted that what was designed at Stanford and Teddington would so rapidly scale up into the world-dominating system on which business, government and individuals increasingly conducted their work and went about their lives. But the speed of its rise meant that inherent weaknesses were embedded in a system which quickly became too large and too essential to roll back and redesign, or to junk and start again (although some still talk of this today). It was just not possible to retrofit security. There is always a balance between openness, speed and innovation on the one side and security on the other. A more secure internet may not have taken off in the same way. But no one even tried. And it was not long before some of the weaknesses were exposed. The first to exploit this inherent insecurity were not the spies but pranksters and explorers within the community itself.

In May 1962, a group of young people clustered in a room at MIT's Science Open House. It was one of the regular occasions when students got to show off what they had been working on. A screen was perched high on a cabinet so that as many people as possible could watch what it was showing. The previous year a small group of young men at the university had realised you could do something a little more fun than intended with the cathode ray tube display on which SAGE and those big grey computers had shown the flight paths of Soviet bombers as bright dots against a dark screen and which had, in part, been designed at MIT. You could programme the computer to move the lights around the dark screen like a bouncing ball. It seemed that you could programme computers to do all kinds of things if you knew what you were doing, and the students had next created a game called 'Spacewar!'. Now, in May 1962, they were making their spaceships fire off torpedoes in battle against a simulation of the night sky. Those involved in modifying code to do new things and showing off to their friends called themselves 'hackers'. MIT was where hacking emerged in its original sense – a playful, rather than illegal, activity.[18] It was all about demonstrating what you could do, revelling in the power, beauty and simplicity of computer code. But as the years went by, a more anti-authoritarian and mischievous part of the culture would come to the fore in which people wanted to demonstrate how they could find vulnerabilities in computers and get into places where they were not supposed to be. It was all about the challenge. The more the people running the systems tried to stop people doing mischievous things, the more these hackers wanted to prove they were better and could get in. Through most of the 1960s this was fun and games for the students at MIT – perhaps at most trying to get round the time limit on how long you could use a machine, or the rental charges – but as computers were connected up to the outside world over phone lines, people outside this community realised they could dial into MIT's computers and display their skills. That would include deliberately crashing the system. That annoyed the hell out of the MIT students and staff, who would lose their work. 'So the people at MIT at the time that were responsible for these computer systems – they were hackers themselves – they said "what's the best way to do this?"', recalls Herb Lin, a student at the time who dabbled in hacking. 'You'd think they would put in big computer security systems – passwords and firewalls. Instead they changed the system to implement a command called "CRASH" and said in a list of system instructions what it did.' The note said: 'If you execute this command "CRASH" you will crash the system and everybody will lose their work and be really mad at you – don't do it.' The result was that no one bothered any more to crash the system. The point was that it had all been about the challenge of showing your peers that you were clever enough to do it. But once it became possible to cause a crash by just entering a command, then you were not showing off to anyone any more. For Lin, it was an example of the psychology of hacking and of how human rather than technical responses were sometimes the solution.[19]

The more connected a computer becomes, the more powerful it may prove to be for the user, but also the more vulnerable. Connections with other computers immediately introduce an element of risk. This was clear from the start, but became increasingly clear as computers and communications merged over the coming decades. 'I can design for you a perfectly secure computer but you would look at it and call it a brick,' says Herb Lin. 'It is a useless computer – it doesn't take any information in or take any information out but it is perfectly secure. What you really want is a functional computer that is secure. The problem is that you have to put information into it to make it useful . . . If I can fool you into putting bad information into it . . . then I win the game.' The more useful a computer is – by being easy for it to talk to other computers and trusting them when it does, rather than demanding complicated security checks – the more vulnerable it is. And the human factor remains the key. One of the key differences between humans and computers is trust. Connecting with another person (whether socially, professionally or intimately) can be richly rewarding but also dangerous, so people have learnt to make judgements on whom to trust. These are subtle and learnt by individuals and societies. Computers are not so good at such judgements when they are asked to connect with other computers. They can be given rules – do not connect with another computer coming from this location or which is bearing this piece of code – but those rules can easily be subverted by a clever person at the other end. Computers lack the ability to make the kinds of judgements which humans can. Put simply, machines are not so good at telling the good guy from the bad guy when the bad guy can disguise himself. 'Deception is what all the threats to cyber security are built on,' argues Lin. Technology can help you make judgements about what is safe, but since a clever human is normally behind the bad code that is being sent over the internet, it takes a very human understanding to spot it and defeat it.

Hackers might be tricksters. But could they also be spies? What opportunities might all these vulnerabilities offer them? Again, Roger Schell was one of the first to wonder about just that. It was 1979 and a KGB officer was addressing a select group of Soviet officials with not just secrecy but an added sense of excitement. 'Comrades,' he began, 'I will brief you about the most significant breakthrough in intelligence collection since the "breaking" of the "unbreakable" Japanese and German ciphers in World War II.' This breakthrough was, he explained, 'the penetration of the security of American computers'. Schell imagined the KGB man then explaining that there was virtually no major American national defence secret that was not stored on a computer somewhere. And almost all of those computers were accessible, at least in theory, to the KGB's prying eyes. What was so wonderful, he said, was that it was no longer the case that they needed to wait for information to be communicated so that it could be intercepted in transit and decoded (as Bletchley had been forced to do). Instead, the data was at rest rather than in motion, just sitting there waiting to be accessed. All this could be done, he said, with 'virtually no risk to our agents'. Schell was just imagining the scene when he wrote an article. He was trying to make a point: that computer security was the 'Achilles' heel' of an electronic air force. Technology, he explained, had always been a mixed blessing, bringing opportunity but with it vulnerability. Everyone could see the opportunities, but he had seen enough of the dark side of the penetration of computer systems to know about the dangers. What was different about computers, Schell understood even more clearly in 1979 than when he had first thought about the problem as a young captain in the 1960s, was that they offered the chance to do much more than intercept data and gather intelligence. If you could get into a system you could actually change and modify the code. In others words, you could make computers do things they were not supposed to do. Fire a missile the wrong way. Switch off a radar defence system. This was more than just traditional espionage: it was raising the possibility of sabotage and subversion.[20] Funding for Schell's proposed solution had just been cut, which made his anger all the more clear. Higher-ups seemed worried about cost, or pointed to the fact that no enemy had actually yet been found inside a system carrying out espionage. Schell's warning was stark. If America did not appreciate these dangers, then it would be like Germany or Japan in the Second World War, complacently thinking its machines (then Enigma, now computers) were safe when in fact the enemy was already in.

The vast, heavily fortified complex buried within Cheyenne Mountain in Colorado had been studied by the Anderson Committee back in 1972 and was the place where fears of the consequences of computer security first took root in both government and public consciousness. It was home to NORAD – the North American Aerospace Defense Command. Data from sensors located all over the world flowed into NORAD to look for missile launches and other signs of attack. Computers sifted and analysed all that data to help the commander know if America was under attack. Even in 1972, NORAD had what was described as a 'highly sophisticated wall-size display system' with twelve-by-sixteen-foot screens showing what was going on and fifteen individual consoles for the staff working at the command post. It was noted by Anderson that there might be some excess computer time available, which 'may be available to devote to war-gaming and other Top Secret activities'. There would also be two remote terminals in secure areas that would be connected into Cheyenne. The terminals were secure but the incoming communications lines, in some cases, were not secure, it was noted, and there were plans to connect more systems together.

Hollywood took the twin fears of nuclear Armageddon and the power of computers and merged them with concern over teenage rebelliousness in the 1983 film *WarGames*. In this hugely popular and influential movie, a young hacker nearly starts off thermonuclear war by getting inside the NORAD systems and playing what he thinks is a game without realising his simulation is being played out for real. The film had an impact on President Reagan, who began asking questions about what might be possible. The media – fascinated by this new creature, the hacker – started to seek them out. A group of sixteen-to-twenty-year-olds called 414 (after the Milwaukee area code) managed to break into supposedly secure systems like Los Alamos Lab. That led to an FBI investigation and one of the group appearing on the cover of *Newsweek* magazine. These rebellious teenagers were attracting an oddly glamorous mystique.[21] People started raising questions about how fictional the Hollywood scenario really was. Amid public concern, six Bills about securing computer crime were introduced into the US House of Representatives in the aftermath and in 1984 a top-secret National Security Decision Directive was signed to improve the security of sensitive systems. Over the next few years there was a major debate as to how far the NSA should be involved in protecting government information. There were many in Congress who disliked a foreign intelligence branch of the military taking on a domestic role, and eventually it was settled that it would only protect military networks and other national security data but not the rest of government. Already at issue was a central question: what did it mean to defend a country in this strange world of computer networks which seemed to exist in a different space? Amid all the concern after *WarGames*, a real-life general was interviewed on ABC TV. He tried to reassure people that computers could not start a war. 'Man is in the loop,' he said. 'Man makes decisions. At NORAD, computers don't make decisions.'[22]

But the *WarGames* scenario was not entirely fantasy. It was, at least in part, based on real events. On the morning of 9 November 1979 the Honeywell computers inside Cheyenne Mountain flashed a signal. The operational duty officers at NORAD were watching their worst fears come true. Soviet missiles had been launched off the West Coast by Soviet submarines. More missiles appeared on those vast screens, this time coming from the Soviet Union itself. What looked like a wave of destruction was about to descend on the US. Bomber crews were scrambled and missile crews told to ready for launch. For precisely six terrifying minutes it looked like the Third World War was under way. The National Security Adviser was woken to be told the end was nigh.[23] But senior officers knew there was something strange about it. A surprise attack at a time of no real tension? When they contacted the radar stations that had been providing the data on which the computers worked, those stations relayed back that they had not seen any missile launches. It quickly became clear that this was not real. It was fortunate the systems were not totally automated. People had to exercise their judgement and they had separate communications channels to the source of the data. It turned out that a technician had fed an incorrect tape containing a training exercise into NORAD's computers. A false alarm. But also a warning.[24]

CHAPTER FIVE

SPY-HUNTING

A jumble of random letters streamed out of a Russian radio station and made its way to the furthest reaches of the world. In attic rooms in Ruislip, in suburban London, and garages in Rochester, New York, undercover Soviet spies would listen in at an agreed time. They would pull out their one-time pad from its hiding place and use pen and paper to decode the message from their KGB controllers providing their latest instructions. Venona had only given Western spy-hunters a glimpse of past communications, and they lay awake fearing what treachery was being planned from Moscow. Could computers help them discover the truth?

Catching spies through their signals or communications had been done for centuries. During the First World War, in the ditch of the drained moat of the Tower of London just after dawn on 30 July 1915, two men, Janssen and Roos, who had come to Britain as cigar salesmen, were shot. Intensive monitoring had discovered telegrams heading to an address in the Netherlands believed to be a front for the German Secret Service. References to orders for Havana cigars were interpreted as code referring to British military vessels. Locals did not smoke cigars in the quantities the men were ordering, nor the type, experts would testify at their short trial.[1]

In the Cold War, hunting for spies would slowly come to rely on data provided in new ways, eventually by computers. At the start of the 1960s, the NSA sent over to GCHQ a huge swathe of the raw Soviet radio traffic it was collecting. On the surface it looked like gibberish. A young GCHQ analyst (and future Director) processed thousands of these broadcasts from hand onto a computer. With this done, the computer could then run what was called 'cluster analysis' to isolate similarities in the traffic. This was used to look for patterns which might help identify who was sending and receiving the messages. The content might not be readable, but perhaps the data about the communications (what today is sometimes called metadata) might help identify someone who sought to hide.[2] The correlation of fragments of information with signals was the work of the real-life George Smileys, John le Carré's fictional spy-hunter. These techniques would lead to Gordon Lonsdale, a Canadian jukebox salesman who was really Konon Molody, a KGB 'illegal' working under deep cover, and his contacts Peter and Helen Kroger, posing as antiquarian booksellers while they sent back to Moscow secrets provided by British traitors. Laborious detective work has always been at the heart of a successful counter-intelligence agency, much more than glamorous gadgets like guns and bugging devices. A domestic security service (like MI5 or the FBI) rises or falls on its ability to collect and collate information about those it is interested in. Stacks of paper folders were built up and an officer would take a trip to the 'Registry' to request a file before setting off on their investigation. This might then involve looking for traces people have left – following a paper trail of birth certificates, passports and employments records and the like – and then trying to piece together whether someone was who they said they were. Identity – and the hunt for false identities – was at the heart of this. A search engine in the days of Smiley involved a person's two feet and a lot of paper. In the early Cold War, technical means of collecting communications signals were a part of spy-hunting for MI5. One way of getting hold of them was to target the embassies of other countries in London where foreign spies might be running agents. That was usually done by bugging and could offer a wealth of intelligence. It might even offer a way around those hard-to-break codes that other countries used.

Hugh Alexander, Bletchley veteran, former British chess champion and now a senior GCHQ official, was briefing a new MI5 scientist named Peter Wright in Cheltenham. Wright recalled Alexander gesturing over to a row of huts being built behind the main complex and explaining the challenge his growing organisation faced. 'Our problem is that our theories are running beyond our computer capacity,' Alexander said. 'So many ciphers we could crack – we know how to crack them. We just don't have sufficiently powerful computers to do the job. We'll get them soon, of course, but in the meantime, any help may give us a short cut.' A top target was the Hagelin machine, one of the cipher machines most commonly used by developing countries. By pretending to be a telephone engineer, Wright got into the Egyptian Embassy in London where he managed to install a microphone inside the handset of a phone in close proximity to a Hagelin machine. This was 1956, the year of the Suez Crisis, when Britain, France and Israel colluded to attack Egypt. The secret microphone broadcast the clicks of the cipher machine which, Wright had worked out, could be picked up and used to work out the initial key setting with which the machine had enciphered a code. As a result, all the messages of the embassy could be read.[3] It was a clever way round the need for a computer to break a code. As they explored these techniques (including against the French Embassy in London), Wright and others in the spy world were learning about the first known technical vulnerability of electronic communications.

Hidden microphones were being found in American and British embassies in places like Moscow, concealed in all kinds of clever places. But there were more than just audio bugs to pick up conversations or even the clicks of a cipher machine. There were other gadgets attached (this was kept secret at the time) which were there to pick up tiny pulses of electromagnetic radiation – radio waves. As a result of this discovery, those at the heart of the secret state would become deeply worried that their computers were being spied on by the Russians in the decades ahead. But not by what we would think of as modern computer espionage. All electrical devices – whether crypto machines or computers – emitted tiny amounts of radio frequency when the different switches and buttons were deployed. These would not carry far: perhaps only a few hundred feet. But an adversary who could get close enough with an antenna might be able to pick them up. And because each different switch or key press emitted a subtly different signal, you could potentially decipher what was being written on an electrical device, including a crypto machine enciphering a secret message. This would allow you to read the content of a message before it was encrypted – bypassing even the best code in the world. In modern parlance, you were going for the endpoint rather than the communication in transit. And this trick also applied to computers.

This 'compromising radiation' was given the cover name 'Tempest' and, decades on, aspects of it still remain classified.[4] GCHQ vans with special antennae ran tests outside government offices. Special masking was developed to try to prevent signals escaping. Strict standards were put in place for the most sensitive computers in London for fear that the Soviets might be parked up outside collecting the signals. When MI6 eventually put its records on a computer, the room would be heavily shielded.[5] This slowed the adoption of personal computer terminals for many spies and senior officials in London all the way through until the 1980s. Trying to shield the emissions was one of the earliest forms of electronic or computer security (although it extended to all mechanical machines) and remains a technique used today.

In America and Britain, there was a paradox about computer security. The agencies with the most advanced computers – GCHQ and the NSA – were often behind others in understanding the risks posed by computer networks. This was because, while their computing power was massive, they were so scared of penetration that they shielded themselves from the outside world. When one computer expert left MIT for the NSA in 1967, he was stunned by what he found in the basement.[6] The NSA had 100 computers occupying almost five acres of floor space.[7] The agency would grow its computing power by 50 per cent every year over ten years from 1963 to 1973 to keep pace with the demand of work.[8] This capacity would increasingly come from the private sector, which the NSA realised was pulling ahead.[9] IBM remained the big player, providing around two-thirds of general-purpose mainframes in America and half around the world by 1979.[10] A small new company called CDC developed a super-fast circuit and became a favourite for building special-purpose machines for the spies, its engineers being given the highest security clearances. One of CDC's key players, a mercurial engineer called Seymour Cray, would go on to form his own eponymous company and build giants like the CRAY 1, machines so powerful they were called 'supercomputers'. Cryptanalysis remained the primary task, taking up about 40 per cent of NSA computer usage, with the 'High Compute Complex' housing machines like the CDC 7600. Meanwhile, the Central Data Processing Complex was the successor to the punch-cards first used in the 1930s and forty years on was still growing at a considerable pace, with plans for a new IBM system called Oak which offered 169 billion bytes of storage.[11]

Britain was doing its best to keep up. Most of the British government had to buy British when it came to computers. One place that was given an exemption was GCHQ, since it clearly needed the fastest and best devices for code-breaking and so by the 1960s it was relying on American machines. The IBM team had their own office and workshop at GCHQ and resisted any other technicians getting a good poke around their systems for fear the technology would be stolen. 'Some of the big IBM machines could gobble up work faster than our old donkey engines,' one GCHQ veteran recalled.[12] At one point in the 1960s, GCHQ was being pushed to use British machines from ICL for some less sensitive jobs. Under duress, they agreed to try them for administrative tasks. But when they attempted to run the payroll through the British machine it took what seemed to be days rather than minutes, leading to much grumbling from staff about their salaries arriving late. IBM machines would become the backbone of general computing at GCHQ, supplemented by supercomputers from companies like Cray for cryptanalysis. Experts at GCHQ did contribute to the chip design and the maths manoeuvres these specialist machines were being built to undertake since that remained, as in Turing's day, a particular field of expertise. But, unlike the days of Colossus, it was now married to American private-sector engineering. These supercomputers were very expensive and GCHQ had to carefully pitch its justifications to the Treasury for the investment they wanted in a new machine for cryptanalysis every couple of years. GCHQ officials would say that if the government did not cough up the sometimes tens of millions of pounds needed, they would forfeit the capability to keep doing what they were being asked to do.[13] The Cray supercomputers are remembered by staff as being not just costly but also beautiful – works of art in their own way, with expansive water-cooling pipes running around the central tower system that were covered with foam cushions, which provided a perfect seating area. And so when one of the old Crays was due to be thrown out, the engineering staff instead gutted it and then put the shell in the tea room so that people could sit on it and enjoy their morning cuppa. A nice cup of tea still had its role in British computing.

Security fears meant spy computers were not allowed to talk to strangers. GCHQ's and the NSA's computers remained – almost without exception – air-gapped and separate from the internet and the outside world. That meant they did not worry that much about being spied on or subverted, unlike the US Air Force. In the 1970s, GCHQ's only real network connection was to the NSA. This explains why, as in the US, the military started worrying about computer security first. Britain's Ministry of Defence had communication lines which stretched internationally to places like Germany, where forces would be deployed, and dealt with everything from command and control to air defence. A team in Malvern (where wartime work on radar had been pioneered) started working with Don Davies' NPL in Teddington from the mid 1970s, trying to understand the security implications of networking. Over at the NSA, internal networking began with a 1972 NSA system called RYE, through which 200 remote terminals could access four main machines.[14] The problems of 'computer support' were also emerging, as evidenced by a humorous account in a 1975 NSA in-house staff journal entitled 'The Yawn of the Computer Age – Or When your Terminal is Terminal', which moaned about the awful excuses when computer systems went down a few times a day and suggested a competition for the best twenty-second rant in response.[15]

The security fear for the NSA and GCHQ throughout much of the Cold War was not an attack over the network but human spies within the agencies who could give away secrets and devastate capabilities (two NSA staff, William Martin and Bernon Mitchell, who went to the USSR in 1960 being among the most damaging, along with Geoffrey Prime from GCHQ in the 1980s). The danger was perceived as the targeting of staff by foreign intelligence agencies who would then be able to access classified material through the poorly secured computers, rather than people entering into the network electronically.

One NSA official who was involved in a two-year project examining the systems used by the NSA, the Pentagon and other federal agencies was shocked by what he found in the early 1980s. It had been claimed that the security of systems was 'ironclad and invulnerable'.[16] But across government, passwords were predictable, often taped to terminals and sometimes shared with unauthorised persons. 'At best, such practises can be labelled sloppy; at worst, they are an outright invitation to compromise.' The official thought it amazing that staff would undertake a panoply of physical security measures like locked doors, sign-in lists and escort requirements to get access to an office where sensitive data was being held, but then happily put the same sensitive data on an unprotected computer file space accessible to lots of other people. The official ended with a stark warning: 'The commercial computer world is replete with incidents of embezzlement, intrusion, deception, thievery and sabotage. Can we honestly expect less of a threat to our computers?' 'The development of computer networks was going much faster in the US,' says Victor Sheymov, who worked on communications security for the KGB until he defected to the US in 1980. 'So the exposure to security risks was much higher in the US. In the Soviet Union, it was less of a concern. They were very conservative about putting sensitive stuff on them.'[17]

Stealing secrets over computer networks was not yet something anyone in GCHQ or the NSA had considered in the middle years of the Cold War. Gathering intelligence relied on intercepting Soviet radio traffic, and particularly planting dishes around the world to intercept the new method of transmission – microwave satellite communications. But it was the oldest of methods, dating back to the First World War – cable-tapping – that thrust GCHQ and the NSA into an uncomfortable spotlight.

Britain's exposé came first. On the afternoon of 16 February 1967, a man called Robert Lawson went to see the legendary *Daily Express* Defence Correspondent, Chapman Pincher, a well-connected journalist well versed in breaking scoops on spies. While working in cable telegraph offices, Lawson had discovered that the telex traffic of all companies was being collected every day by a van belonging to the Ministry of Public Works and then driven off to the Admiralty building before being returned forty-eight hours later. The Ministry of Defence denied this when Pincher asked if it was true. That denial was possible because it was not them doing the examining – it was of course GCHQ (which was looking at international rather than domestic traffic). But word that Pincher was asking questions spread panic across Whitehall. In true British style, the Foreign Secretary called up the editor of the newspaper, who was dining at the Garrick Club, and asked him to stop the story (he had to talk in guarded terms because it was an insecure phone line). It was too late. There in black and white the next morning was the front page: 'Cable Vetting Sensation, Security Check on Private Messages out of Britain.' It went on: 'This "Big Brother" intrusion into privacy, which ranks with telephone tapping and the opening of letters, was disclosed last night.'

Cable messages going in and out of the country were secretly being copied and sifted by the security services. The article said that this was conducted under a special warrant, signed by the Secretary of State under Section 4 of the Official Secrets Act and regularly reviewed. A clause had been inserted into the 1920 Act allowing ministers to sign warrants demanding co-operation from cable companies. 'It is important to leave this part of our activity to the deepest possible obscurity,' the Foreign Secretary noted at the time.[18] But such was the obscurity that when officials looked into this in the 1960s after Pincher's exposé, they were embarrassed to discover the entire practice relied on the legal authority of Home Office warrants which had been signed in 1921 and which had never actually been renewed since. What was more, no one could find a copy of the original warrant to check what it authorised. The Home Secretary of the day was totally unaware of the authorisation process until he read it in Chapman Pincher's article. Officials suddenly realised that the whole system might not be legal at all. A huge storm gathered, with questions asked in Parliament and carefully worded denials issued in response. But it soon passed. And the practice returned to the darker recesses from which it had briefly been dragged.[19]

The NSA had the CIA's penchant for risky covert action and assassination to thank for its exposure in the mid 1970s. The CIA was taking heat for plots against Cuba's Castro and Congo's Lumumba, two countries that Washington feared might fall to Communism and where it had gone as far as plotting the assassination of the two leaders. But even more toxic was the revelation that what was supposed to be a foreign intelligence agency had been spying within America on Americans. Congress established a committee chaired by Senator Frank Church to investigate. Two of the younger investigators on the committee were given the daunting task of seeing if they could unearth anything about the NSA. Thirty-year-old Britt Snider asked the Congressional Research Service for everything they had on the organisation, whose name was often joked within government to stand for 'No Such Agency'. He was given a one-paragraph description from a government document and a *Rolling Stone* magazine article that was patently full of errors. When he asked retired NSA staffers about any abuses, they only gave him a litany of complaints about the way parking spaces were allocated.[20] But within a set of files about the CIA, he and his colleague unearthed two small references which suggested that the CIA had provided an office in New York for the NSA to copy telegrams. It seemed as if the CIA had asked the NSA to monitor the communications of some Americans involved in the anti-war movement. They had stumbled upon Shamrock.

It was a Saturday afternoon when Britt Snider knocked on the door of Louis Tordella at his suburban home in Maryland. Although technically the number two, the lanky Tordella had effectively been running the NSA from 1958 until he retired in 1974, the year before Snider's visit. He greeted the young investigator politely. But he was clearly uncomfortable at Snider's questions about Shamrock and asked him what he knew. When Snider had finished, Tordella sighed a long sigh and talked into the early evening. He explained that Shamrock was essentially a continuation of the Second World War practice of copying all international telegraphic traffic to look for intelligence. Every day a courier took reels of magnetic tape from three telegraph companies from New York down to the NSA which would then be processed, looking for anything of interest linked to an intelligence target (like a foreign government) or simply because it was encrypted. Tordella implied that many senior officials in the government may not have known the details and said that even within the NSA only one person below him had any responsibility for the programme (and that individual was only the second person ever to hold that job). The material was mainly used for regular foreign intelligence purposes, Tordella said. But when he was asked if it was used to spy on the communications of Americans, Tordella replied 'Not per se,' but suggested it might have happened in a few cases. In fact, it later transpired that hundreds of American citizens were on an NSA watchlist which Shamrock traffic was matched against. This programme, known as Minaret, had originally been drawn up to look for people with links to Cuba, then those who threatened the President, but later expanded. Presidents Lyndon Johnson and then Nixon were convinced a foreign Communist hand was behind the race riots, anti-Vietnam protests and other unrest that was creating confrontation on campuses and sending American inner cities up in flames. As a result the CIA and NSA were all turning towards domestic surveillance.

Technology and computing were also offering something new. The shift of the source material from paper to magnetic tapes in 1963 had made a big difference to the NSA. It allowed the agency to put the data through its new high-end computer system – Harvest – and do something we only associate with modern computers. The machine could work through all the traffic, searching for keywords such as addressees or senders or other items of interest – keyword matching. It could scan 7 million messages for any of 7,000 key words in under four hours. This was a steep change in capability as the whole process of collection and processing became increasingly automated.[21] The volume of traffic was too large for people to comb through, as they had done in First World War sorting offices. Now a computer could hunt through the material looking for the match of a name and address from the blacklist and flag it to a person.

But was it legal for the NSA to read the telegrams of American citizens? 'You'll have to ask the lawyers,' Tordella replied. He went on to explain that what worried him was losing the co-operation of the companies. They did it for patriotic reasons, he explained, and feared that exposure would make them reluctant. When the three companies were approached some were unapologetic, others brought in a phalanx of corporate lawyers for a meeting with Snider. The President personally – and unsuccessfully – called members of the Church Committee to lobby them against making the role of the companies public.

After the report was published, the NSA would 'find' more documents. These showed that when the Second World War was over, the army feared losing its relationships with companies that provided cable traffic. 'If we broke them off and then had to go start them again, you see, that would be just like building the world all over again,' said Frank Rowlett, who could recall the informal system that went back to Friedman's in the 1930s.[22] Companies had been told that the Attorney General had said the operation was 'not illegal' if there was a foreign intelligence purpose. However, an NSA memo later revealed that there may not have been such an opinion.[23]

Shamrock was said to have been halted by the time of the Church Committee, but in reality it simply evolved with more formal, but still secret, arrangements with companies. When President Carter expressed surprise in the late 1970s that a similar system was still being run, the NSA Director made a point of removing all the intelligence derived from that source from White House briefings. It was soon reauthorised. Out of the controversy came a new 1978 law to protect the privacy of Americans – the Foreign Intelligence Surveillance Act – in which a special group of judges would have to approve warrants for the electronic surveillance of US persons. A quarter of a century later, an updated version of the Shamrock programme would be instituted in the wake of a terrorist attack on the United States. It would prove even more controversial when revealed. The common thread of these programmes was the search for persons unknown and the hidden hand behind them by examining communications, as had been done in the First World War. They might be spies, they might be terrorists, they might be activists communicating with foreign powers. Computers were slowly transforming the way this could be done by looking at communications and also, increasingly, other forms of data.

In America, a CIA veteran used to tell people that when he started at Langley, the CIA headquarters, there was one computer in the basement and one lawyer on the top floor, but in the subsequent years each had spawned many offspring until, he feared, eventually they would simply meet in the middle. The filing cabinets holding all the data for spy-hunters were slowly being superseded by computers. The FBI began looking at computerising its own (mainly criminal) files from the late 1960s – although its computer systems would prove desperately backwards until well into the twenty-first century.

In the mid 1970s, MI5 in Britain started to hook up into computer databases as part of its efforts to find Soviet spies and also keep tabs on 'subversion'. A 'Computer Working Party' looked into how to connect their computers directly to the National Insurance system in Newcastle so that people's records and identity could be checked almost in real time rather than asking an undercover officer based up there to check paper files and ferry the records down the country.[24] At this point, the information that was needed was still government data, not private data (communications were in private hands through phone companies, but they were subject to warrants allowing a pair of copper wires to be attached in the local exchanges by the men in grey overcoats). Old hands like MI5 technical expert Peter Wright, who saw plots and subversion everywhere, felt wistful about the past as they witnessed the rise of computers and what they saw as the 'relegation' of the individual when it came to this kind of detective work. 'From now on, we were to be data processors, scanning tens of thousands of names at the press of a button,' Wright would say. 'The fun has gone,' some said.

The process of computerisation was still slow. Although the index for its legendary Registry of files began to be computerised in the 1970s, it was only in the mid 1990s that the MI5 files themselves were digitised, ending the days of trolleys of paperwork being wheeled around. Computers and databases held out the promise of being able to collate and systematise information about a place, a person or a subject. Like all technology, this could be used for good or for ill, depending on where you sat. Definitions of subversion could be elastic and 'counter-subversion' and spy-hunting could shade into domestic surveillance. In Britain there was never quite the controversy over mass surveillance that the US saw in the 1970s, but there were certainly some groups – like peace activists, anti-nuclear campaigners and trade unionists – who felt they were being targeted (again, often on the pretext of looking for a foreign, secret Communist hand behind them), and even former heads of MI5 concede that too many files were opened on too many people.[25]

East Germany's domestic spy service – the Stasi – went to the extreme of opening a file on pretty much everyone in its desire to maintain Communist control. But it took a different approach when it came to computers. The head of the Stasi, Markus Wolf, resisted pleas by 'bright young things' to computerise the organisation's endless library of files, insisting they remain on paper cards.[26] On these cards were the endless personal details of much of the country's population, recorded by hand. They were based on individual informers and bugged conversations, husbands spying on wives and agents listening in to phone calls to create a police state which knew what everyone was doing. Wolf's fear was that if they were computerised, the records could then be accessed too easily, potentially revealing the names of agents. He insisted that the most sensitive details were not even put down on paper but kept in his and a few other people's heads. 'Time and time again, experts tried to persuade me that a computerised system was fail-safe. It always sounded convincing until a few weeks or months later, when a newspaper report would appear about some twelve-year-old hacking into a military computer from his bedroom. I have never trusted computers.' As a result, the Ministry of State Security files went on for miles. Literally. About one hundred kilometres in all. That was what could happen when the concept of 'domestic security' was stretched but not yet computerised. What if the two concepts of authoritarian domestic control and computers were combined?

In the 1970s, the CIA, NSA and FBI had edged into monitoring domestic dissent and anti-war activity as well as finding Soviet spies. This was nothing like the scale or the brutality of the Stasi, but still proved controversial when exposed. Files were created on something like a hundred thousand Americans, their names being scanned by the computers which churned through the telegraph traffic. The Church Committee raised an important question. The power of intelligence agencies was growing in secret. It was designed to collect foreign intelligence. But what if that changed and was combined

with advancing technology? Senator Frank Church issued a powerful warning that outlined both the utility and dangers of combining
computers and spies. It bears quoting in full:

> In the need to develop a capacity to know what potential ene-
> mies are doing, the United States government has perfected a
> technological capability that enables us to monitor the messages
> that go through the air. Now, that is necessary and important
> to the United States as we look abroad at enemies or potential
> enemies. We must know, at the same time, that capability at
> any time could be turned around on the American people and
> no American would have any privacy left, such is the capability
> to monitor everything. Telephone conversations, telegrams, it
> doesn't matter. There would be no place to hide. If this gov-
> ernment ever became a tyranny, if a dictator ever took charge
> in this country, the technological capacity that the intelligence
> community has given the government could enable it to impose
> total tyranny, and there would be no way to fight back, because
> the most careful effort to combine together in resistance to the
> government, no matter how privately it was done, is within
> the reach of the government to know. Such is the capability of
> this technology. I don't want to see this country ever go across
> the bridge. I know the capability that is there to make tyranny
> total in America, and we must see to it that this agency and all
> agencies that possess this technology operate within the law and
> under proper supervision, so that we never cross over that abyss.
> That is the abyss from which there is no return.

CHAPTER SIX

CRYPTO WARS

A conference room inside the sprawling Stanford University campus in northern California was an inconspicuous place for the opening salvo of a war. And the combatants sitting across a table in 1976 were not exactly the type you would normally find on a battlefield. On one side were a pair of twenty-something West Coast-based researchers – Martin Hellman and Whitfield Diffie. Diffie's long hair was just one sign that the two were definitely part of the 1970s and not part of the establishment. On the other side were three men in suits who had come over from the East Coast. Two came from the NSA. The other said he came from the National Bureau of Standards, but in fact had just transferred from the NSA. The most senior of the NSA men (technically retired, but still very much involved) was Arthur Levenson. More than thirty years before, he had been one of the select Americans who had worked at Bletchley Park during the Second World War. He had gone on to become one of America's leading code-breakers, running 'A Group' which targeted the Soviet Union, and most recently the Machine Processing Organisation, which ran the NSA's computers. The NSA had ballooned in size, not just to deal with the Soviets but also to support the ill-fated military campaign in Vietnam. But now the American national security elite were on the defensive. There had been the disaster of Vietnam, with Saigon falling the previous year to the Communists. There had been the Church Committee raising fears of what the spies might be up to at home. And Washington as a whole was reeling from the Watergate scandal that had forced President Nixon from the White House after his abuse of institutions had been revealed. A lot of people outside the secret world had come to the conclusion that power was not to be trusted. And sitting across the table from Levenson and his NSA colleagues were a pair of young upstarts determined to lay siege to the NSA's innermost secrets.

That morning session at Stanford was recorded on cassette tape (by Paul Baran, the RAND analyst who had worked on packet-switching concepts more than a decade earlier). The young academics asked if they could and NSA offered no objection. 'It's appropriate that they approve their own wiretapping,' joked Diffie. On the tape the scratch of chalk is audible as Hellman goes into battle by writing up numbers on a blackboard to make his case. At dispute that day was the seemingly obscure topic of whether a proposed government encryption standard had been made deliberately too weak. The argument began with maths. But really it was about much more. It was the start of a forty-year war which continues to the present between the state and computer experts about privacy and national security.

There was a culture clash at that meeting. Not only between the uptight East Coast suits – the men in black – and the West Coast hippie academics, but also over ideology. To the NSA, cryptography was their domain, vital in protecting the country, and had to be kept secret. And now suddenly these two outsiders were asking questions, causing trouble and endangering national security. Diffie and Hellman saw themselves as fighting the NSA's presumed monopoly because they believed people – and not just the state – had the right to a high degree of privacy and security for their communications. 'My view at the time was I was Luke Skywalker in Star Wars,' Hellman recalls, 'and NSA was Darth Vadar.' In other words, he was the hero who was going to take on the might of the evil empire almost single-handed.[1]

At stake was an issue that was going to become a defining one for security in the computer age and the source of intense friction between spies and the outside world. How secure was information going to be on computers and could the state be trusted to secure it? The meeting at Stanford was the beginning of a struggle over encryption that is still gathering pace forty years later. For centuries, diplomats, generals and spies had worried about keeping their secrets, but it was less of a concern for everybody else. By the 1970s, however, computers had moved out of the basement of the NSA and were becoming more common in government use. They were holding more and more information in one place in the form of databases. Some of this was sensitive, but not what was traditionally thought of as secret or highly classified data – for instance, people's tax and health records. The private sector was also increasingly holding sensitive information on computers it wanted to protect. The obvious answer was for the near-magical power of encryption to move outside the secret realm where it had resided.

In the late 1960s, a naturalised German engineer for IBM called Horst Feistel came up with a computer key generator to encrypt data used in government databases. This would be a physical device with a computer programme in it which would generate a key to scramble a message – a kind of electronic version of a mechanical Enigma machine. It was christened 'Lucifer', a strange tribute to its power, perhaps. At first there had been no interest but then the first customer stepped forward. As with the Enigma machine, it was a bank. Lloyds Bank in London was building a network of cash machines that would spew out money after a computer contacted another to validate the request. It was obviously important to be sure that this process was secure, otherwise it would be a magnet for thieves. Previous commercial ciphers had proved easy to crack: in 1963, Donald Davies of the NPL in Teddington had broken the first magnetic stripe card for an ATM in a matter of hours.

IBM realised a lot was riding on Lucifer when it came to public and commercial confidence as well as the reputation of the company. To test their key, they asked a team of people to see if they could break it. This was one of the earliest cases of commercial 'white hat hacking' or penetration testing – asking someone to test whether you are as secure as you hope you are. After seven weeks of trying, the hackers failed to get into Lucifer.[2] Next up was the NSA. It could see that as IBM and others wrestled with ways to break this code, they were inevitably moving towards forbidden territory – the techniques it used to attack foreign encryption systems, which it considered its deepest secrets. The NSA wanted to ensure it had some control over the process, and so suggested that it both test Lucifer to make sure it was secure and oversee its distribution. It was a way of keeping control. A few lines in a government publication called the *Federal*

Register in March 1975 announced that a regulatory body called the National Bureau of Standards (NBS, now called NIST) was going to set a new encryption standard for the government to secure its information. This was theoretically open to competition. The victor would inevitably become the default security standard for the wider world, including business. IBM would be the only real candidate and so discussions began about adapting Lucifer.

This is where Whit Diffie and Martin Hellman entered the fray. Hellman was one of the tiny band of people outside government who had become fascinated with cryptography. His colleagues told him he was mad to work in that field, but he persisted. He had worked with Feistel briefly at IBM before going to Stanford in 1971. He had grown up in the Bronx but California offered more freedom. At conferences, people with badges suggesting they worked for the 'US Government' would ask him if he was interested in doing some consulting for them. But when he asked if he would be free to publish and they said no, he decided he did not want to work for what was clearly the NSA. There was a bit of him that relished the challenge of taking on the big guys as well. At Stanford, a colleague had recommended he talk to another young man who shared his fascination. When he met Whit Diffie, Martin Hellman immediately knew he had found an intellectual soulmate.

With his long hair and immaculately tailored London suits, Diffie is a free-thinking, idiosyncratic character who, in many ways, epitomised the counter-cultural challenge to the establishment. Conventional academic work had not suited him, but he had ended up at MIT working on computer programming in the 1960s. In part he was looking at the issues Alan Turing had studied: how to define intelligence in machines. But while doing his research on computers he became interested in how to keep his own information private – partly the result of games played between fellow computer researchers who tried to find out what each other was doing. This was the fundamental impulse behind encryption: stopping other people prying into your data. But details of the way in which codes could do that were a kind of forbidden knowledge, kept in the confines of the secret world out of reach of ordinary mortals. That intrigued but also angered Diffie. Diffie and Hellman's first meeting went on for hours as they headed back to Hellman's house to keep talking late into the night. The two men could see that the growing marriage of computers, data and communication brought with it the need for the average citizen to protect information like medical and banking records. The US was the most computerised country in the world. It had the most to lose by insecure computer encryption.

When the NBS issued details of the proposed federal Data Encryption Standard (DES) – a modified version of Lucifer – Hellman and Diffie began poring over the details. They thought they saw something fishy. First of all, the design of the 'S Box' which housed the mathematical algorithm to encode the data was secret. They started wondering whether there was something the government did not want people to see. A relative had told Hellman that the NSA had secretly 'colonised' the NBS. It seemed to the young Stanford men as if this dark, secret, malevolent force was working behind the scenes in the shadows. Could they have installed a 'trapdoor' in the new system for their own ends?

'We came up with the idea of trapdoors from the Hardy Boys or other mystery books I'd read as a kid,' Hellman says, citing a series of children's adventure novels. 'There was some tomb I think they were stuck in, with a million bricks. And if you press on the right brick a door opens and you'll survive, otherwise you die of thirst in this tomb. The designer knows which brick to push and can survive.'[3] To anyone else, the room would look secure. But the person who knows where the trapdoor is simply has to open it to get in.

There was something else, though, which was much more obvious. The strength of the encryption seemed much lower than it needed to be. The original Lucifer key was 128 bits. The version being proposed had been reduced much further, first down to 64 and now to 56 bits. A way to think about 56-bit encryption is to imagine 56 switches, each of which could be on or off. That may not sound like a lot, but the possible permutations are 2 to the power of 56, which comes out at 70 quadrillion. An increase of one bit doubles the strength. So a 57-bit code, not a 112-bit code, is double the strength of a 56-bit code. So choosing a 56-bit code made it a lot weaker than a 64-bit code, let alone a 128-bit code. Why?

Seventy quadrillion still sounds like a lot, but Diffie and Hellman thought that the way computers were heading they would be able to cycle through those possibilities in a realistic time frame by doing a brute-force attack. That might not be possible for criminals yet, but Diffie and Hellman thought computing power was growing so fast that it might not be long before they could. And someone with a lot of supercomputers in their basement might be able to break such a code right now. Who could that be? Everyone suspected but few knew for sure that the NSA had built hugely powerful cryptanalytic computers to do just that kind of calculation.

So that morning at Stanford, Hellman laid straight in. 'The first thing we'd like to hit is the key size of the proposed standard,' he told the suits. 'We feel it's too small, both today – we feel it's vulnerable to attack by an agency such as NSA, although quite secure against commercial assault . . . And we feel that, looking fifteen years down the line, and extrapolating the decreasing cost of computation, the standard would be insecure against attack by almost anyone.'

The meeting began politely. But that did not last long. 'Levenson is a Brooklyn boy and Marty is a Bronx guy,' Diffie recalls of how his colleague began to get riled by the men in suits. 'There were two abrasive up-front New York personalities. They tried to brush off his [Hellman's] analysis, [saying] this is amateur incompetence. That was what started the fight.'[4] Hellman said it might eventually be possible to break the key with brute force in a day. Levenson countered that it would take ninety-one years. Hellman said that was rubbish. The two New York men started to go at each other. It became clear that, as well as worrying about a weak key being vulnerable to all kinds of people in the future as computing power improved, Hellman was saying he was worried about the NSA as a threat right now.

The political and cultural context had changed since the days of the Second World War and early Cold War. In the wake of Shamrock and Watergate, a new generation saw the government as not necessarily the protector of the people and their liberty but a potential threat to those liberties. American culture was changing in profound ways, and Diffie and Hellman shared an anti-authoritarian outlook.

'You think NSA today is looking at anybody's income taxes?' asks Levenson incredulously.

'Honestly, we don't think so, but that's ... In the climate of Watergate, we think that's something we hardly take for granted,' Hellman replies. The row intensifies as the argument moves from the maths to the politics.

'I spent my career there, and I never read anybody's income tax return,' the NSA man Levenson exclaims.

Hellman interrupts him. 'But if there's a request from the Executive Branch to do that, what would you do?' he asks.

'If the Executive Branch comes and tells me as a civil servant to shoot Marty Hellman, I don't know what I'd do,' Levenson retorts as a joke.

'You would!' exclaims Hellman to laughter.

There was a trade-off over national security, both sides agreed: a balance between being able to crack codes and securing information. 'We feel we've struck that balance,' argues Levenson.

'Here's the deal,' replies Hellman. 'I don't think that the National Security Agency is the appropriate organ of the government to decide where that balance lies.'

The question of whether the NSA was best placed to judge the balance between offence and defence was a key question then and remains one now. As the meeting came to an end, Hellman suggests the NSA is his enemy. For Levenson, a man who sees himself as having worked to protect his country from the dark days of the Second World War through the Cold War, this questioning of his patriotism is too much.

'I feel a little funny that you regard me as an enemy,' he replies.

Hellman says that is not quite right but Levenson persists. 'Well, that's what you're saying – that NSA is the enemy.'

'Not you specifically,' Hellman replies, not backing away from the idea that the NSA as a whole is a potential enemy. At that point the meeting is concluded and the tape switched off.

Hellman saw himself at the time as a (self-appointed) security officer for the public, taking on the NSA on their behalf; but is more reflective now. 'My view back then was they were not interested in national security, they were interested in job security. Now I look at it very differently. I do think they were concerned. They had some legitimate concerns and I should have taken those into account. I still would have taken the position that I did but I would have fought it more fairly. Now, they didn't fight fairly either. In fact they fought less fairly – they were telling me I was wrong when I was right.'

Former NSA and NBS officials from the time are still sensitive about the whole affair four decades on and choose their words carefully when they speak about it. They say the secrecy around the 'S Box' design was not because of a trapdoor but actually to protect a specific code-breaking technique – known as differential cryptanalysis – which the outside world had not yet discovered. Most critics now accept there was not a trapdoor. However, the reduced key size remains more mysterious and suspicious. Former officials say it had to be reduced to fit on a single commercially produced chip to make it practicable – in other words, it had to be good crypto but also cheap enough crypto. They also say that since it was for US use only and not for export, making it too strong was not so much of an issue. But the key sections covering DES remain blacked out in a recently declassified NSA official history. And that history makes it clear there was an internal disagreement leading to a 'compromise' on 56 bit which suggests some people might have wanted it even weaker.[5]

Within the NSA, the decision to get involved with the federal standard had been contentious. Both the NSA and GCHQ have a dual mission: to gather intelligence from communications of adversaries, but also to protect their own country's communications from those adversaries. This had been a lesson learnt the hard way in the Second World War. If you knew what you could do to your enemy, you could stop them doing similar things to you (and vice versa). The relationship between these two functions was fairly straightforward for many years. Your enemy used a machine like Enigma that you tried to unlock, while learning from this process to make your equivalent machine as secure as possible. But what if both your people and your adversaries started using the same encryption techniques? How weak or strong did you want it? This new emerging world of encryption posed a challenge, because the system you were securing was one that the public and not just the government might be using and which could also spread internationally. Did you want a system used by your own public that might be totally secure for them if that meant those abroad you were trying to spy on would also use it and be secure themselves? Offence and defence were previously complementary. Now there was an inherent tension. Some, like the Second World War veteran Frank Rowlett, had always argued that in the long run it was more important to secure your own communications than to exploit those of the enemy, but the NSA and GCHQ still saw themselves primarily as intelligence-producing agencies.[6]

An official report into the controversy said the key size was 'more than adequate for all commercial applications for which the DES was intended' over the intended time span.[7] The phrase was very carefully chosen. The system was designed so that it was secure enough to keep criminals and states out – apart from those with the highest level of capability. That meant the NSA and unfortunately also the Soviet Union. Making a code un-crackable to the Soviets would have the unfortunate by-product of also keeping the NSA out. And that was not what was wanted. DES was used for many years for everything from banking to police communications in the US, and 'by the early 1990s, it had become the most widely used encryption algorithm in the world', an NSA official history remarks.[8] Though technically restricted to the US, it was used in thirty-three countries. A rather useful outcome if you know you can break it.

Diffie and Hellman lost that first battle. Looking back, they both now agree the standard was not as bad as they feared. But at the time, this tactical defeat in the opening battle would, strangely, lead to the two men's much more important strategic victory against their opponent. After that morning at Stanford they were fired up.

The two began reflecting on how to secure information from the prying eyes of the state. If DES was weak, then maybe they could find something stronger. They began thinking about the trapdoors, the type they feared were hidden in the chip. Diffie had a radical idea. What if there was a way of using the concept to provide security rather than vulnerability? One night he came up with an answer: if you could put a secure trapdoor in your own message, then it could be scrambled as much as you like and only you would know how to get back in and read it. So what if, rather than using the same key to lock and unlock a message, you had two keys, one which was public and which you gave to anyone who wanted to send you a message? This would be used to scramble the message. But then there was a second, private key that you kept yourself – the trapdoor key – which was the only way of unscrambling a message. This was utterly revolutionary.

When Hellman explained the idea to Horst Feistel at IBM he said it was absurd. The whole trick of encryption – as everyone had known for centuries – was to keep your keys secret. So the two German operators using an Enigma machine would both need to use the same key to lock and unlock a message. But this was hard work. The central problem of encryption for centuries had been key distribution. To share a secret with someone you both needed the same key, but how do you distribute that key secretly in the first place? One-time pads might be secure, but getting a single pad to both sender and recipient totally securely was hard work. Even swapping Enigma or Tunny settings was difficult in wartime or if you worked across the world. Changing keys regularly was also laborious. What was called 'public key' offered a different model. For Diffie, the motivation had also been ideological – part of his distrust of authority. 'I started all of this with a very anti-establishment viewpoint,' he explains. 'I never understood what's called a key distribution centre, which is a pre-public key technique for distributing keys, because it involves trusting a central organisation centre.'

Public key meant you did not need to courier a key secretly. Everyone who wanted to send a message to you could use the same published key. This also meant you could change it much more easily. With an Enigma machine you needed to keep the workings of the machine secret as well as the key. But if the public were going to use cryptography, then keeping the machine secret was clearly not possible. The secrecy had to reside in the key, not the machine. Public key was going to offer encryption to the masses and not just the state. It also offered another benefit. It meant you could authenticate a message. People had been wondering how you could prove who sent a message in the digital world when it consisted of ones and zeros which could be manipulated. A signature worked in person but not electronically. If I was to agree a contract remotely and electronically, how would you know I would not back out later claiming it had been faked?

With public key you could scramble a message with your private key and then send it to someone else. They could then use your public key to read it. In doing so they would know for sure that the message had come from you as only the combination of the two keys would make the message readable. This made it the digital equivalent of a signature. The system opened up the possibility of contractual exchange and commerce in a digital world. At the time, this seemed abstract. But internet commerce today would be impossible without these two advances. If you wanted to send payment to a bank securely, could you really wait for a courier to deliver an electronic key every time? And how would the bank know it was you and not someone else manipulating ones and zeros to give your name? The modern world of secure electronic communications is built, in no small part, on what is known as public key cryptography. In turn this is based on the breakthrough known as the Diffie-Hellman key exchange.

When the two men published a paper entitled 'New Directions in Cryptography', most of the world, as it had done with Alan Turing's 'On Computable Numbers', did not quite grasp what was being proposed. Diffie and Hellman's paper found its way to the academic Ron Rivest at MIT. Along with two colleagues, Leonard Adelman and Adi Shamir, he began working on applications for public key. In technical terms what was needed to make a trapdoor was a mathematical one-way function, something that was very easy to do one-way but nigh impossible to reverse. Multiplying two large, randomly chosen prime numbers proved the breakthrough since it is almost impossible to work out which two numbers have been used if you are presented with the product. They published an article in *Scientific American* that made a huge stir – not least because it offered $100 to anyone who could crack a message. Out of the three men's work would come the company bearing their initials, RSA – still a major force in the computer security world. But, just as importantly, out of the magazine article came a flood of letters from the public. A revolution was under way. Cryptography was going public. The old era of secret code making and breaking was dying.

*

Thousands of miles away in Britain, one mathematician could be forgiven for feeling rather frustrated as he read the article in *Scientific American*. 'I thought "gosh, yes, I've seen this before",' Clifford Cocks recalls with a wry smile. He had seen it before because he had also developed the same mathematical scheme as RSA. But a few years earlier. And he had to watch as the rest of the world lauded this new discovery which he had in fact worked out in his first few months in his office. The problem for Cocks was that his office was GCHQ and that meant his work had all been utterly secret.[9]

Cocks had joined GCHQ as a young mathematician in 1973 after a university tutor at Oxford had suggested he might be interested in some fascinating, but rather secret, work – the fabled 'tap on the shoulder' which was always known to operate for MI6 but also applied to mathematicians for its sister agency. On starting, Cocks had been presented with a range of mathematical problems. One in particular caught his eye. It was something that a colleague, James Ellis, had been working on. Ellis, born in Australia but brought up in London's East End, was, like Whit Diffie, mildly eccentric but also brilliant. He had found a home for his talents at GCHQ. 'James was something of a maverick. If he was given a problem, he would never try and solve it directly,' says Cocks.

In the 1960s, the agency was struggling with distributing the vast number of encryption keys used by the armed forces. As things like secure telephones were being pushed to more and more people, the strain of developing and distributing a growing number of keys securely was becoming immense. Was there a way of overcoming the need to courier these around the world securely so that both sides would have the same key?

A Second World War report into secure speech provided a burst of inspiration for Ellis. A 1944 paper suggested that the person who wanted to receive a message could send out seemingly random noise. The person sending the message would then add their voice message to the random noise they had received and send it back. Anyone intercepting the message would only hear the combination of the two noises. Only the person who had put the random noise on in the first place would know what it was and be able to take it off and hear the message they had been sent. The key concept was that the recipient, not just the sender, took part in the encryption process. Ellis says that one night in bed in 1969 he came up with a theoretical proof for what he called 'non-secret encryption' in which someone added a random key to a digital message which only they would know how to remove. Another analogy was the idea of sending someone a box with the padlock open. They put the message in and closed the lock. They never needed the actual key and nor did the courier (who could not open it on the way since only the sender had the key). He was then faced with the challenge of making it work in practice and finding a mathematical function to generate the right type of key. For a few years Ellis was stuck. Within GCHQ this was viewed as an interesting challenge but not a top priority: one camp thought it could not possibly work and it was just that no one had figured out why; another camp thought it might work but could not see how.

When the maths team introduced their twenty-two-year-old new recruit to the puzzle, Clifford Cocks was intrigued. By a stroke of luck he had been working on number theory. 'I just happened to be the right person in the right place at the right time,' he says modestly. What that meant was that the very night he was told about the problem, he had the solution. 'I thought of it that night. I wasn't able to write anything down as you weren't allowed to write anything down. So I just hoped that when I went to sleep I would remember it when I woke up in the morning.' Cocks realised that multiplying two large prime numbers would be easy to do one-way but hard to reverse. Just as Ellis had come up with something almost identical to the Diffie-Hellman concept (although without the idea of a digital signature), now Cocks had come up with his version of the RSA solution. The next morning, having fortunately not forgotten it, he wrote it up and took it to his boss. 'This is the most important cryptographic discovery of the century,' Cocks's boss exclaimed, and ran off down the corridor.[10] Another colleague, Malcolm Williamson, would also come up with a further innovation, but among the top brass at GCHQ there was nervousness about this idea. It just seemed wrong. Too easy. Too different to what had been done before. There had to be a flaw, one which could then be exploited like a backdoor by the Russians. Within the secret world, investing in such a scheme involved too much risk, some thought. The technology was also not quite there yet to be able to build such devices cheaply. And so the idea remained just that.

A few years later, the British team began to hear about a revolution coming out of America. And they had to bite their lips and remain silent. Williamson talked about trying to get the US patents blocked but he was told not to bother. And so, despite having technically come up with the idea (or a very similar one) first, the British team had to sit and watch as all the credit, the glory – and the money – went to the Americans, who began to develop public key systems. Britain – and GCHQ – had missed a huge opportunity to be pioneers. When Clifford Cocks was asked whether he was frustrated by this, there was a telling pause before a careful answer: 'I just thought of it as something I had done and I had moved on. And that was just how life was.' Ellis had been frustrated by the lack of recognition and wrote a paper to try and set the record straight, but this paper itself was then classified. He tried year after year to get it cleared by GCHQ. Finally it was released in December 1997. James Ellis had died a month earlier. Secrecy has a price.

In the summer of 1976, Whit Diffie went to see Arthur Levenson of the NSA – one of his opponents at the Stanford meeting. Diffie told him about the key exchange theory. At the time, Diffie was surprised by the fact that Levenson was not overcome with awe at the originality of the theory. Looking back now, Diffie believes it was because the NSA already knew about it. This may be because the NSA had been told of the work at GCHQ or it may be, as some former NSA officials suggest, that they themselves had also come up with a form of public key exchange in secret.

Since the first murmurings of new work on codes outside the US, the NSA had reacted with a mixture of panic and denial. Surely everyone who wanted to work on codes would want to do so inside the NSA? Concern really began to escalate with the publication of the *Scientific American* article and the realisation that a lot of people were starting to get interested in the field. This needed to be stopped. 'NSA hunted diligently for a way to stop cryptography from going public,' records a declassified official history of the agency.[11]

For the NSA, public key was a nightmare. They might be able to beat something like a 56-bit DES key with their supercomputers working away on the permutations for a while. But if the key could be changed every day, then no sooner would they break into a code than it would be gone. And the fear was that everyone would be able to use this system.

In 1977 one of the more influential Directors of the NSA took the helm. Bobby Ray Inman was the son of a Texas gas station owner. He was something of an outsider: not from the traditional navy elite, and also unusual in that he had always wanted to be Director of the NSA. He was a workaholic with a photographic memory who started work at 4 a.m. (in his eighties he starts at 5 a.m.). He immersed himself deeply in the technical aspects of the business and worked with an intensity that left his staff often unable to keep up. 'He appeared perpetually calm, but in reality was about as stable as high voltage across an air gap,' an NSA historian noted.[12]

Inman had a distaste for human intelligence and also for commercial espionage, both of which he viewed as somehow unclean. It was on his watch that a rule was laid down: no use of the intelligence machinery for the advancement of the commercial interests of American companies. Before that intelligence had been passed to companies, but Inman insisted, over objections, that it be stopped. President Carter had been hostile to the work of the entire intelligence community, including the NSA. But when White House officials realised the cost in not renewing the successor to Shamrock, they changed their stance. Inman also finally won a long-running battle with the CIA over their attempts to muscle in on signals intelligence. His approach gained him much-needed allies in Congress.

But it was the battle over cryptography that caused some of his biggest headaches. His staff were seriously worried. First of all he inherited the mistrust over DES. 'The NSA was actually engaging in a highly complex balancing act in working out how secure to make non-classified communications,' Inman argues. The aim was to make it secure enough that most other states could not break it but not so strong that, if it was exported, the NSA itself would not be able to get into it. Export was the real fear for it. 'Finding the balance,' Inman explains. 'You wanted it strong enough that it could not be broken easily.' And now public key raised even more challenges. Again, the fears that staff expressed to the new Director were about export. 'The great worry was that this effort would produce cryptographic systems that they couldn't break and it wasn't just worry about drug dealers and the rest of that. It was that they could be picked up by foreign countries. And that they would reach a level of encryption that NSA would be unable to successfully attack.'[13]

The day Inman took the reins, one member of staff took events into his own hands by picking up his pen. In July 1977, a letter arrived at the Institute of Electrical and Electronic Engineers, a prestigious computing body. It raised concerns over cryptography research and drew attention to specific paragraphs in the International Traffic in Arms Regulation Code that made exporting certain military technologies illegal. It sounded like a warning and it soon emerged that the author worked for the NSA. Inman would later recall ruefully that it had been sent without him knowing anything about it and not as official policy. However, the perception grew that the NSA was going to fight against the new ideas. Academic grants to people working on cryptology were questioned. Secrecy orders were put out on people trying to patent new encryption technology, but this again led to the thing the NSA hated most – press reports.

Martin Hellman received darker warnings. 'Silicon Valley is sometimes called Spook Valley because not only is silicon made here but there is a huge presence here by the intelligence community,' he explains. 'And some of my friends in Spook Valley told me I was stepping on some very dangerous toes. And by the way I wasn't just stepping on NSA's toes, I was stepping on GCHQ's toes, I was stepping on the KGB, GRU's toes. And so they warned me that not only was I potentially going to suffer legal action but my life might be in danger.'[14] Codes suddenly looked dangerous. But also kind of sexy and anti-authoritarian.

So, how to respond? Some in the agency said the best thing would be to take the usual NSA position and say absolutely nothing. They argued that any public discussion of cryptography would heighten awareness and lead other countries – especially in the developing world – to buy more secure devices than the weak ones NSA was regularly penetrating. Others talked of getting new legislation passed to increase government controls. At first Inman pursued this path, seeking a new law; but there was no appetite for it in Congress and so a third option was taken by default – to talk.

Inman decided to deal direct. While on the West Coast, he arranged a parley. 'Since I was already on the [West] coast I decided to drive down to Stanford and get acquainted with Professor Marty Hellman,' Inman recalled with a chuckle decades later. 'I think he was somewhat surprised, maybe impressed, that Mohammed had come to the mountain.' Decades later, too, Hellman could recall the events, starting with a phone call:

> I get this call and it says Admiral Inman is coming to the Bay area and would like to meet with you if that's possible. So here I've got Darth Vadar in mind – at least in my mind at the time – wants to meet with me. But of course I didn't know – being Luke Skywalker – that he was actually in some sense my father. But I said I'd be happy to and we met in my office here at Stanford . . . And I'll never forget. His first words jokingly were, 'It's nice to see you don't have horns.' Because that's what he was getting from everybody at NSA and from their perspective . . . And so I looked back at Admiral Inman and I said, 'Same here,' because they'd seemed like the devil to me.[15]

Out of this unusual meeting came something of a dialogue: a group of academics including Hellman agreed a voluntary system in which they could submit planned papers to the NSA, who would explain if there were any sensitive issues. However there was no veto over publication. 'Part of it was persuading the counter-culture types that we in fact did not have evil intent,' Inman says of his initiative, 'and that we genuinely were focused on national security – how you protect it – and we made the presumption that, notwithstanding their political views might be different, they still were patriotic.'

Inman did something else unusual for an NSA Director. He spoke in public. In front of a friendly audience of private-sector contractors in early 1979, he argued that collecting secret intelligence required secrecy in order to protect intelligence sources of the 'utmost fragility and sensitivity'. The NSA had until recently 'enjoyed the luxury of relative obscurity', but wider interest in the security of communications had led to a 'novel encounter' between the NSA and the outside world which had not been a happy one.

Inman acknowledged that, as more information was being held on computers, people worried that it was vulnerable to being stolen or accessed in an unauthorised fashion. 'The public has become increasingly aware of the danger that automated data processing systems, if not adequately protected, can be exploited for fraudulent and illegal purposes. Moreover, the vast amount of personal information stored in and handled by automated data systems, both private and governmental, has given rise to serious concerns about individual privacy.'

But there was a 'tension' between the intelligence and national security interests of government on the one hand and the public and private sector on the other, he told the audience. If it was not resolved, Inman argued, there was a risk of serious damage to national security. 'There is a very real and critical danger that unrestrained public discussion of cryptologic matters will seriously damage the ability of the government to conduct signals intelligence and the ability of this government to protect national security information from hostile exploitation.' These were the dark warnings spoken by spies from the Second World War to the present. Any discussion was dangerous. The world might go dark.

Inman's speech was half a warning to the outside world and half an acknowledgement that the two sides were going to have to talk. Looking back now, just as Hellman can see the other side of the argument, Inman too acknowledges public key was not the end of the world. At the time, the NSA ended up realising it could not stifle academic interest in cryptography. Free speech was too strong a principle. And people wanted their information secure. The NSA could not stop cryptography, but it could try to control it. There was another problem, though. For Americans to use these systems was just about tolerable; but what if the technology was exported? Export was the big issue, but the concern was not the Soviet Union. It would always be building its own systems that the agency was still struggling to deal with. In practice it was other, less advanced countries which often provided the best intelligence on the Soviet

Union. One of the things the NSA had learnt was that by listening to traffic from allies of the USSR you could find out a lot about the main enemy – for instance, arms exports or who was visiting where or planning what. 'You may not be able to get the Soviet traffic, but you could get what was happening on the other end and get understanding,' Inman explains. The fear was that now these countries would adopt more secure systems, which would deny a valuable stream of intelligence.

With computers and information spreading, people sensed there was money to be made in security. After early struggles, RSA began to sell its product. Companies like Lotus Notes began to use encryption in their product in which employees collaborated over a network. The NSA and others in government worked on companies to ensure exported systems were weaker. So the Lotus Notes system sold abroad had only a 32-bit encryption standard.[16]

In Britain, companies were also pressed only to export weaker versions of cryptography. They would be visited by GCHQ and told 'you will be wanting this' when it came to a particular encryption algorithm. 'Not really,' the company might reply, noting it was weaker. 'Yes, you will, if you want to export,' was the reply, as a person involved at the time remembers. (One bank got into trouble when it was discovered it had shipped a strong domestic crypto system to a branch in Moscow.)

How far did the US go in ensuring they could break commercial systems used abroad? Over the decades, whispers have occasionally surfaced out of the spy world that the US has placed backdoors in exported commercial products. This, it was said, was based on a mixture of appeals for patriotism and commercial self-interest to the companies, including the fear of not getting export approvals in the future. 'It is not unheard-of for NSA to offer preferential export treatment to a company if it builds a backdoor into its equipment,' one person with long experience was reported as saying soon after the end of the Cold War. 'I've seen it. I've been in the room.' There was also talk of efforts to weaken encryption systems built by foreign companies, tampering with the algorithms so they were either weak enough to break or had a trapdoor which could be exploited (or a combination of the two). Engineers at foreign companies talked in hushed tones of mysterious visits by unnamed figures which led to subtle changes. Those companies have always strongly denied the allegations.[17] The likelihood is that NSA teams ensured that weaknesses were left in export versions which only they knew about and would allow someone who knew the right statistical or cryptanalytic attack to break in.

A new set of combatants stepped onto the battlefield of the crypto wars in the latter part of the 1980s. Joining the academics and the businessmen arrayed against the NSA were a ragtag militia of privacy activists who raised the banner of encryption as an ideological cause. Hellman and Diffie had been influenced by the culture of the 1970s, but this new generation took it a step further. A leading figure was Phil Zimmermann. Zimmermann's work intersected two transformative trends. One was a belief in privacy as a means of empowering people against the state. The other was the arrival of the personal computer.

As a ten-year-old in Florida, Zimmermann read one of those children's books on codes and secret writing that taught you how to make invisible ink out of lemon juice. By the time he was at university he learnt how to have the same result in masking your communications by writing computer code. He would also read the 1977 *Scientific American* article outlining how public key might work. Zimmermann was from the post-Watergate generation that was deeply suspicious of government. He was a leading anti-nuclear campaigner and saw cryptography as a means to empower fellow activists to protect their secrets from the government. And by the early 1980s personal computers were also just emerging which might allow ordinary people to do that. Zimmermann wanted to spread cryptography to the masses. In his bedroom, he wrote a system modestly called 'Pretty Good Privacy' or PGP, which for the first time made encryption relatively easy for anyone. Previously it was controlled through hardware – a physical machine – that someone needed to buy and whose distribution could be regulated by a company or the government. But now encryption was moving into the realm of software, a set of instructions that could easily be distributed, perhaps on a disk or even downloaded from the internet.

That meant that when PGP arrived in 1991, it scared the hell out of the government. The FBI had come to rely on wiretaps of criminals – for instance, organised crime groups like the mafia. Suddenly they worried encryption could put that at risk. They began to lobby hard, holding briefings with government officials, issuing terrifying warnings of what would happen if they went 'dark'. Senator Joe Biden proposed legislation to ensure that government could access unencrypted communications. This possibility in turn created uproar in the crypto community.

'I suggest you begin to stock up on crypto gear while you can still get it,' one person warned his fellow believers on an internet bulletin board.[18] The language is redolent of what you hear from gun owners and lobby groups when gun-control legislation is proposed: 'load up on guns before the government takes them away'. The analogy is imperfect but still illuminating, since there is a strong libertarian streak to the crypto community. The state is perceived as the threat to individual freedom and it is the patriotic duty of the individual to limit its power. Resistance is seen as heroic. In America there is a type of revolutionary, idealistic zeal to many of the crypto advocates. Zimmermann had a motto: 'When crypto is outlawed, only outlaws will have crypto.' It was the Wild West out there and everyone should have a gun in the form of crypto (or at least, to be more accurate, the means to defend themselves).

Zimmermann took the danger posed by legislation seriously and started to push out his PGP software before it could be passed. He decided to release PGP for free, telling people to distribute it far and wide by placing it on servers to be downloaded. And they did. And of course not just from within the US but around the world. PGP was killing the system of export control on cryptography. Biden ended up withdrawing the clause from the Bill, but PGP was now out there (much to the annoyance of the commercial companies hoping to sell their products). Zimmermann himself began to feel the heat. Some officials from US Customs got in touch.[19] The reason it was Customs men who initially came to see Zimmermann was that cryptography had been designated by the US as a 'munition', exporting it abroad being the equivalent of being an international arms dealer. Zimmermann was put under criminal investigation for three years although never charged. It was a miserable experience at the time, but looking back he recognises it as the moment that made his career. The press coverage enhanced PGP's popularity. Zimmermann began to get messages from activists around the world who explained that PGP had become a vital tool for them to communicate and organise while protecting themselves against authoritarian or repressive regimes.

A culture was growing up of people who valued privacy and saw the state as the enemy. 'What do I have to hide?' Tim May, one of the key ideologues, asked rhetorically. 'None of your business,' was his response. At its extreme, the crypto community extended from libertarianism into out-and-out anarchism and a desire actually to destroy the state altogether, seeing it as the enemy of freedom. A few advocates saw crypto as offering the chance to challenge the whole power structure of society and to empower individuals.[20]

The cypherpunks had arrived. This was the description used for themselves by activists who were evangelical about spreading cryptography. They were a spin-off from a wider cyberpunk movement which focused on the power of computers and the internet. The playful aspect of early hacking was now being joined by an ideology. Information should be free, some activists argued, believing everything should be open and there should be no 'secrets'. Others said privacy was a fundamental, political right. Those two ideas are sometimes expressed in the same breath but are far from easy to reconcile.

When people entered the online space, they did not have to say who they were and could even create a false identity or profile. Because of the way the internet had been established, the only thing that might trace you was an Internet Protocol address where data would come and go, but this did not necessarily associate with an individual name and could also be masked. People like David Chaum looked at ways of trying to ensure anonymity in communications: this meant going beyond just encrypting the content of a message and actually being able to hide who was sending it, perhaps by remailing it around a network so that the point of origin was hidden. This was useful for all kinds of people. These included those who simply wanted to post abusive messages, but also those seeking to hide from spies – activists, libertarians, whistle-blowers – and of course sometimes spies themselves, for whom multiple identities and working undercover are an essential part of their trade.

If you were a dissident living in fear of a knock on the door from the authorities, anonymous communication might be all that stood between you and a dark prison cell or even death. The US Naval Research Laboratory helped develop a system called TOR (The Onion Router) to provide a means for people living under oppressive regimes to communicate and organise without fear. But, as with any technology, anonymity can be used for good or for ill. One internet forum created something called the 'Assassination Politics Plan' in which people could anonymously place bets on the timing of the death of important people: the implicit idea was that if a large enough pool of cash existed, it would create an incentive for someone to carry out the act themselves in order to claim the pot of money. Such a site would only be possible with anonymity.[21] The same tools that would protect people from being imprisoned by their government could offer a safe haven for those with darker purposes. That duality was at the heart of the internet and a contest between its supporters and the state.

The cyberpunk ideology spread not just in the US but around the world, to a hacker community that could speak to each other and form a virtual community on the new technology of internet bulletin boards (making hackers perhaps the second virtual community after spies). These message forums would be filled with a mix of technological tricks and personal abuse but also carried, for some, a political edge. An Australian hacker who went by the name of Mendax was one of the earliest people to sign up on the cypherpunk mailing list in the early 1990s. He equated the battle for cryptography with the right to bear arms to prevent totalitarianism.[22] His real name was Julian Assange. As a young man he lived and breathed computers; like many others he was most comfortable and felt he was who he was supposed to be when he was online. He hacked into NASA and Lockheed Martin in the early 1990s, doing no damage but exploring. For him, encryption in those days was something almost mystical:

The universe believes in encryption. It is easier to encrypt information than it is to decrypt it. We saw we could use this strange property to create the laws of a new world. To abstract away our new platonic realm from its base underpinnings of satellites, undersea cables and their controllers. To fortify our space behind a cryptographic veil.

A world in which people could communicate and organise anonymously without fear of government was the cypherpunks' dream but the authorities' nightmare. They saw their intelligence targets become some of the early adopters of the new technology. Previously only states could encrypt, and only the best, like the Russians, could do it at a high level. But now, Inman says, they began to see international drugs dealers encrypt their voice communications.

When Stewart Baker was brought into the NSA as General Counsel in the early 1990s, the head of the agency, Mike McConnell, sent another member of staff to brief him on the crypto problem. 'He handed me a bottle of aspirin and said after this briefing you are going to need them all.'[23] Encryption was spreading faster and faster. 'It turns out that the biggest, most enthusiastic market for strong encryption are people who have a lot to hide ... so as soon as you start making new forms of encryption, especially stronger forms of encryption, available you find that they are misused by criminals and foreign spies and the like,' Baker recalls.

One NSA expert, Clint Brooks, came up with what he thought was a solution. It was called the Clipper Chip. It could be put in phones and potentially other devices to provide secure, encrypted communication – something the general public did not have at the time. But the keys to unlock it would be held by a third party – in escrow – and they could be retrieved by the authorities when they got permission, for instance a search warrant from a judge. This was a backdoor. But not a secret one. A public one.

Stewart Baker wrote an article in *Wired* magazine extolling the virtue of Clipper and attacking what he called 'romantic high-tech anarchism'. Baker attacked the spread of PGP. 'Some argue that widespread availability of this encryption will help Latvian freedom fighters today and American freedom fighters tomorrow. Well, not quite,' Baker wrote. 'Rather one of the earliest users of PGP was a high-tech paedophile in Santa Clara, California. He used PGP to encrypt files that, police suspect, include a diary of his contacts.' There might be kids who needed help, but Baker quoted a policeman saying, 'Thanks to this encryption, we'll never reach them.'[24] Now familiar battle lines were being drawn between privacy activists on one side and authorities talking about criminals and paedophiles on the other. Little would change about the terms of this debate in the years to come.

The FBI wanted guaranteed access. They made sure they had that to phones (which were not encrypted). The Communications Assistance for Law Enforcement Act of 1994 ensured all phone companies would build in the facility to intercept communications on the production of a warrant – effectively a kind of backdoor for law enforcement in the network. But the problem was that this kind of system did not exist to deal with the two emerging worlds of internet communications and encryption.

In the 1990s battlefield of the crypto wars, the NSA would suffer a painful defeat over Clipper Chip. It inspired a fierce counter-attack from the privacy community. It was not helped when an outsider called Matt Blaze spotted some potentially serious security vulnerabilities in the system. NSA veterans admit they made mistakes too, including not letting the FBI lead the charge. They also say they failed to explain to people that Clipper could actually have provided more security for everyone against other people listening in by deploying encryption and leaving only the government with the possibility of getting in. But this ignores the fact that critics saw the government as the problem. Could they be trusted with the keys? This was the Big Brother Chip, they said. Those activists also had a new ally who had just taken to the battlefield.

The burgeoning tech community did not like Clipper Chip at all. They had a strong West Coast anti-government libertarian streak: they believed they were the future and government was the past. The idea that technology promised to liberate the individual became a staple of Silicon Valley culture. For a while, the punks and businessmen would be allies: many of the new corporations were started by people who came up through the hacker culture (like Steve Jobs) who rejected suits and ties. It was an alliance of convenience which would not last for ever, though. The businessmen were motivated by money as well as ideology in rejecting encryption and state control. How were they going to export new technologies around the world with a US government backdoor in them? they asked. This was a major problem the government had not really addressed. If you shipped technology with encryption to the French, did you keep the key or did you give it to the French government? European countries did not like the idea anyway, as it meant the Americans might have the key. And what about less savoury governments who might want to spy in an oppressive way on their citizens? One person described the mess as the 'Bosnia of telecommunications', a reference to the unfolding disaster in the Balkans at the time. The business lobbyists from Silicon Valley were emerging as a new force. The Cold War is over, they said. The world is globalising, they argued. Did the government really want America to lose its edge and its exports due to this whole encryption nonsense? Who were these spies anyway and did we really need them?

For the cypherpunks, cyberspace was a place where they lived, not just a tool to communicate. It was a new frontier which, like the West in the America of the nineteenth century, offered those heading that way freedom from the past and from authority. A leading group battling for privacy rights online would even call itself the 'Electronic Frontier Foundation'. This new frontier was a place where you could go for a variety of motives: to be free from the strictures of traditional authority, to escape the past and also to try to strike gold and get rich. The latter applied to the tech industry, where new fortunes would be made as with the barons of the industrialising nineteenth century who had built the railroads and struck oil. But the internet was a frontier that business drove into and colonised, not just in America but around the world. The notion of globalisation was intimately bound up with the spread of the web. Like the telegraph more than a century before, it promised to draw the world closer together, make it easier for people to talk and trade, to create a global village. Tech utopianism was often disdainful of the state (and its spies) and believed that they were leftover relics of the past whose importance would quickly fade. That did not quite turn out to be the case.

For spies and governments, this spread of commercial technology – whether hardware, software or encryption – was making the balance between defending your own systems and getting inside your opponent's much trickier. If you were using commercial systems, did you want them to be totally secure? Yes, to protect your own communications, but what if your targets were using them? Was it more important to secure them or be able to break them? If you spotted a weakness did you close it or keep it open? In the NSA this was called the 'equities' problem and a board was created to try to resolve the tensions.

Security was about more than just scrambling a message with code. One former NSA Director recalls an old crypto hand walking into his office soon after he took over.

'You know you are responsible for the integrity of the US command and control system?' the mathematician said to his new boss. That was the system that gave the orders from the President downwards for the launch of nuclear weapons. He then turned to the door.

'Before you leave, let's talk through this,' the Director replied.

'What's the most important thing about encryption?' asked the mathematician.

'Scrambled text,' his boss replied, thinking that stopping the other side reading the message was the main point of encryption.

'Wrong,' the visitor said. 'The single most important thing is attribution.' Only one person can give the order to fire nuclear weapons – the President – so you have to know it is him and not someone else. 'What's the second most important thing about encryption?' he asked.

'Scrambled text,' the Director said again.

'No. It's the integrity of the data.' If the message is somehow corrupted and you get a digit wrong, you could send a nuclear missile against the wrong target.

'How many of these do we have to go through?' the Director asked.

'Five,' the man replied.

'Third one is non-repudiation. If you did it, you can't repudiate that you did it and say it was not you who gave the order,' the crypto man explained. 'Fourth?'

'Scrambled text?'

'Wrong. Infinity.' This is now called availability. The system has to keep working in the same way for infinity or always be available – it cannot go down. 'Fifth?' asked the visitor.

'Scrambled text for secrecy,' the Director replied.

'You got it,' he said.

This was the level of assurance and security demanded by government for its most important communications. It had to offer confidence as well as confidentiality. And as the world would move online, the public and people like banks wanted assurance too. Security was about more than just secrecy. If health records telling doctors your blood group are going to be kept in a database or online, then security is a concern; but if you are rushed to hospital after an accident, being sure of the *integrity* of those records (and that the blood group is correct and could not have been wilfully or accidentally altered) and the *availability* of those records so the system is always there for a doctor to refer to, are likely to be even higher in your priority list. Providing this in the world of the internet might require encryption, but even that was not always enough to deal with issues of attribution – knowing who sent a message – because of the many pathways the internet offered for obscuring identity and maintaining anonymity. People liked the anonymous nature of the internet for some purposes – web-browsing, perhaps – but it was not always practical for others, like buying things or banking.

By the late 1990s, the internet was arriving for ordinary people as companies emerged who would allow you to do things like buy a book on the web. Life was moving online, and with it piles of data from people and about people. Everyone would want this secured. The general public were starting to ask questions about security and vulnerability that in the past only the generals (and then the banks) had asked. Encryption was moving centre-stage.

In 1997 Robert Morris, a computer expert who had recently retired from the NSA, gave a talk to MIT students about protecting information. Morris advised the students not to even think about using email for financial transactions since it was not encrypted. When asked what the most secure way to send a message was, Morris had an answer: 'Probably the US mail.'[25] His take on information security might have seemed alarmist to the students, but it reflected time spent in the world of spies. 'If people think I see wires and mikes everywhere, it's because I do,' he told them. 'To protect information, one has to be paranoid.' He explained the tensions over the NSA helping secure information and still being able to break it, acknowledging that the same argument the National Rifle Association made about gun control might apply: that government regulation only stopped ordinary people protecting themselves.

The spread of the internet and the debate over encryption reached the highest levels of the Clinton administration in the late 1990s. The new President and especially his Vice-President Al Gore made a big play of being from a new technologically aware generation in favour of building an 'information superhighway'. So what was it supposed to do with the internet? The US government had been a key sponsor and had retained controls at various levels, including over encryption and things like the domain names which people used for websites. But now many thought it was time to let go. The first tech boom was beginning. Companies were taking over. People were starting to buy and sell things online. Globalisation was in vogue. Security threats had receded since the end of the Cold War. The consensus was emerging that the best thing the US government could do was to deregulate and leave American companies to get rich, and for the internet to spread peace, freedom and prosperity around the world, as many predicted it would.

This 'letting go' ranged from setting up an arm's-length body to distribute global internet address names to relaxing encryption controls. 'The theory was, at the end of the day in terms of economic growth, in terms of our ability to spread democracy around the world – which ultimately is the guarantee of our national security – it was better to let it go,' says James Lewis, an official involved in the discussions at the time. The debate over encryption was particularly tough and raged for years within the administration between the national security people like the head of the FBI and NSA in one corner and the business, commerce-facing officials in the other. Eventually, though, even the most senior intelligence officials could see they were losing. The decision was that it was better for the American economy and national security if you let people make their networks more secure and created a safe environment through online commerce through encryption. Intelligence-gathering was not over, it would just have to find new ways of working. Making money was the priority, but allied to an ideology that technology would aid globalisation and bring peace and harmony and therefore security. 'The US internet policy is still largely a creature of the 1990s and the beliefs of the decline of the state and of a global economy. It all turned out to be wrong, you know the triumph of market democracy was nice while it lasted but it's very much a creature of the California culture of the 1970s and 1980s and then the Washington foreign policies of the 1990s,' argues Lewis.[26]

And so by the end of the 1990s not only was Clipper Chip dead but export controls on cryptography were being lifted. And so the ragtag Rebel Alliance had defeated the Empire and blown up the Death Star. The crypto wars were over. Weren't they?

Back in 1976 at that meeting at Stanford there had been an afternoon session that had not been recorded. It was even more candid than the heated opening meeting. At one point, Diffie spoke with Arthur Levenson about what the spread of encryption meant for the future of communications – or signals – intelligence. 'I went through what I thought was the economics of the matter: roughly speaking, encryption was falling in cost and the value of using it was rising and so at some point signals intelligence would go away,' Diffie told the NSA veteran. 'Whit, we've heard these arguments before,' Levenson, whose memory stretched back to Bletchley Park, replied. 'After World War Two people said people know how to make good rotor machines and business would dry up.' They had been wrong, Levenson told Diffie. It had not turned out that way in the end. Forty years on, as Diffie recalls the conversation, he shakes his head with a rueful smile. 'I was clearly mistaken,' he says. 'Utterly.'

Diffie slowly came to understand a central paradox that does much to explain the last seventy years of the battle between securing and exploiting communications. 'The sources are fragile' when it comes to signals intelligence, he says. 'But the phenomenon is robust.' In other words, code-breakers are always fighting to keep up with new technologies and devices and new means of securing communications in the form of encryption, but even as they lose their fragile hold on some of those sources, they still manage to gather intelligence. Why? 'It took me a while to realise that the growth of communications has outrun the protection of communications for all human history,' Diffie explains. 'If you were to write out a mathematical expression for the value of signals intelligence, the first term is how much the opponents are communicating.' In other words, intelligence collection remained robust because, even as specific sources became easier or harder to read, the number of people communicating was rising and the amount each individual communicated was likewise growing as part of an information revolution that was only just beginning in the 1970s. In the years after Diffie's conversation with Levenson, the volume of information that was out there in the ether began to grow faster and faster like an exploding star. And as that happened, people began to understand for the first time that in this new world they were under attack. Robert Morris would be one of the first to understand that, thanks to a painful lesson.

CHAPTER SEVEN

ATTACK

Robert H. Morris's phone at the NSA did not stop ringing on the morning of 3 November 1988. Morris was Chief Scientist at the agency's National Computer Security Center (NCSC) and, with his long, scraggly beard, a man who seemed to blend the past and present, every inch the brilliant eccentric, equally at home poking holes in a computer system or building a sheep pen in his back garden based on designs from medieval manuscripts. The three chalkboards in his office were filled with what he told visitors were Russian riddles and mathematical equations. But he also had a Dockmaster computer terminal – the first (and at that time the only) NSA system which connected to the outside world. It was used to share unclassified computer security information with contractors and gave the NSA user a most unusual thing: a public email address and an insight into the early internet. Morris logged on to Dockmaster at 6.30 a.m. that November morning as it became clear something bad was happening online. At 6.45 a.m. he received the first call. It came from Cliff Stoll, a computer expert-cum-sleuth at Lawrence Berkeley Laboratory in California. 'Hi Bob,' said Stoll. 'We've got troubles.' Stoll had no idea how true that was for Morris.

A few hours earlier, Stoll had been woken by a phone call. His glow-in-the-dark alarm clock told him it was the inhuman hour of 2.25 a.m. A friend from an NASA computer lab was on the other end of the line. 'No apologies for waking you up,' the friend said excitedly, 'our computers are under attack.' Stoll stumbled over to his Macintosh. But it didn't start. After five minutes still nothing. When eventually he connected to a bulletin board he found a message from another Berkeley user that had just been posted. It read: 'We are under attack.' The internet was going down.

In the tight confines of the community, there was blind panic.

All across America machines were grinding to a halt, sometimes slowing down and stumbling like late-night drunks, sometimes falling over entirely, crashing to the ground in a catatonic state. This was only happening to machines connected to the internet and it seemed to be spreading. This suggested that something was travelling through the internet itself and *infecting* the machines it came across. The small group of computer experts would never forget the next few days and the lesson they taught about the vulnerability of the computer systems on which they depended.

It was called a 'worm' (although some thought 'virus' was a better description: the difference is based on the biological analogy that a virus works by infecting a host, whether a person or a computer programme, while a worm lives and replicates by itself). Whatever name people preferred, it had been unleashed at 5 p.m. on America's East Coast the evening before Stoll's call to Morris. It took an hour to infect its first machine. But then it picked up pace. Half an hour later it hit the RAND Corporation, which carried out defence research for the government. By 9 p.m. it had headed over to the West Coast and Stanford. The internet had been designed so that responsibility for securing computers was devolved to each 'host' or institution, like a university. The result was that no one was in charge. By 11.45 p.m. the Army Ballistics Research Laboratory was hit. Fearing an attack by some kind of unknown enemy, it cut itself off from the internet.

No one knew where the worm was coming from or what it was doing to their systems other than slowing them down. Was it destroying data or stealing it? Would their machines ever work again? It was utter confusion. Just after midnight, Princeton University's main computer crashed. Just after 1 a.m. Lawrence Livermore National Laboratory was struck. Desperate researchers began capturing the worm and dissecting its code to understand what it was doing. By 11.30 in the morning, the US military closed down any links between its own internal MilNet system and the ARPANET to prevent cross-infection. It was an epidemic.

The handwritten operations logs by the desperate computer staff trying to respond reveal, minute by minute, the frantic attempts to control the spread. Shutting Cray supercomputers on and off. Disconnecting terminals from the network and searching through code. Working through the night, some had figured out ways to stop the spread and were sending instructions. But these messages were themselves becoming clogged in the system and not getting through. In all, 6,000 machines had been infected. That may not sound a lot but in 1988 the entire internet consisted only of 60,000 connected machines (up from twenty-five machines on ARPANET in 1973). There was a simple equation here: the easier you made it for machines to talk to each other, the easier you made it for something bad to spread among them. (Britain was barely touched by the event since it had very limited connections with the US ARPANET.)[1] A network had been constructed to share, but sharing meant sharing whatever bad things got into the system. All of this meant that one-tenth of the entire internet had been taken down in hours.

'Whoever wrote the virus must be laughing, but it's going to mean a rough day for everyone,' Stoll told Morris on the phone that morning.[2]

'Any ideas who started it?' Morris asked.

'Nope,' replied Stoll.

'Don't worry about it. I'll look into it and see what I can do,' replied Morris. The call ended and Stoll stared back at his computer. Within a few moments he had fallen asleep in his bathrobe, head slumped on the keyboard. But as Bob Morris tried to find out who had attacked the internet, he would have no idea how close to home the source of the problem lay.

A chance find in a Manhattan junk shop had led Bob Morris into the classified nerve centre of America's computer defences. Twenty-five years after leaving Harvard, he wrote a message in the anniversary notebook for his class. 'A long time ago,' he wrote, 'I promised myself that I would learn to read Greek, learn in some detail how the planets move in their orbits, and how to decipher secret codes. I have gone a long way toward keeping all three promises.'[3] From the 1960s he was working at Bell Labs, at the leading edge of computer development. He had been part of the team that had designed the first modem, which could then move data at the rate of forty-eight bits per second, and was one of the first people to have a remote computer terminal in his home through which he could dial into the office. In the early 1960s, colleagues played a game called Darwin. They would design code which would be unleashed in a computer and fight the other person's code to the death. Morris came along and wrote a simple programme that adapted to its opponent's behaviour and wiped the field.[4]

Morris helped develop the UNIX operating system for computers, and from the late 1960s one of his tasks had been developing the process for encrypting passwords for it. This had got him interested in cryptography. Scouring a junk shop in New York around 1970, he came across a vintage Hagelin machine – similar in many ways to an Enigma. He transferred some of the techniques to UNIX and also began looking at mathematical and statistical attacks to try to break it – similar to the kind of Bletchley maths of the war. Morris, along with Dennis Richie, a colleague at Bell Labs, and Jim Reeds, who had worked on a similar attack, planned to publish a paper outlining their methods in a journal. The three agreed they should probably talk to the NSA. Morris and Richie then received a visit from a man whom Morris called 'a retired gentleman from Virginia'. 'He was quite a charmer,' recalled Dennis Richie. Over lunch, the man explained that the problem was that the method the three had discovered to attack the Hagelin machine was also applicable to systems still used by other governments, and so while there was nothing that could be done to stop publication, it might not be wise.[5]

There had been no explicit threat. That was not needed. 'I suspect we all had the feeling that we'd shaken a velvet-gloved hand on friendly terms and sensed that there was steel underneath,' Richie noted. But the man also made an offer. Their work had been good but the work inside the NSA was better. Would they be interested in consulting? Then they could be privy to the highest level of research. This was the era when the NSA was fighting to hold onto its monopoly on encryption and still hoped to co-opt those on the outside. In the end the paper was quietly shelved and a few years later Morris was not just consulting but working inside the NSA for the National Computer Security Center.

The Center had been set up as a result of the NSA's increasing use of commercially produced software: this meant there needed to be some mechanism for the spies and the outside world to talk about security issues and for the products to be vetted and certified as secure enough to be used for secret work. Roger Schell had been the driving force until he left in 1984. The NCSC did not have an easy time at the start. The communications security people in the NSA did not like talking, since their work was highly classified and they did not want anything leaking out to the new Center's industry contacts. Meanwhile, industry was wary of any kind of security demands that might slow down its systems or raise costs.

Now, in 1988, the Center was watching the first internet crisis.[6] The teams at the various universities and labs had worked out how to kill the virus, but the hunt was on for the perpetrator. TV news reports were talking colourfully of a 'dark genius' behind the attack.[7] On 4 November, two days after it all began, a reporter from the *New York Times* working on the story put a call into the NSA. Bob Morris called him back. The reporter had been talking to Cliff Stoll, who had worked out the likely computer username of the perpetrator and then what his actual name might be. The reporter could tell that Morris seemed to know a lot about what was going on and so he tried out the name he had heard from Stoll. Morris confirmed the name was indeed correct. The reporter then noticed something odd. 'It's a funny coincidence,' the reporter said, almost as an aside, 'you both have the same name.' Without missing a beat, Morris replied: 'That's no coincidence. He's my son.'[8]

'I had a feeling this kind of thing would come to an end the day he found out about girls,' his father said later, deadpan. 'Girls are more of a challenge.'[9] But not every child has his father bring home an Enigma machine from the office. The young Robert T. Morris had immersed himself in the world of computers his father had introduced him to. As a teenager he was one of a couple of dozen children of the staff at Bell Labs who had been given their own computer accounts. They had begun communicating online with each other, forming perhaps the first teenage social network.[10] Morris junior had enjoyed finding holes in computer security. When he was sixteen, he explained to a reporter (introduced to him by his father) how you could get into computer systems. The magazine described 'a quiet, polite young man with soft brown hair and rosy

cheeks' who 'has broken into password files, read supposedly private computer mail, and has broken into computers that are linked
together in networks'.[11] Morris junior even gave a lecture at the NSA
on computer security, one part of which was entitled 'How not to
get caught'. He was a shy boy who loved reading computer manuals
but also enjoyed playing pranks on friends and colleagues.

Morris junior had ended up first at Harvard and then at Cornell
University's Graduate School. In October 1988 he had started work
on a 'worm' – the term for a self-replicating piece of code which
was first coined in an influential science-fiction novel, *The Shock-
wave Rider* by John Brunner, published in 1975. From the early 1960s,
hackers had developed 'rabbits', given the name no doubt because
they bred rapidly and could bring a single computer to its knees by
flooding it with commands. By the mid 1970s, as computers were
being networked, the first worms designed to spread from machine
to machine were being developed. At Xerox they designed one, not
as something destructive but in order to move around a network
looking for free space which could be used.[12] In Brunner's book the
worm was a weapon. A totalitarian government exercised power
through computers, and so the resistance released a 'tapeworm'
which replicated itself to destroy the network. Was the Morris
Worm a Brunner weapon or a Xerox test? Perhaps the answer lies
somewhere in between. It was made to copy itself from machine to
machine through the internet, exploiting security flaws in UNIX. A
brute-force 'dictionary' attack exploited the fact that people used obvious passwords (something his father had specifically worked on).
The worm actually did nothing other than spread from machine to
machine exploiting flaws. It did no damage in itself. It was an experi-
ment to see how far something could spread.

The younger Morris designed his worm so it would spread fast
and be hard to detect and to kill. The worm was only removed
when a computer was shut down – but that only happened properly once every week or two in those days. Morris understood
that, as it spread, the worm risked clogging up the system. So
he took a precaution. Every time a worm spread to a machine it
would check whether there was already a worm present. If there
was, there would be an electronic toss of a coin and the worm that

lost would commit suicide. However, Morris knew this was also a weakness. A clever person could spot this and fool the worm into thinking there was already a copy on a machine, leading it to kill itself. So he made another adjustment. A certain proportion of worms would be 'immortal' worms which would not be killed by a coin-toss and would duplicate themselves even if there was another worm on a computer. Morris's catastrophic error, though, was over how many worms he would make immortal. He picked one in seven.

And so, on the evening of 7 November, Morris went to MIT (to hide his tracks), released his worm and went off for dinner. A few hours later he came back and noticed his machine was slow. Something had gone wrong. The one-in-seven immortal worms were spreading so fast they were bringing computers to a grinding halt. Here was a crucial lesson about unleashing attacks over this new interconnected medium: however clever you thought you were, you could never be quite sure how far your code might travel and what exactly it might do. The law of unintended consequences acted with added force in cyberspace.

Morris panicked. Within hours he had contacted a friend at Harvard to discuss a solution. They sent an anonymous message instructing people how to kill the worm. But by then it was too late. The network was already clogged under the traffic and the message did not get through for another twenty-four hours. Universities, medical research facilities and the military found their machines crashing. Once they eventually worked out a fix to stop the spread, the tiny world of the internet began to calm down. But then the hunt had begun for the culprit.

Morris junior knew there was no escape. Close to midnight on the evening after he had released his worm, Morris had called his father and told him what had happened. The next step was finding a lawyer. Within days, his name was on the front page of the *New York Times* and the national media were camped out on the driveway while his mother fed him chicken soup inside. Bob Morris acknowledged that his son's behaviour 'is not a career plus' but, ever the proud father, he also pointed out just how clever a programme his son had written. Morris senior said the action would make people more aware of vulnerabilities, but others warned that imitators would adapt the code to make it even more dangerous.

On the morning of Tuesday, 8 November the internet was largely back up and the fifty experts involved in picking up the pieces gathered to meet with officials from the National Computer Security Center. It was clear that the worm had only been stopped thanks to an informal network of people who knew each other and were able to swap notes quickly in an emergency. Everyone knew this was not going to be sustainable as the internet grew. There needed to be something more formal to deal with any future crisis. This led to the creation of the Computer Emergency Response Team, or CERT – a group to fire-fight any crisis. Today almost every country has a CERT. This was not the first virus or worm ever to spread and it would not be the last: others like Michelangelo and Conficker had significant impact on the way people saw computer security. But the 'Morris Worm' was the first to make the national news because of the havoc it caused, and the first to make the wider world – and not just the small community of computer experts – understand the potential vulnerabilities. Computers were vulnerable, but so was the internet itself.

The worm unleashed by Bob Morris's son caused shockwaves inside his father's agency, the NSA, as well as Britain's GCHQ, where experts realised they had not been paying sufficient attention. From the early 1980s they had been aware of computer hackers developing viruses to spread on early personal computers like the Apple II and first IBM PCs. An internal Top Secret NSA paper in late 1985 had outlined the possibility of something malicious self-propagating through 'viral infection', or another problem called 'denial of service' in which the computer's processing power would be used up by a virus so that it could not perform other functions.[13] But until the Morris Worm this was all theoretical and nothing to do with the secret agencies, partly because they were not connected to networks, and so could not grasp the damage that could be wreaked through them. 'That was the real wake-up call. I remember hearing about it and essentially understanding what happened, and saying, wow, we have missed it – in terms of being something we really need to be paying attention to,' one official later reflected.[14]

From the mid 1980s, computer viruses began proliferating. Many were designed to show off or send a message – like the 'Peace Virus', which popped up a message calling for universal peace to all Macintosh users (the following year saw the 'Worms Against Nuclear Killers' virus – the joke was in the name). There were the first signs of real-world consequences as well, though: two hospitals reported virus infections in 1989 which affected image displays for cardiac studies and patient diagnosis. Companies began offering virus-scanning products and then hackers started to hide their work better and even pose as virus-scanning companies to get people to download infected products.[15]

The NSA had its first experience of a hacker in 1986 with a string of incidents involving Dockmaster, a system which could be accessed externally via dial-in modems with the right password and user ID. In October someone noticed that his login was already active as he entered the system. A Trojan horse had been used to grab user passwords and then dial in. Soon after, another user noticed that details about when he had last logged in were wrong. An intruder had been in the system for two hours and fourteen minutes over several days, potentially allowing them to grab over 2 billion bytes of information. Earlier that year, one of America's most notorious hackers seems to have been the culprit in another intrusion.[16]

Kevin Mitnick's forte was employing a mix of technical knowledge and social engineering – meaning manipulating people to get them to help you. This might involve learning the jargon and structures of a particular organisation, so that when he called someone up and said he was an in-house engineer who needed help, he would sound as convincing as possible – techniques which are the forerunners of modern highly targeted email attacks.[17] He phoned a guest user of Dockmaster at the NSA and claimed he was from the NSA and was issuing new passwords. He asked if he could please get the name of the user's old password?[18] Mitnick's hacking had grown out of 'phone phreaking' in the early 1970s, in which people worked out that you could imitate the sounds that controlled the electro-mechanical telephone switches to get yourself free calls. A hacker called Captain Crunch took his name from a type of cereal after he discovered that a free whistle found in the bottom of the packet was exactly the right tone to control an AT&T telephone switch. Stephen Wozniak got to know Captain Crunch and soon he and a friend, Steven Jobs, began building boxes to do the same thing, offering them to students, part of a community of people who mixed computer hacking with phone phreaking and building their own hardware. Wozniak and Jobs decided to build a basic low-cost computer they could sell to a burgeoning subculture of people like them, and so the two started a company called Apple in the late 1970s, imbued with the hacker ethic. Computers were beginning to get personal.

Computers were also taking over the telephone switching process as telecoms companies began moving towards fully electronic exchanges to connect up calls. This was the next stage of telecommunications and computers merging (a process still under way). For hackers like Mitnick, the challenge was getting into the computers that controlled phone lines. Mitnick wrote his own programme to steal people's passwords by creating a fake login page on their computer. His motive was not to destroy but to poke fun, to embarrass, to reveal lapses and in so doing show his own superiority. The thrill of attacking a tougher target was, Mitnick says, like getting to the higher level of a video game. He began sharing tips and tricks, tools and weaknesses with a community of other hackers during the 1980s, developing tricks like routing his access through other countries (including the UK) to keep law enforcement off his back. At times the mythology surrounding him exceeded the truth. A *Time* magazine article claimed that 'putting a phone in Mitnick's hands is like giving a gun to a hit man'. And one Federal Prosecutor said of Mitnick to a judge: 'He can whistle into a telephone and launch a nuclear missile from NORAD.' It was a claim clearly designed to play into the fears arising from the movie *WarGames*, but was also patently untrue and indicative of a lack of understanding. An older generation were realising that a group of drop-out kids knew more about the technology that made society tick than they did.[19] Hacking was moving from a quirky subculture into the wider realms of social fear and becoming criminalised.

Robert Morris junior was the first person to be caught and

prosecuted, his motives never being entirely clear. In the late 1980s there had been a lot of debate within the computer community about how hard he should be punished. Some felt it was vital to be tough to establish deterrence or else others might follow suit. But one of the problems was trying to work out what damage he had actually caused. Some said it was nearly $100 million, and yet within a few hours systems had been patched and nothing had been destroyed, so others said the damage had been close to zero. This pointed to a problem that remains today when it comes to breaches in cyber security: how do you calculate the damage? Morris became the first person convicted under the Computer Fraud and Abuse Act, receiving three years' probation, a $10,000 fine and 400 hours of community service. He went on to earn his Ph.D. and become a serious and respected computer scientist and inventor, putting his past well behind him. If he had not done it, someone else would have done; but Morris was the first to reveal what chaos could be wreaked over computer networks.

Robert Morris, his father, gave the occasional talk on what was then called computer security after he retired from the NSA. He would enjoy explaining a truth that his years of technical understanding had brought. 'There are three golden rules to ensure computer security,' he would say. 'They are: do not own a computer; do not power it on; and do not use it.'

He had learnt the hard way. His son's worm had highlighted a number of key lessons: the ability to exploit vulnerabilities (both within programmes and also of humans, for instance weak passwords) – and to do so remotely; the ability of viruses and worms to propagate rapidly but also in ways even the author might not be able to predict; and the ability to wreak significant harm by bringing a system juddering to a halt. There was a final lesson, reflected in his golden rules: computers were inherently insecure. The internet had come under attack. But from one of its own, and from within the tight confines of America's computer community.

That same year there were also darker signs – signs that other countries might be able to exploit computers to spy on America. Cliff Stoll, the astronomer-cum-amateur computer sleuth who had called Bob Morris early that November morning to tell him there was something wrong, wrote a report on the impact of the Morris Worm. Thousands of miles away soon afterwards, a young man handed over a copy of that report to the KGB.

CHAPTER EIGHT

ENTER THE KGB

Berlin in the mid 1980s was a divided city on the front line of the Cold War – a wall cut off the Communist East from the West, with border guards ready to shoot and kill those trying to escape while spies plied their trade. It was here that a young West German man handed Cliff Stoll's report on the Morris Worm to an older man. The young man, a sometime croupier at casinos, fronted for a loose group of oddball hackers. He thought the older man would be interested in Stoll's damage assessment. He knew it might provide useful ideas about replicating the kind of attack on computers that the internet had just witnessed. The older man who received the report was known to him as Sergei and worked for the KGB.[1] The two were engaged in the first proven case of state espionage over computer networks. What they did not know was that Stoll was on their trail.

Cliff Stoll was not the kind of person you would expect to lead an international cyber counter-espionage operation. He was a wild-haired, slightly manic, left-wing Californian astronomer who had turned to being a computer systems administrator when his academic funding ran out. The role of systems administrator is like being the caretaker of an apartment building – making sure everything is working right, that no one is causing any problems and everyone is playing by the house rules, keeping the place clean and dealing with any problems or faults by repairing or patching them up. Stoll was like a caretaker who stumbled on a Cold War spy ring operating out of his building – but in electronic form.

The paltry sum of seventy-five cents set Cliff Stoll off on the spy trail.[2] A record was kept of who had used the system and for how long so they could be properly billed for their time. Someone in 1986 had been using his system and not paying. Stoll wanted to find out who they were. A few days later Stoll had got a message from a user of an obscure computer named Dockmaster based in Maryland saying that someone had tried to break into their system over the weekend and it looked like it was coming from Cliff Stoll's lab. What is Dockmaster? Stoll had wondered at the time. Probably some bank, he and his colleagues had decided. Stoll started to dig deeper. Who was the mysterious user behind the strange logins who used the name Hunter? Was it a student? The first trails ended up being dead ends or false leads. Whoever was playing around with the systems was managing to cover their tracks, using the anonymity of computer systems to route their phone connections through different locations in the US to hide where they really were. This created what is known today as the 'attribution problem': when everything is ones and zeros it is easy to route them through different network points and mask who you really are and where you are coming from.

Stoll became a man on a mission. He slumbered in a sleeping bag by his computer and waited until an alarm told him there was a suspicious connection. He realised that if he was going to catch the perpetrator, it was better to watch him and learn rather than shut him off. This was the beginning of counter-espionage work on the web. It is the same type of counter-espionage activity that spy-catchers in places like MI5 and the FBI carried out in the Cold War: put your suspect under surveillance, follow him, learn about him, build a case, wait for a misstep.

Stoll realised the attacker was smart. He had adopted privileges on the computer which allowed him to change the way programmes operated. But he was not destroying things. He was exploring. And he was also careful. When he had hacked in, he would period-ically issue a command to tell him who else was logged in. He was adopting what spies would call counter-surveillance measures, the equivalent of stopping and tying your shoelaces to look back over your shoulder and see if anyone was following you. Stoll realised he had to hide himself. This was electronic hide and seek. When he approached them, the local FBI seemed uninterested. 'You want us to investigate someone who has stolen seventy-five cents' worth of computing time?' they asked incredulously. Stoll watched as the hacker jumped from his system over to MilNet – the military part of the internet which had been hived off from the civilian part in 1983. He was heading for air force systems, contractors working on satellites and part of the US Army missile complex that was online, using a Trojan horse programme to fool people into giving away their passwords.

At the US Air Force Stoll's call found its way to Jim Christy, who worked as an investigator at the Air Force Office of Special Investigations at Bolling Air Force Base in Washington DC. Even though in 1978 the air force had set up what was thought to have been the first law enforcement team dedicated to computer crime, Christy found higher-ups uninterested in Stoll's case. 'I had to meet with the CI [counter-intelligence] guys who spelt computer with a K. I spent four hours trying to convince them that it was a national security issue.'[3] Since the counter-intelligence people were dismissive, Christy began running the case as a 'time and attendance' fraud case. Slowly, a small community grew up who shared an interest in solving the mystery.

The hacker seemed to be searching in particular for information about NORAD, and one particular programme. 'Star Wars'– or, to give it its proper name, the Strategic Defense Initiative [SDI] – was a signature programme of Ronald Reagan's presidency. In the 1980s, the danger of nuclear oblivion still loomed. Star Wars aimed to alter the balance of the Cold War by shooting down Soviet missiles with lasers in space. This would protect America, but also undermine the concept of mutually assured nuclear destruction. This was part of a new strategy to push the Soviet Union harder — arming rebels in Afghanistan and upping the rhetoric about an 'evil empire'. The project embodied Reagan's optimistic faith in technology. As science, it was hokum and never came close to working. But, as part of a broader campaign of putting psychological pressure on the enemy, it worked a treat. It made the Soviets paranoid that they were falling behind in technology (which they were) and undermined their morale. The Soviets were desperate to find out what they could about Star Wars and – if possible – steal the technology. The CIA had already taken advantage of the Soviet hunger for technology earlier in the decade in one of the most noteworthy, but under-appreciated, intelligence operations of the late Cold War.

President Mitterrand of France took President Reagan aside for a discreet conversation on the sidelines of a meeting of world leaders in July 1981. Mitterrand surprised Reagan by revealing that France had managed something that America's CIA had struggled to achieve: a deep-level penetration of the KGB. The mole, Colonel Vladimir Vetrov, was providing thousands of documents – snapped with a Minox camera or photocopied. The information outlined how 'Line X' of the KGB was tasked with stealing the most advanced Western technology. The Soviets estimated that stealing documents on the US F-18 fighter jet saved five years of development and tens of millions of dollars, as they could copy elements like the fire-control radar for their own latest-generation fighters. The target list also included high-end computers, which Communist countries were banned from importing.[4] Computers, like fighter jets, were becoming a key component of national security and therefore a target for espionage.

The Soviet Union had fallen well behind in the computer race by the 1980s, fifteen years by some estimates. At the start of the Cold War it had been close to the pace, starting to build its own stored programme computer in 1948. And the country was as good if not better than the West at cryptanalysis, with a rich tradition of codes and brilliant mathematicians. Some of the work was based at Marfino Sharashka, a secret research and development laboratory in the Soviet gulag labour camps described by Alexander Solzhenitsyn in *The First Circle*. 'Everything stemmed from there; and the methods were barbaric, of course, but ultimately a pretty serious cryptography tradition was created whose fruits we are enjoying to this day,' a senior Russian electronics expert commented recently.[5] The USSR invested in cryptanalytic special-purpose machines but struggled with general-purpose digital computers. In the 1970s, their latest RYAD system was a reverse-engineered (and poorer-functioning) version of an IBM computer.[6] A centrally planned economy stifled the kind of innovation and risk-taking needed in high technology.

From the late 1960s the USSR had become increasingly worried about lagging behind in something clearly important to both economic growth and military might, as well as a symbol of superpower status. That led the KGB to focus on stealing Western computer technology to catch up. It had a long-standing agent (a French citizen who had been born in Russia, codenamed ALVAR) in a senior position inside IBM's European headquarters in Paris from the 1950s through to his retirement in the late 1970s, who received the Order of the Red Banner for his efforts.[7] ALVAR seems to have passed on details of computer networking and its security which were then copied by the Soviet Defence Ministry. Other agents were also in place in companies. In Communist East Germany, the spy service expended considerable energy in trying to place spies inside the West German wing of IBM. This meant that the top East German electronics company soon 'became so heavily dependent on surreptitiously acquiring IBM's technological advances that it was, in effect, a sort of illegal subsidiary of that company,' wrote East Germany's top spymaster Markus Wolf.[8] This form of spying – transferring technology – was crucial to any power worried that its opponent was ahead.

Vetrov's role was to sit at the meeting point between the demands of Soviet state industry and the work of the KGB in fulfilling its need for a wide array of Western high-tech equipment. The Russian had crashed his car in Paris and made a panicked call to a man who was a source for the French domestic security service, the DST. That led to Vetrov being recruited as a spy. He was a classic disaffected agent, egotistical, unstable, unhappy with his career and colleagues, who drank too much and had a messy personal life. But he had top-level access. The French DST, unused to running agents abroad, turned not to their own overseas intelligence agency but to Britain's MI6 for advice on how to run Vetrov in Moscow since they had more experience with agents.

Vetrov had been given the codename Farewell and revealed a gold mine, including the full requirements list for Soviet technical intelligence, which ran to hundreds of pages and listed every piece of technology the Soviets were seeking from the West and where they hoped to get them from. Vetrov also provided the names of the undercover Soviet agents tasked with undertaking this massive acquisition. British intelligence advised the French that, to fully exploit this, the Americans should be brought in – leading to Mitterrand's disclosure to Reagan.

In Washington, an adviser to the President on technology called Gus Weiss saw a possibility. Immediately recognisable for his lack of hair and known by some as Dr Strangeweiss, in tribute to the scientist in the iconic Cold War film *Dr Strangelove*, Weiss came up with a plan to exploit this intelligence treasure trove to support Regan's strategy of putting pressure on the Soviets. This could be done by making use of the Line X shopping list. 'I met with Director of Central Intelligence William Casey on an afternoon in January 1982. I proposed using the Farewell material to feed or play back the products sought by Line X, but these would come from our own sources and would have been "improved", that is, designed so that on arrival in the Soviet Union they would appear genuine but would later fail,' Weiss later wrote in a document now declassified by the CIA. The theory was that even if the Soviets began to see what was happening, they would no longer be able to trust any of the technology they acquired, rendering their whole effort useless.

The Pentagon introduced misleading information on stealth aircraft, space defence and tactical aircraft. The Soviets ended up using a design for a space shuttle rejected by NASA. Computers were part of this operation. 'Contrived computer chips found their way into Soviet military equipment, flawed turbines were installed on a gas pipeline, and defective plans disrupted the output of chemical plants and a tractor factory,' Weiss claimed. There have been widely repeated allegations that a massive explosion on a Siberian pipeline was the result of a virus introduced into the exported software controlling valves and turbines (a so-called 'logic bomb'). US officials with knowledge of the classified record offer no confirmation (although it may be that it remains classified), and so this use of computer code to carry out physical sabotage – often cited as the first case – remains unproven. Russian experts also say they are doubtful of this story. Vetrov himself ended up stabbing a man in a park as part of a bizarre incident in which he may have been trying to kill his mistress. The result was that his treachery was uncovered and he met the traditional Soviet fate of execution. But he had a significant impact in helping the West understand Soviet technological insecurity and in increasing that insecurity by raising suspicions about whether stolen technology could be trusted – by sabotaging the hardware.

In his nightly vigils by his computer screen, Cliff Stoll had stumbled on the Soviet Union learning how to spy over computer networks in the later part of the 1980s as it sought to use a new means – what we now call cyber espionage – to steal technology and catch up. The USSR was spying through computer networks but it could not yet be spied on. The whole Soviet Bloc simply did not have the widespread use of computers among companies, contractors and laboratories that the US enjoyed. And what computers it did have were not yet linked up to the early internet, which was still US-dominated (Russia continues to have its own classified network not connected to the internet). This meant the Soviet systems were not accessible in the way American systems were. And the places that were using computers and were networked-up were precisely the places that engaged in the high-tech research that the Soviets were most hungry for.

In 1979 Roger Schell had predicted a KGB man explaining to colleagues what riches lay on American networks. Now, a few years later, he was being proved right. What was crucial about this new world of hacking compared to the old world of intercepting communications was that it offered access to the mass of data stored on computers in their databases (known as 'data at rest'), rather than just the narrower field of 'data in motion' which was being sent from one place to another by someone. If you could get inside these systems, you could get access to the trove of information kept in databases and then extract it. This was at the heart of the operation Cliff Stoll had stumbled across.

Stoll watched as the hacker began looking for pathways into the CIA and into Dockmaster (which Stoll now realised was linked to the NSA). Those agencies began to get interested in Stoll's work. The CIA began following the case daily. They and the NSA encouraged Stoll to keep at it although, thanks to the country's bureaucratic turf wars, neither wanted to tell him or the other agency what they knew. So Stoll was left to his own devices.

The hacker was also patient and persistent. He wasn't picking locks so much as wandering about seeing who had left their windows open through poor security and then using that as a chance to get in and rummage around. He was different from vandals who were noisily trashing things. Stoll began to get to know his adversary and his personality through his online behaviour, just as the people intercepting radio messages in the Second World War could often tell who was tapping out a Morse code signal from the way they hit the keys (known as 'fingerprinting'). This hacker was confident and arrogant, Stoll thought. He also noticed he operated at unusual times – normally early afternoon. Most hackers worked late at night. What if he was in Europe?

Stoll gave a talk to the National Computer Security Center about what he was learning. He kept being interrupted by a bearded guy who asked him about astronomy. Afterwards the man introduced himself as Bob Morris and took him to meet the Assistant Director of the NSA. Stoll wondered what he, as a long-haired astronomer without either a tie or a security clearance, was doing briefing the NSA but nevertheless explained what he had found. The Assistant Director explained that the NSA was having difficulty convincing senior officials that computer security really was a problem and wanted Stoll to brief some of them.

'Can't you just tell them?' Stoll asked.

'We've been telling them for years,' Morris said. 'But this is the first documented case.'

Cliff Stoll's day-to-day logbook, the NSA man explained, was the first hard evidence that America's enemies were spying on it over computer networks. Stoll was becoming less and less sure that the people he was talking to were 'the war-mongering puppets of Wall Street' that he had expected (leading some of his Berkeley friends to think he had gone a bit native with the spooks). Stoll's motivation was anger at someone getting inside his network and then using it to jump off and do the same to others, even if they did belong to the military and do strange things with satellites and missiles. For him, this was a community being attacked, not a computer being penetrated. There was a feeling of vulnerability and violation that came from someone else being on your system.

Stoll's girlfriend came up with a clever idea. Why not create a file of what looked like 'Star Wars' material to lure the hacker out into a place where he might spend long enough for his phone to be traced? This was the electronic equivalent of the good old spy trick of a 'honey trap'. Put something too good to be resisted (whether a secret file or a pretty woman or man) and wait for your target to do the rest. People were behind cyber attacks and they were only human after all. The hacker fell for the trap. When he came across the files full of jargon which seemed to be about the programme he spent an hour looking at them. That was enough time to trace him to Hanover.

In Germany a rich subculture of hacking was developing. It over-lapped with a youth culture that resented authority and the older generation – it rejected the politics of the Cold War and those of corporations it saw as serving the state. This was not the violent anti-authoritarianism of the Red Army Faction, which carried out physical attacks, but it was a subculture which encouraged resistance and found common purpose in the American libertarian, cyberpunk generation. The German Chaos Computer Club, a mildly anarchic collective founded in 1984, was at the leading edge, with many experimenting with viruses. Politics and hacking fused in Germany in a way that never quite occurred in Britain. It was fuelled by fiction, including the book *Neuromancer* by William Gibson, published in 1984. In it all computers linked into a global network which cre-ated an artificial three-dimensional place through which you could navigate. Gibson coined the term 'cyberspace' to describe it. A well-thumbed copy of *Neuromancer* was in Robert T. Morris's room. Also inspired by *Neuromancer* was the group of hackers from Hanover that Stoll had stumbled upon.

The member of the group whom Stoll had been watching was called Markus Hess. Hess enjoyed reading spy thrillers, and from his quiet suburban background he seemed to enjoy being part of his own spy story. He had been inspired by seeing the film *WarGames* on German TV and wanted to imitate the character in it by getting into NORAD. Hacking offered empowerment, a chance to reach out from your bedroom and get inside the most powerful organ-isations in the world. But Hess's skills had come to be recognised by others.

One of them was a wacky, drugged-up individual who went by

the name Hagbard Celine. His real name was Karl Koch, but he had become convinced that a secret group called the Illuminati were controlling the world through computer systems. Subverting the systems was his act of resistance. Selling what he and his hacker friends could steal from the systems was also a way of funding his cocaine habit. Another hacker – known as Pengo, because he was addicted to a computer game in which a penguin of that name pushed round blocks of ice – was another key player in the motley crew that gathered. Pengo had started by hacking into the computer lab at CERN where Tim Berners-Lee was starting to think about organising information but, like his kindred spirits in the gang, the Americans were the most tempting target. In early 1986, the men had decided to offer their skills to the Soviets – both sides seemed as bad as each other so why not make some money off one of them? They tried to justify it by saying that they would be helping the less technologically advanced side catch up with the other, evening things out. So they called it Project Equalizer.[9]

In September, Peter Carl, the sometime croupier, had driven over to Berlin and then taken the subway to the East of the city where he walked into the Soviet Trade Mission. Like Cold War defectors, he had simply walked up to the glass partition and asked to speak to the KGB. When Sergei Markov appeared, Carl proposed a business deal. Sergei did not understand much about hacking but seemed interested to learn more. Carl said he and his friends could offer a package of secret information. His price (which you can almost hear him saying in an Austin Powers voice) was one million German marks. Sergei told him to bring some material. After that, they could talk about a price. At the next meeting, over coffee in an apartment on Leipzigerstrasse, the men handed over some disks of information they thought would be interesting. But then Sergei pulled out his wish-list: he wanted high-level engineering and operating systems software of the type the West had banned from being exported to the Eastern Bloc. He also wanted computer-aided design software for making chips. It was a shopping list for technology transfer. Sergei told them after receiving the first package that it was not worth a million dollars (and the hackers did indeed sometimes pass off freely available software and pretended it was classified), but he explained there were things the Soviets wanted – the Line X requirements list. He wanted material on nuclear weapons and the Star Wars SDI initiative. Sergei was very clear about what he wanted: information from US military computers plus code compilers and source code.

The hackers noticed that Sergei had a very precise catalogue of what he wanted in terms of databases and software. Intriguingly, some of the items had been crossed out, even though the men knew they had not provided these items. 'You have competition,' the KGB man remarked.[10] We know the details of the gang that Cliff Stoll tracked, but this suggests there may have been others.

The phone-call trace led German authorities to track down the gang Stoll had stumbled across. Hess was arrested in 1989 when he returned to his apartment after an early-morning swim. The men at his door told him he was under suspicion of espionage. Pengo wrote a strange post on an international online computer forum saying he had been motivated by the hacker ethic and not by geopolitics. The trial proved difficult. Explaining cyber attacks to a judge – and getting the right evidence into court – was not easy. Stoll went over to give evidence. Hess ended up in jail. Hagbard Celine's charred body was found out in the forest. Suicide. Probably. Although it was hard to be absolutely sure. How much damage was done by the group? It is hard to be certain what the Soviets did with the fresh information they received. Robert Morris wrote a memo for the NSA on the subject that suggested the Russians had got ripped off in terms of what they had got for the money.

But by the time the trial took place, the world had changed. A new form of espionage had arrived just at the moment when the old game of spies was ending. The KGB, the CIA, MI6 and everyone else was taken by surprise in November 1989 when the residents of Berlin first began to chip away at the wall that divided their city with hand tools, and then tore it down as East Germany's willpower to enforce the dividing line collapsed. Soon afterwards the Soviet Union, unable to compete economically and technologically with the West, its legitimacy hollowed out, died its own death. The Cold

War was over. But spying was not going to disappear, nor was the new world of computer-based espionage that had just emerged. In the Hanover case, hackers were becoming spies. Eventually spies would become hackers.

CHAPTER NINE

OUT OF THE COLD
AND INTO CYBERSPACE

August 1991 was a month in which the world changed. Tanks moved into Red Square as the KGB and the old Communist elite launched one last desperate and ill-fated attempt to prevent the demise of the Soviet Union. The coup lasted only a matter of days, and by the end of the month the statue of the founder of the secret police was being torn down in Moscow and the Cold War was truly over. The other event – much less newsworthy at the time – occurred in a quiet part of Switzerland, not far from Lake Geneva. There, at the laboratories belonging to CERN, which carried out advanced nuclear and particle research, the first ever website was put online thanks to the ideas of Tim Berners-Lee. A note at the top of the page explained to visitors that it was 'a wide-area hypermedia information retrieval initiative aiming to give universal access to a large universe of documents.'[1] Following the end of the Cold War, a New World Order was promised in which the single remaining superpower would act as global cop to ensure peace. And out of that peace would come prosperity driven by international trade, globalisation and technology – symbolised by the World Wide Web. Silicon Valley was growing in power and influence, increasingly decoupled from supporting the military and instead focusing more on consumers and commerce in America but also around the world. Spying and all that cloak-and-dagger stuff seemed a little passé.

It was in these years that GCHQ and the NSA realised they could learn something from their old, now departed sparring partner the KGB. Cliff Stoll's Hanover case had shown that it might have been possible to gather intelligence over computer networks. Now, Western spies realised that others were connecting online and they could find their targets on the internet. And once they started looking, they found the doors were unlocked. 'It was like Christmas,' says one former senior intelligence official. 'There were so many open ports [access points] and networks.' This was an era in which nothing was encrypted and everything was easy to get to.

So who were the first targets for Western computer espionage? At that time, there were only a limited number of users of networked computers in countries other than the US and the UK. But one very interesting community was online. The clue is in the fact that CERN put up the first website. That was the community of scientists. Now CERN, with its particle physics work, would not have been of significant interest. So which scientists were? The answer was scientists working on nuclear, chemical and biological weapons programmes in states of concern. These scientists were communicating with their colleagues within their country at labs and universities and sharing details of their work. And, of course, at this time none of them would have had much idea about computer security and the possibility of espionage over computer networks. It was simply not something people in those countries would have been thinking of – which left them open to spying. One of Britain's first targets was the nuclear programme of Pakistan, whose scientists needed computing power for modelling and research as well as communications, leaving themselves vulnerable to GCHQ's spies.

A similar shift to online spying came in the US around the same time. 'NSA, in 92, 93, 94, started moving its activities onto the internet,' says the former US cyber tsar Richard Clarke. 'And found that it was remarkably easier to collect diplomatic transmissions and other information when people used the internet than if they were doing encrypted things over the air.' One individual in the US, an Italian-American, was found to be particularly proficient at hacking and was hired to lead a team at the NSA, according to a former US official. Britain was keen not to leave the Americans to this game alone. One Briton describes GCHQ's mindset as that of offering tailored Savile Row suits while the American product was more like mass-produced off-the-peg Marks and Spencer's suits. The British thought that, in cyber espionage just as in human intelligence with MI6 and the CIA, they could offer quality rather than quantity of their product. German spies also seem to have begun carrying out computer espionage around this period, perhaps even a couple of years earlier (although not quite on the scale that some have claimed).[2]

In Russia there was an awareness of vulnerability. That was evident to Roger Schell, who had been one of the first to understand computer security in the US Air Force in the 1960s, when he visited Russia just after the Cold War ended. He was working for a private company trying to sell software, and an admiral in the Kremlin in charge of evaluating security said their primary concern was what the translator described as 'undisclosed functionality'– in other words, hidden vulnerabilities in hardware and software. Schell was peppered with questions at a conference about trapdoors and Trojan horses. The Russians had learnt to fear technology as a tool for sabotage. The Farewell operation had also emphasised the way in which you could sow confusion and doubt in your opponent's mind and keep them off-balance by making them unsure of what was genuine and what had been doctored, thereby forcing them to discard genuine material as well as fake. Of course, if you were an American spy after the Farewell case, you would have two questions. Who else can we do this to? And if we are doing it to others, might someone else do it to us? This was the new world of computer espionage – one defined by opportunity and anxiety.

When Saddam Hussein's Iraq invaded Kuwait in the hot summer of 1990, America gathered its forces to drive him out. Strangely, one of the people summoned for duty in that mission was Bob Morris of the National Computer Security Center at the NSA. He was sent to work with the Joint Chiefs of Staff. This was a highly unusual assignment for a scraggly-bearded computer programmer. He was, it seems, involved in finding ways to target the Iraqi defence system, but the exact details remain classified. The First Gulf War is often portrayed as the first modern information war – a world of smart bombs and CNN, of deception and the lightning annihilation of a vast but static army. It certainly was a stunning display of firepower: America and its allies defeated what was the fourth-largest army in the world in 100 hours. Burnt-out tanks, obliterated from the air, littered the road from Basra to Baghdad, which became known as the Highway of Death. Accounts from the time, though, contrast with the now established view that the deployment of data was seamless. In fact, they make it clear that the use of computers was messy and improvised. Packet-switched computer networks were deployed but were not always compatible, meaning that floppy disks had to be carried back and forth, with people desperately trying to send data down slow lines.

The use of what was called 'computer intrusion' as part of the effort to destroy Iraqi radar and communications remains classified, but other techniques – such as clouds of anti-radiation missiles despatched from aircraft and even carbon-fibre ribbons dropped on electric power stations to short-circuit them – seem to have been far more important. But, taken as a whole, this 'electronic warfare' was highly significant. The centralised Iraq system was largely knocked out, although just enough communications were left to ensure that a deception operation regarding coalition plans of attack could be played on the Iraqis, which could be monitored to ensure they had bought it (much in the way Bletchley Park had been used to check that the diversionary plan over D-Day had been bought). It may have helped that modified communications equipment had reportedly been sold to the Iraqis in the 1980s which could be monitored.[3] Even if computer attacks were limited, there were signs that the military was now beginning to understand the vital need to integrate information into warfare as part of the 'Revolution in Military Affairs'. Bandwidth and connectivity were going to be vital for warfare.

Every change in communications technology had profound implications for warfare and the point at which warfare overlaps with espionage. The telegraph had allowed mobilisation and communication in a way previously unknown (leading to the cutting of cables in the First World War); radio had allowed a central hub to direct navy ships and others at sea. Now connected computers offered something new – but no one was yet quite sure what. US thinkers and strategists began to talk openly about 'information warfare' in the 1990s, even when it was not clear how computers might be employed. Because other countries were barely hooked up to the internet, the big puzzle for military thinkers was how they would get access to enemy systems.[4] In one publication, senior officials talked about limiting an enemy's mobility by targeting oil refineries. It says that this could be done through penetrating the automated control systems for the refineries in advance so that they could be turned off 'at a moment of our choosing'. This was, the article noted, 'a classic example of strategic attack'.[5]

On 3 March 1997, the Secretary of Defense officially gave the NSA the authority not just to spy ('computer network exploitation') but also to develop what were called 'computer network attack' techniques – meaning operations to disrupt, deny, degrade or destroy information resident in computers and computer networks, or the computers and networks themselves.[6] It was recognised that new weapons were needed. Alongside tanks, missiles and guns there would be viruses, worms, logic bombs, trapdoors and Trojan horses. 'They are publicly available, very powerful, and, if effectively executed, extremely destructive to any society's information infrastructure,' wrote William Black, one of the NSA's leading figures, in a once secret but now declassified internal article. The potential to attack another state over computer networks was already understood, for instance by the 'digital coercion option' of targeting the information infrastructure of an adversary. 'Such information infrastructures are expected to be primarily computer-controlled, operated by the commercial-civilian sector (unprotected), and those primary infrastructures upon which military forces almost totally depend. For IW [information warfare] purposes, access to these computer-controlled infrastructures can permit the degradation, disruption or destruction of the network and/or the functions they serve. As a result, the "computers" become the intelligence "targets" of highest priority.' It was recognised that 'shaping cyberspace is a long-term activity which will require a serious continuity of effort'.

America has a remarkably open national security culture, and after a while the US realised that talking about this capability might not be such a clever idea. The Pentagon changed the terminology from 'information warfare' to the less martial-sounding 'information operations'. But word was out. Others were watching the American military closely and apprehensively, especially in China. 'We believe very strongly that the Chinese went to school on us in the First Gulf War,' says Michael Hayden, who took over the NSA at the end of the 1990s after a career in the US Air Force. 'We no longer launched waves of bombers that darkened the sun but two or three bombers with two or three bombs that had an incredible ability to put those weapons precisely where you wanted them to be . . . So the Chinese became very aware that we were an information independent – an information systems dependent – military.'[7]

America's opponents considered their choices. They could not directly confront its strength on the battlefield. One option was to make the most of the 'asymmetric' imbalance of power. This could be done in various ways. If you were a terrorist group like Al Qaeda, perhaps the best option was to think low-tech, like hijacking a plane and flying it into a building. If you were another country who feared American power, though, then perhaps you could take a different path. Perhaps you could get hold of American technology to boost your own national strength – whether in the military or in the economy at large. Perhaps you could also try to exploit America's dependence on technology to try and find its weak spot – its technology – and then use it against it.

During the First Gulf War, something happened that alarmed the Pentagon. A group of Dutch hackers had rooted around US military systems. And so, at almost exactly the same moment that it realised its huge computing lead gave it the knowledge and power to spy on and attack others, America was becoming aware that, precisely because it was the most wired nation on earth, it was also the most vulnerable to being spied on and attacked by others over computer networks. This created a strange, schizophrenic mindset which persists in talk of 'America the vulnerable' in cyberspace at the same time as it is arguably the most aggressive player, not just in developing but deploying espionage and attack techniques. It has been America's supremacy in cyber espionage that has made it so paranoid. It knows what can be done. What if other states could use cyberspace to coerce Americans and prevent the deployment of American power?

'Information Warfare poses the greatest threat to the national security of the United States,' warned one NSA official in the second half of the 1990s, just as the US was gearing up its own massive capabilities.[8] Computer networks seemed to offer small groups – perhaps even individuals – the chance to become empowered against big

states. That was precisely what the hackers loved about it. The US
military also offered – in the jargon of present-day hackers – a large
attack surface. In others words, there were a lot of places to get in.
By the mid 1990s the US Department of Defense had a vast, sprawling network of over 2 million computers and 10,000 local networks,
100 long-distance networks, 200 command centres and sixteen central computer processing facilities. Another 2 million non-Defense
users did business with the Department.[9] The first cyber defence
work was begun, to do things like detect anomalies in computer
behaviour and monitor for intrusions. Computer forensics – trying
to trace the origins of an attack for either law enforcement or intelligence purposes – was becoming a discipline, but one with huge
problems. The military network was increasingly under attack from
those seeking to steal its secrets. But how could you know who they
were?

Rome Laboratory is a long way from the Colosseum in Italy, instead sitting in a quiet town of that name in upstate New York. The
air force laboratory was home to the most advanced research on
command and control, including sensitive projects like artificial intelligence and radar guidance. So when, on 28 March 1994, a systems
administrator saw something amiss, they feared the worst. A person
was logged on who was on holiday. A hacker was at work. They had
then logged in and put something called a sniffer on the system.
Sniffers were used to secretly monitor what information people
were typing in to capture their passwords.

The US Air Force and Pentagon were called and Jim Christy, who
had worked with Stoll a few years back, got on the case with his
team at Bolling Air Force Base in Washington DC. The US Air Force
computer investigative team in the mid 1990s included people who
would go on to play a key role in American computer security – like
Howard Schmidt, later White House cyber tsar, and Kevin Mandia,
who would form his own company called Mandiant. The team
slept under their desks as they began looking into Rome Labs. They
found the labs' computer systems had been accessed 150 times by
two people who had hidden their path over the previous five days,
weaving their way over international phone switches in Bogota in

South America before capturing the details of 100 users. Trojan horses were installed and the hackers took control of the labs' network, taking all thirty-three subnetworks offline for several days. Users' emails were deleted. They had downloaded sensitive (unclassified) air tasking order research data. By masquerading as a trusted user at Rome Labs they also used their access as a jumping-off point to target defence contractors and other bases across the US. Government and commercial systems were so intertwined as to both be vulnerable to a single attack.[10]

The investigators got to work. They isolated the hackers onto one part of the network that could be watched, putting their own sniffers back on the attackers. They tried to trace the phone calls through which the attacks came, but air force officials discovered they were using a form of phone phreaking to route their attacks. They were like ghosts – intangible and elusive – running through a maze.

The main hacker went by the name Datastream. But he was not alone. One of the most disturbing things was that the authorities could watch Datastream try to attack a site and fail. Then he would go into a private internet chat room that they could not monitor and communicate with another hacker calling himself Kuji for half an hour. After Datastream came out, he would then attack the same site he had tried before, but this time he would succeed. It looked like Kuji was mentoring the other hacker. They feared this mysterious character might be working for a foreign country interested in finding research data. Were they dealing with Russian spies? Chinese spies? Germans? Or even American hackers paid off or being manipulated by a foreign intelligence agency?

The air force investigators had got as far as they could technically. So Christy and his team turned to good old-fashioned detective work. They turned to snouts – a network of informants who surfed the net. This was the equivalent in the cop show of the detective going to some dark street corner and asking the local hood if he could put his ears to the ground and find out who was behind the stick-up at the bank the other week. These informers were often people who had been investigated and arrested and had agreed to work for the detectives secretly to mitigate their sentences. Could they find out who Datastream and Kuji were? It worked. On 5 April, a week after the attack had first been noticed, one of the informants said he had previously been in email conversation with someone called Datastream Cowboy. His location was a surprise. 'He runs a bulletin board out of London. Here's his number,' Christy recalls the informer saying. This was an important lesson of the Rome Labs case: the attribution problems of the internet might be overcome with detective work and the employment of people rather than just machines.

Scotland Yard's Computer Crime Unit was only a few years old and was housed in Holborn police station in London. It had grown out of a team which dealt with company fraud. Comprising half a dozen officers, it was not well resourced and had to rely on equipment being donated by companies and universities. One of the team was Mark Morris, who had become interested in computers after putting case notes onto a system while based at the high-security Paddington Green station. The team he joined were trying to gauge the new problem, occasionally posing as hackers to go undercover and gather intelligence. They had already dealt with a few cases – notably one involving a man who had written a nasty piece of work called the AIDS virus. People downloaded a programme off a disk that offered details on how likely it was that they would catch AIDS (at a time when there was much fear but little understanding of the virus). Except there was another kind of virus hidden on a disk – a Trojan horse programme which would encrypt people's own information so that they could not read it. They would then be told they had to pay $189 to an account in Panama to release the information again. It was an early form of what is known as Ransomware – holding a computer to ransom. Scotland Yard investigated and found the author was a Harvard-trained biologist called Dr Joseph Popp who was arrested and extradited to Britain. He claimed he was trying to raise money for AIDS research but was declared unfit for trial because he walked around with a cardboard box on his head.

Rome Labs was now going to be the first major international case for the Scotland Yard unit. Morris went out to the US to talk to the air force team. The difference in resources was notable. 'Compared to us, it was like Star Trek,' he recalls. 'They had so much more money and understanding of the threat.'[11] At one point he was taken to a US Air Force base and onto a lift which went deep underground into a bunker reinforced against nuclear attack. As the lift doors opened, corridors stretched so far you could not see the end. It was, he thought, like a scene out of a James Bond film. 'The Americans were apoplectic,' Morris recalls of their attitude to the Rome Labs attackers. 'Their reaction was that these people need to be found and locked up for the rest of their lives.' They remained convinced that a foreign intelligence agency might be behind this – a bit like the Cliff Stoll case. The British team traced back the phone calls to a house in the north London suburb of Colindale. They then asked British Telecom to start a pen register on the phone lines that recorded all the numbers dialled. They revealed someone phone-phreaking. Scotland Yard could see that every time Rome Labs got hit, the occupier of the house was making free calls out of the UK into South America, through Europe, Mexico and Hawaii, often ending up at Rome Labs, and then onwards to contractors in California and Texas. They held their breath as they saw him penetrate something called the Korean Atomic Research Institute (it was not clear whether it was North or South Korea at the time). The US and North Korea were at that moment negotiating over the North Korean nuclear programme. If the North Koreans saw an attack coming via Rome Labs they might well assume the US was behind it. Fortunately it emerged the facility was South Korean. But what was going on?

Scotland Yard went in on 12 May. That evening Morris and other officers waited in four unmarked cars outside the Colindale house until Rome Labs confirmed that the attacker was in the system. BT also confirmed the phone line was in use through South America. One officer knocked on the door and pretended to be a courier. As it was opened, they surged in. The officers checked the house but, to their despair, initially found nothing. Then they realised that two houses had been knocked together and there was a door between them. They moved into the second building and made their way up to a bedroom in the attic. At the far end of the room the police could see a teenager on his chair with his back to them, engrossed in his home computer. Like any other kid on his machine, he was oblivious to what was happening around him and had no idea anyone else was in the room. 'He didn't hear us come right up behind him,' says Morris.

The boy was just logging out of the system. A policeman walked up quietly behind him, lifted his hands from the keyboard and then pulled back his chair. It was only at that point that the sixteen-year-old grasped what was happening and collapsed in tears. 'We scared the living daylights out of him,' says Morris. 'The look on his face was one of horror. He just curled up in a little ball on the floor.' This was the person one US official report would hyperbolically describe as the 'No. 1 threat to US security'. Like many a hacker, he had believed he was anonymous and secure.

Richard Pryce, a talented music student, was armed with a basic desktop computer. He admitted to breaking into Rome Labs and other locations and taking documents. He said he looked for the word 'missile', only to find out about artificial intelligence. He explained that he had started as a phone phreak. He had then begun to browse the hacker forums on the internet. He had first got into the American system at Rome Labs thanks to someone using an easy password that was the name of their pet ferret. A mystery remained. Who was Kuji? He had offered help and received files, but Pryce said he did not know who he was. For the next two years, the identity of the apparent mastermind remained unknown. 'The fear was that he could be a spy working for a hostile foreign power,' Morris said at the time.[12] Morris worked on the files seized from Pryce in his own time at the weekend. Finally, he found something that looked like a phone number. That led him to Kuji.

So if Datastream Cowboy was a schoolboy, then was Kuji a master spy? No. He was Mathew Bevan from Cardiff, only just out of his teens. The police turned up at his workplace, the computer department of the Admiral Insurance Company. A manager asked him to come and look at a boss's computer. 'There were seven people in the office, your typical men in black,' Bevan later recalled. When police got to his house they saw it was filled with posters and videos from the US TV series The X-Files, all about the hunt for aliens. The show's tagline was 'The Truth is Out There'. Bevan had once seen some strange lights when going from Cardiff to Newport and he believed the truth about UFOs was out there on computer systems. He admitted only to looking for a mysterious 'Hangar 18' at Wright-Patterson Air Force Base, a less famous version of 'Area 51' where Ufologists believed alien spacecraft wreckage was secretly kept. To Mark Morris's amusement, when he asked his US Air Force colleagues what actually was in Hangar 18, they told him that if they did have any UFO material that was where it would be, but other than that they could not say anything. Bevan had engaged in his clandestine pursuit 'because he could' – the traditional hacker motive. 'I did it for the pure adrenaline buzz of hacking a secret system,' he said. Bevan told the police he started getting strange phone calls from people claiming to be interested in him. In one case a caller knew not just his current phone number but also a new one he had not even started using yet. They gave an Asian name and what seemed to be a Chinese phone number.[13] He may not have been a superspy, but his skills might have attracted them. No evidence was found of any foreign intelligence agency involvement in the original case, but did someone want to piggyback off his skills once they were exposed?

When Pryce and Bevan appeared at Bow Street Magistrates' Court, it was the first time Datastream Cowboy and Kuji had actually met.[14] The US estimated the cost of their investigation was half a million dollars. The supporters of the young Britons argued that they had not done any real damage and were just kids larking around. Pryce was fined £1,200 under the Computer Misuse Act after his lawyers said he had not been motivated by gain and had not done any direct damage. The charges against Bevan (conspiracy with 'unknown others') were dropped when it was decided it was no longer in the public interest. US investigators were frustrated. Years of work had not led to much of a return. It was a strange end to what had seemed an international cyber-spying mystery. But there was also a wider point. If kids from Britain could get into Pentagon systems, then how secure were they?

In February 1998, the US was preparing for one of its periodic face-offs with Saddam Hussein over weapons of mass destruction. But, just as thousands of troops were being deployed, the US Air Force realised their systems were being hacked from a gateway in the Middle East.[15] The investigation was codenamed Solar Sunrise.

Because of the political context, Iraq was initially believed to be the culprit. President Clinton was informed. The attackers had only got into non-classified networks and seemed to be gathering rather than destroying data, but the kind of information that might flow through here would include the logistics behind troop deployments, among other things. The fear was that they could use the access to disrupt the support infrastructure needed to organise a military campaign somewhere like Iraq. Are we under attack? people asked. What constitutes an attack in cyberspace anyway? Does stealing information constitute one or is that traditional espionage? What if, by gaining entry into the system, you understand how to switch things off and stop a military attack – is that in itself an attack? A 1995 RAND study had asked what would happen if another state attacked and crippled US critical networks as the country planned an intervention in the Middle East. The simulation forced the attackers to drastically reconsider their options and the report contained a dire warning: 'The US homeland may no longer provide a sanctuary from outside attack.'[16] Was this now happening?

Senior officials briefed NATO allies to prepare themselves for their own attack. But then, as the cyber sleuths started tracing the origins of the hack, they came to a startling conclusion. The enemy was not the Iraqis. Nor even another government. It was two sixteen-year-olds in northern California who went by the online monikers of Stimpy and Makeveli. When the FBI raided their houses, they found the typical hacker set-up of a computer, Pepsi cans and half-eaten cheeseburgers. An eighteen-year-old Israeli hacker was also involved. He had been the leader of the group and was the reason a Middle East gateway had shown up. 'We had only rudimentary capacities even to monitor our own systems at that stage, so the attacks looked more serious than they turned out to have been,' says John Hamre, Deputy Secretary of Defense at the time.[17] It was a false alarm. But only of sorts. The fact that a couple of American kids had been 'mentored' by a foreigner to hack into American government systems suggested that it might be all too easy for someone clever to manipulate naïve people at home. And maybe next time it would not be an Israeli out to show off, but someone with a darker purpose.

It was the tragedy of a very traditional attack – a huge fertiliser bomb that destroyed the federal building in Oklahoma City, killing 168 people – that finally made cyber security an issue at the highest level of the White House. In the wake of the attack, President Clinton established a commission to look at protecting critical national infrastructure. This meant major buildings as well as bridges, dams and also areas like power, transport and telecoms. The commission reported back fifteen months later, in October 1997, with a startling conclusion. Rather than focus on physical attack, it spent most of its time talking about cyber threats. It found that critical infrastructure was increasingly being controlled by software and was vulnerable to attack. 'A satchel of dynamite and a truckload of fertilizer and diesel fuel are known terrorist tools. Today, the right command sent over a network to a power generating station's control computer could be just as devastating as a backpack full of explosives, and the perpetrator would be more difficult to identify and apprehend,' the report began, before calling for the government to formulate a clear policy to deal with this danger. 'That all sounded nice,' said Richard Clarke, the plain-speaking, bureaucratic warrior who worked on counter-terrorism issues at the White House. 'It didn't seem to have anything to do with me until President Clinton said we need someone to worry about this and since you worry about other forms of protection and other threats you get to worry about this as well. And I had to start learning then what cyber space was and how it worked.'[18] Clarke knew little about the subject and was shocked by what he found.

It was just as the commission was completing its report that the US government ran its most significant exercise to date. Organised by the Joint Chiefs of Staff and codenamed Eligible Receiver, the 1997 test involved a two-pronged attack. First, power companies in selected cities were subject to a simulated cyber attack. The resulting power outages were made to appear random and unrelated, leading to confusion at the FBI as to which team to put on it.[19] Senior defence officials realised they might not even know when a cyber attack was going on. 'The first three days of Eligible Receiver, nobody believed we were under cyber attack,' John Hamre later recalled. 'But back then, it was so novel and unpredictable that the game players just genuinely didn't know that that's what was going on. And the red force, the attacking force, did a very good job of masking their attack profile, so that it didn't look like it was cyber warfare.'[20] This part was just a scenario and simulation, but still scary. 'We didn't really let them take down the power system in the country,' said Hamre, 'but we made them prove they knew how to do it.'[21]

The second prong of the attack involved setting the NSA's hackers loose on the Pentagon. They were not allowed to use any inside information but only hacker techniques off the internet, and remained constrained by US laws. The results were still devastating. 'By the second day of a five-day exercise we had to call it off because the team of internet hackers from NSA had gotten into the Pentagon and gotten into the command and control network. In fact they had gotten onto the floor computers of the National Military Command Center,' recalls Richard Clarke. 'The very heart of the Pentagon. They could have sent secret-level orders to troops from the National Military Command Center. It was very shocking to me and particularly shocking to the Pentagon.' It was assessed that the attacks would have disrupted operations at select military bases, undermining the ability to deploy forces. At one point in the exercise, the Pentagon went to the National Security Council at the White House with a plan of how it wanted to respond. At this point, the Department of Justice explained that the actions proposed would be illegal since they involved operating through domestic computer networks. The admirals and generals were flabbergasted that some lawyer was telling them what they could not do to defend themselves. Suddenly cyber was not just on the NSA's agenda but that of the Pentagon as well. 'You can trace back to Eligible Receiver 97 the Pentagon's waking-up on this issue and realising not only that they had to do a lot to defend their own networks but that here was an opportunity to do it to other people as well,' says Clarke.

There is an alternative view of this exercise, though. Some of those whose job it was to defend against the attacks dispute the narrative. In fact, they say it was something of a fix. They say that their teams had spotted some of the reconnaissance for the intrusions in the exercise months earlier. They could see the vulnerabilities that were going to be exploited. But they had been told to ignore what they saw. One of them even renamed the whole thing 'Inarticulate Deceiver' because he thought it was so dishonest. In his view, the whole exercise was a carefully crafted marketing ploy, artificially designed to ensure that offence won and defence failed so that senior Pentagon officials could be convinced that there was a problem. 'You had to get a briefing saying the sky was falling in,' says one of those involved now, although he does not dispute that it served a useful purpose. This was not the first time that those concerned about cyber security would be accused of exaggerating the threat for effect. (In 1990 Britain's Ministry of Defence had only two people looking at computer vulnerabilities – mainly in terms of password controls and the like – but there was no real understanding of hacking. Britain had its first scenario-planning for a potential cyber attack in 1995, partly as it had been pushed by the US to take the issue seriously, and American pressure over an ally's vulnerability rather than genuine domestic concern continued to be the main driver.)

In 1998 in the US, fears over computer security and espionage merging – which had been present since the 1960s – became real. Just as Solar Sunrise ended, another, far more sophisticated, intrusion was detected in defence computers. Someone with a poetic, astrological bent was clearly behind the naming of investigations since this one was codenamed Moonlight Maze. Most of the details remain classified to this day. It was on a totally different scale. It was described by an official report as 'low and slow', a stealthy, highly advanced and long-term campaign only discovered by chance. Pentagon cyber-watchers did not see anyone penetrate their system, but when they happened to check, they noticed that information was going out from Wright-Patterson Air Force Base and other military labs. And not just a little, but a lot – and mainly technical defence information. Desperate attempts began to work out who had got in and how, but the teams struggled to grapple with the problem. One investigator watched a document whisked away from a print queue in front of his very eyes. They began to try to trick the attackers into revealing who they might be and what they were up to. Eventually, it emerged that someone had managed to install a backdoor (called LOKI) which provided access and which was so stealthy it might never have been seen and was almost impossible to conduct forensics on. In all, the attack had gone on for three years and was a persistent attempt to steal technical information from military and related scientific networks, largely by first penetrating academic and scientific supercomputer centres. One of the search routines the attackers kept using was for the word 'secret'.

These were no dial-up bedroom hackers doing a bit of freelance work for spies, like Markus Hess a decade earlier. 'It was very sophisticated. This was not two kids from Cloverdale, California using an automated technique to bust into as many computers as they could after school,' John Hamre recalled a few years later. 'They had strong operational doctrine on how they operated. These were people who not only had strong computer skills, but they also had very strong security skills. It suggested that it was potentially perpetrators who came more out of an intelligence background.'[22] The scale, sophistication and duration of the attack suggested it was a state.

The attackers were not using home computers. They might have used sophisticated techniques to mask their identity, but it was clear they were using powerful machines – the kind only a state was likely to employ. 'They took huge amounts of information,' says Hamre, who believed it was an attempt to steal military-related intellectual property. The fear was that this might be more than espionage – more like reconnaissance finding weaknesses and shaping the battlefield ready for war. 'We were quite worried that an opponent of this skill could insinuate surreptitious code inside machines,' says Hamre. In other words, that the opponent could leave something behind that could sabotage a system at the press of a button – the kind of 'logic bomb' that people had been talking about for years. A new task force was inevitably established by the Pentagon which looked at 'defense in depth' rather than just trying to hold off an attacker at the perimeter, which risked being a 'Maginot Line' that, once breached, was useless.

The US has never publicly confirmed who was responsible. However, insiders say they have no doubt it was Russia. The compilers that brought together the code to carry out the attack were in the Russian language. The attackers were operating from 8 a.m. to 5 p.m. Russian time and never worked on Russian holidays. Attacks could be traced back to high-speed mainframe computers in the

Moscow area – allegedly one linked to the Russian Academy of Sciences. 'Low and slow' is the way the Russians have always done their spying, and so after a brief hiatus in the early 1990s, as the Cold War ended, Russian spies were back in business. The evidence seemed strong and the US made a complaint in 2000 and even sent over FBI teams, but it was met with firm Russian denials and minimal co-operation.[23]

A few years earlier, in the mid 1990s, there had been another sign that the Russians were becoming more aggressive in computer espionage. Many of the people working in the computer industry at that time used to attend a major exhibition called COMDEX. Like most conferences, free gifts were often on offer at various stalls to get people interested. These were often T-shirts with a (perhaps) witty slogan or else a floppy disk with some new software you could download. US Air Force investigators found that not all of these floppy disks were what they seemed, though. Some of them had malicious software hidden inside them. They would secretly install on a machine and then, if the computer connected to the internet (which was not necessarily very often in those days), transmit data to another party. Who was responsible? The best guess was the Russians. A decade later they would use the same trick with USB sticks to get inside US classified networks.

The Cold War may have been over but, with the US and UK beginning to spy at the start of the 1990s and others reciprocating by the end, the era of cyber espionage had truly begun. And, just as Moonlight Maze was unearthed, attention to where the problems of cyber espionage originated from was turning further east than Moscow. A new player had entered the game and was playing for different stakes.

CHAPTER TEN

TITAN RAIN

In 1789, an ambitious textile worker called Samuel Slater left England for a new life in the United States. Slater had risen to become a superintendent at a mill where a British inventor called Richard Arkwright was developing new techniques to use water-power to drive machines as well as to divide up tasks between his workers. Before he left, Slater memorised the design and workings of Arkwright's latest water mill. England had restricted the export of such know-how and Slater had to lie and say he was a farmer. The year after he arrived in the New World, Slater worked with a Quaker merchant in Rhode Island to set up the first American water-powered cotton mill. It was such a success that Slater built up a town called Slatersville to service a growing New England factory system built, in part, on the understanding of mass production Slater had taken from home. The other great pioneer of the factory system at the time was Francis Cabot Lowell, who travelled from Boston to Britain for 'health reasons' but used the trip to tour every factory and memorise every detail. His bags were searched, but that did not matter because he had everything in his head. Slater's and Lowell's actions played an instrumental role in speeding up industrialisation in America as it sought to catch up with Britain. Slater went down in his new country's folklore as the 'father of the American industrial revolution', a phrase coined by Andrew Jackson. In Britain he had another name, 'Slater the Traitor', especially in the town whose workers suddenly felt the force of competition from an upstart, rising power.[1]

Anyone who thinks Britain has a spotless record should pause to consider the case of Robert Fortune. Three-quarters of a century after Slater had done his work, Fortune was one of the few foreigners to trek deep into China. Fortune was a botanist and gardener but also an industrial spy. His mission was to steal a Chinese secret that Britain desperately coveted – its tea. The tea plants of China were known to be the finest in the world.[2] A few years before Fortune set out in the late 1840s, Britain had despatched the Royal Navy to bully China to sell its tea in return for receiving opium. The resulting short, sharp war even won Britain the useful island of Hong Kong as a colony. China jealously guarded its precious tea secrets, but after years of effort – and some danger – Fortune, employed by the British East India Company, managed to smuggle out thousands of tea plants and seeds as well as the know-how about growing them. This was all taken to Darjeeling in British imperial India. In a few years Darjeeling tea was being sold on the global market in much larger quantities and more cheaply than the Chinese tea that had once been so prized. The Chinese tea industry was never the same again.

Call it what you will – intellectual property theft, commercial or industrial espionage or just plain old spying – but stealing business secrets from the competition is not new. It has a pedigree through the ages. As with other forms of espionage, computers have allowed something old to be done in new ways. Initially computers were the targets of commercial espionage rather than the medium through which it took place. The Soviet Union had long sought Western computer know-how, as the Farewell case showed. By the 1980s, as now, the threat of high-tech espionage was seen as coming from an Asian power whose economy was predicted to eclipse that of the West. In 1982, a joint FBI and IBM team set up a fake consulting company. It was supposed to ensnare Soviet spies seeking high-tech equipment.[3] 'We had always assumed that we would be busting foreigners with Russian accents,' one FBI official later recalled. 'I guess we never figured that we would be busting Japanese businessmen.' A group of executives from Hitachi were snared after stuffing confidential information about disk drives into golf bags. This was corporate espionage – one company spying on another. Japan's intelligence service was suspected of working closely with its companies on a vast programme to steal secrets from the car industry and high-tech sectors like computing. But then Japan's economy went into decline and attention moved on.

The other nation that upset the US around the same time was

France. Its intelligence service was said to be targeting high-tech computer companies like IBM and Texas Instruments in the late 1980s, recruiting agents and then passing information to French competitors. One story has it that a French agent inside IBM was unmasked when he drunkenly told an American friend what he was doing. Through the early 1990s, US officials – including the Secretary of State and head of the CIA – told France to dial it down. Congress started to ask US intelligence officials, 'Shouldn't we do this commercial espionage thing too?' The response of the Clinton administration, according to officials who served at the time, was to ask: 'Whom do you want us to support?' In other words, which companies would receive the stolen information? One of the obvious but often overlooked aspects of state-led industrial espionage is that it requires an industrial policy that the espionage is supporting. This is more straightforward in countries with state-owned enterprises in which the line between the state and the company is relatively porous. It can also apply when certain private-sector companies are designated as 'national champions', making it possible for them to be directly supported by the intelligence agencies. In the early 1950s Anglo-Iranian Oil (now BP) was helped out when MI6 and the CIA overthrew a democratically elected Iranian government because it nationalised the Iranian side of the business. But as time went on, in the British and American free-market economies 'national champions' became less obvious as companies grew more globalised and as competition ensured multiple players. This meant, former intelligence officials say, that while the odd bit of useful intelligence might have been passed on from spy to company, there was never any wholesale, official policy or formal channel of spying to support corporations.

The 1980s and 1990s saw industrial espionage targeting information about computers (as the KGB had done extensively in the Cold War), but it was not using computer networks themselves as the means of stealing the information – rather it still used people, who were often bribed to carry out documents. It was only a matter of time, though, before corporate espionage and computer hacking would meet. Computers made espionage easy, cheap and low-risk. Now there was no need to hand over the IBM disk drive in a hotel

room and then fly it over to Asia while worrying about the FBI busting through the door. You could do the whole thing over the internet.
As spies were dismissed with the end of the Cold War, corporate
espionage took off. Intelligence was becoming privatised, with con-
sultants offering 'due diligence' and 'business intelligence', which
they would sometimes use as a cover to steal secrets for their clients.
The private sector began to spy and to hack. But so did states. And
the big – but not the only – player would be a new entrant into the
world of electronic espionage, taking the work of men like Samuel
Slater into the twenty-first century and paying back the West for
Robert Fortune's tea-smuggling.

A local employee in Britain was a little bit curious about why a senior
executive in his company was showing interest in what he was doing.
The employee worked in the local office of the Canadian telecoms
company Nortel. Its roots went back to the nineteenth-century
era of building the telegraph cable system, and by the start of the
twenty-first century it was one of the big players in building modern
communications infrastructure. At its peak, Nortel employed 90,000
people worldwide and made up around a third of the entire value
of the Toronto Stock Exchange. In 2004, the British employee noticed that an executive vice-president of the company seemed to
be downloading from the corporate network a lot of documents
that he had been working on. It was not clear exactly why. So he
thought he would send an email to the executive and see if he had
any questions about his material and if he could help out. The executive responded tersely – 'I don't know what you are talking about'
– saying he had not pulled off any of the British employee's documents from the computer network.

Brian Shields from the IT security team was asked to look into
it. He could see that in the first six months of 2004 alone, over 1,500
documents were taken out using seven different accounts belonging to Nortel executives. These were not any old executives. The
greatest number of documents had been taken using the account
of the company's chief executive. The central security team had not
noticed anything wrong until the British employee had asked his
question. 'Nobody detected it,' says Shields. When he started going

back through the logs of activity, Shields could see it was not new. There were traces as far back as 2000 – as long as the logs had been in existence.[4]

The executives were in Canada, but the logs showed the documents were going back to internet addresses in China – mostly in the Beijing area, but also Shanghai. There was no picking and choosing. It was like a lorry turning up outside in the middle of the night, copying every document in sight and then driving off. Since Nortel was a Canadian company, the Royal Canadian Mounted Police were informed. They asked for all the reports and information. 'We got nothing back,' says Shields. He believes the authorities did not understand what was happening or know what to do.

The attackers had almost certainly used the fact that Nortel had very poor security when employees logged into the system remotely, say from their home computers. All you needed to get in was a username and password. There was no need for what was known as two-factor authentication – the addition of a second password. This might be a constantly changing number randomly generated by a key fob you would carry around (often supplied by the company RSA, which had been founded by Rivest, Adelman and Shamir), which would then be added to your own password and matched by your corporate network in a form of modern encryption. Shields asked one person involved in the security of the firm for his impressions. 'We are the antithesis of security,' that person told him. 'We did nothing from a security standpoint to keep them out,' says Shields. 'I don't consider resetting seven high-level employee passwords as taking any real action.'

Nortel had been trying to get into the growing Chinese market from the late 1990s, seeing it as the next great sales opportunity to drive growth. Chinese rules meant that required setting up manufacturing and software coding operations in the country as a joint venture. This was a deliberate Chinese strategy to bring in as much technology as possible. The concern over espionage became apparent from when Nortel went into a joint venture in the 1990s. One person thought the faxes were being monitored during the negotiations, which meant the Chinese could negotiate down the deal until it reached Nortel's bottom line and minimum position. Shields was also advised by colleagues from other countries that luggage would be searched and laptops examined in hotel rooms.

Shields knew all about the Chinese threat because he also was part of a group that received classified briefings from government. The Network Security Information Exchange brought together the government and the private sector. In the early 2000s this group began to put together the pieces of what was happening in cyberspace and realised something serious was going on. On the corporate side, representatives of sensitive companies in fields like defence and telecoms took part – the likes of Lockheed Martin, Boeing, CISCO and AT&T – but also UK companies like British Telecom.[5] From the government side were a string of people from the three-letter agencies – FBI, CIA, NSA and the like – but also Britain's CPNI (the Centre for the Protection of National Infrastructure, linked to MI5). The group would meet in Washington every other month for a day and a half to share their knowledge of security threats. Those discussions remain classified, but there is little doubt that what came up was the emergence of a large-scale and sophisticated Chinese threat. Government intelligence agencies knew a lot more than they could say. They wanted industry to understand there was a problem but they did not want to reveal everything. If they did, then in classic spy-versus-spy fashion it would tip off the Chinese to what they knew about them and induce them to change their behaviour so that they could no longer be watched.

What were they stealing from Nortel? Research and development information, pricing information, sales plans, customer information, new features, planned enhancements. Every type of business information. It was priceless to a competitor. 'They had untethered access to it all,' Shields says. 'It is the Chinese stealing the technology, making it for less, selling it at cut-throat prices and hands-down winning bids where they compete with Nortel for business,' Shields wrote to his CEO. 'How can you survive when you have information about your intimate goings-on being turned over to competitors?' he asked. Proving that the material was stolen is easy. Proving it went to China is harder but possible (even if not definitively), but proving who received it in China and what they might have done with it? That is almost impossible. At the time, Nortel was losing out in bids against a new Chinese company on the rise. Nortel was coming in at nearly 50 per cent more costly to do the same work as the Chinese firm Huawei.[6] However, there is no evidence to prove any connection between Huawei's rise and the hacking. Correlating one company's demise with another's rise and attributing that to cyber espionage is almost impossible. And Shields does not believe that Huawei itself was attacking Nortel. The attackers were high-end and stealthy. 'Are average hackers going to be that good?' asks Shields rhetorically. 'No, I don't think so. People that are pros are doing this.' He believed the Chinese state was responsible. 'This isn't fair,' says Shields, arguing that the power of a state targeting a company is immoral. 'It is not a level playing field.'

Shields wrote reports on the problem, but he believes they never reached the board or the company's new chief executive. Shields's view was that those immediately above him did not want to own up and tell the world what had happened because investors would lose confidence and the stock price would tank. His proposed solution was severe. Close off people doing sensitive research and development – no internet access, laptops and the like so that you air-gap your systems like intelligence agencies. Make your opponent go back to having to break in or recruit a human spy. 'That kind of attack is much harder and more costly to do. Right now, they can do this in a nice cosy room over there in China as they laugh at how easy it is to take from us what we have worked so hard to make,' he wrote to the CEO. But such advice went against the overwhelming trend in business to move everything online and connect it up all the time. That, of course, made everyone and everything vulnerable – all the time.

Nortel's business was starting to fall apart as it struggled to compete and pay its debts. Cuts in the workforce took their toll. Shields was one of those laid off. He then spent two weeks drafting a fifteen-page letter to the chief executive. 'Let me begin by telling you I am certain the Chinese are inside Nortel's network,' he wrote. 'They have free rein to take whatever they want and have for a long time . . . I firmly believe it is the unfair Chinese competition that is running this company out of existence.' But by then it was already too late.

In January 2009, Nortel – a company which could trace its origins back to the late nineteenth century – filed for bankruptcy. What was left of the business was sold off in pieces, leaving only arguments over the amount executives paid out in the final years compared to what little was left in the pension fund. Why did it collapse? Shields believes cyber espionage was the pivotal reason. 'I personally think it ran the company into the ground. It ran them out of business,' he argues of the collapse. 'I think it was mostly because of the cyber espionage.'

Can that really be true? In reality it is impossible to be sure about the relative role played by industrial cyber espionage. Wider problems plagued Nortel and a major study found that the company failed because of dozens of issues over a long period.[7] Among them was sharper competition from rivals, against which Nortel failed to adapt. Some of that competition was American – notably Cisco, who were developing better products. Some of that foreign competition came from China. Chinese businesses would maintain they succeeded by driving down costs rather than stealing business plans. And so the question that dominates discussion about Chinese corporate cyber espionage remains: how do you prove what is being done?

Britain and America have a fairly set idea of what espionage consists of. This, at least in terms of popular culture, comes from watching James Bond films and reading John le Carré. Espionage involves professional spies doing dangerous and duplicitous things in shady places. But espionage and intelligence are not concepts fixed in stone, and it is a mistake when trying to grasp Chinese cyber espionage to attempt to make it conform to a world it does not inhabit. China operates its own unique system in which what the West thinks of as 'traditional' espionage plays only a partial role.[8]

The modern Chinese Communist Party has three strategic goals: maintain social stability, defend the country and its interests and become a regional and ultimately global power. Maintaining a frenetic pace of economic growth is an imperative for all three. Growth is essential to project power and influence abroad but also to satisfy the growing middle class at home, and in so doing maintain the legitimacy of rule by the Communist Party elite and avoid questions about the lack of democracy. As it came out of the dark period of the Cultural Revolution, China in 1978 adopted a policy of 'four modernisations' with an aim of turning the country into a world power by the twenty-first century. In March 1986 a programme was launched which aimed to end what was seen as a century of humiliation by foreign powers (Britain and the Opium Wars, Japan more recently) by using technology to 'leapfrog' into the future (and over other countries). This meant engaging with the outside world overtly and covertly. Sustaining growth over the long term can be difficult, especially once the easy work has been done and you aspire to move up the value chain. Making cheap things cheaper than everyone else only gets you so far. China needed to innovate. And what if, at the same time as you realise this, it becomes easier to get hold of others' innovation by penetrating their computer systems and obtaining their intellectual property? China was also determined from around 2000 to become an advanced information society – using the application of technology to modernise. So cyber espionage offered a tool for state policy to support modernisation, growth, military power and independence.

One way of understanding Chinese espionage is the 'thousand grains of sand' model. In this account, three different countries want to know what kind of sand is on a beach in a foreign country. One country – let's call it Britain – would send a submarine to the vicinity and then a commando team (probably wearing black tie underneath their wetsuits) to get a sample in the dead of night. Another country – let's call it America – might use the latest technology by pointing their satellites to stare down at the beach and sending sniffer planes to collect samples from the air. A third country – let's call it China – would do things differently. It would send thousands of its own citizens to take a holiday on the beach. And, as one writer puts it, 'at sunset they would all go home and simply shake out their towels; and the Chinese would end up with more sand – and more data – than other nations'.[9] This is also known as the 'vacuum cleaner' approach – use a mass of people to suck up literally everything you can – rather than the surgical approach, supposedly favoured by both Western and Russian intelligence, in which you pinpoint the one or two hard-to-get secrets that you are really after and focus your efforts on those. Critics have suggested that this picture is too crude and that China uses traditional espionage methods as well. In practice it pursues a more 'layered' approach, ranging from the 'vacuum cleaner' to the surgical operation. This, it is said, also fits with the lack of a clear distinction between intelligence and information in China.

We think of intelligence as meaning 'secret intelligence' and involving spies as distinct from 'information', unprocessed data that is not secret. One Western lawyer says the danger is the 'poisoned umbrella fallacy' – that spying always involves danger and death. Spying in Western countries is a specialised, professional discipline which is kept as secret as possible. In China, the distinction between intelligence and information is less clearly defined. This creates an ambiguity that is useful as China built a system to gather information from abroad, often involving its own citizens, as distinct from the Western and Russian model of engaging or blackmailing locals in your target country. When a Chinese person abroad is asked to get hold of something, they might be told that this is useful for China. But they would not think of themselves as a spy. There were many ways to make such requests, from playing on patriotism to more direct forms of pressure on family back home, depending on the situation. There are also the vast numbers of Chinese students coming to the West, especially in engineering and scientific disciplines ('the quiet ones at the back' says one Western intelligence official). These are not spies. They have come to learn. There is nothing illegal in what they do. But they are all part of the plan to absorb as much as possible, which includes making use of academic research publications, product samples, patent documents and technical journals, so-called 'open source' intelligence.

When it comes to technology, you do not always have to 'steal' it by going abroad. You can persuade someone to come to you and hand it over. China did this – around the turn of the millennium – by encouraging foreign firms to set up research and development labs there, with the promise of access to the vast and growing Chinese market. In return they needed to share their technological know-how. Among those who took this path were high-tech companies

like Microsoft, IBM, Nortel, Siemens and Motorola. Access to the
Chinese market was worth any minimal risk, they thought. This
played a part in the wider Chinese strategy of acquiring technology
and then adapting it in a distinctly Chinese way.[10]

Western counter-intelligence or spy-catching services like Britain's MI5 and America's FBI were geared up to spot the tiny traces of
a Soviet agent operating under deep cover. This meant they had no
idea how to deal with large-scale Chinese espionage when they first
saw it, leading to much confusion. In the late 1990s anxieties surfaced
amid fears that top-secret nuclear weapon designs had been stolen
from US nuclear laboratories. A secret report in 1998 warned of an
'acute intelligence threat' to nuclear weapons labs as it emerged that
their unclassified computer systems had been penetrated 324 times
in a matter of months.[11] The few cyber investigators in government
were pulled away from Moonlight Maze and towards Chinese targets. But the potential for cyber espionage was not yet understood,
so the real concern remained about people with Asian backgrounds
working with security clearances. This led to the ill-fated pursuit
of Wen Ho Lee, an ethnic Chinese scientist at Los Alamos Laboratories, wrongly accused of leaking nuclear secrets in the late 1990s
and later cleared. The collapse of that case, followed by the 11 September 2001 attacks, put the issue on the back burner. But it did not
go away.

By the turn of the new century, China had woken up to the possibilities of computer-based espionage (employing it alongside other
techniques like academic and commercial solicitation rather than
instead of them). Many countries and companies conduct commercial espionage. What is different about China? The simple answer is
scale. China has attempted computer-based commercial espionage
on a transformative scale. One American writer rejects the compar-
ison with the way his country stole British technology in the past.
'In a manner of speaking, the United States stole books; China steals
libraries.'[12] In industrial espionage, as in other ways, computers allowed a scaling-up which changed the game. It made it easier to do
and harder to get caught.

The first traces of sustained Chinese cyber espionage were spotted
around the turn of the century. The targets were primarily defence

and high-tech companies of the so-called 'Defense Industrial Base', to harness their intellectual property and technological know-how, not so much for the benefit of Chinese companies as to be able to understand American military expertise and build Chinese versions of those systems. The reality of Chinese penetration was some-times met with denial, even within the most sensitive places that were targeted. People did not want to own up to the fact that they had been taken to the cleaners. The politics and secrecy surrounding intrusions could spell trouble for individuals.

When he was still a student Shawn Carpenter was drawn into the world of computer security by Cliff Stoll's book *The Cuckoo's Egg*, which detailed the investigation of the KGB-linked Hanover hack-ers. But his own experience of tracking hackers would get Carpenter into deeper water than Stoll. In May 2004, as a computer intrusion expert at Sandia National Laboratories, which dealt with America's nuclear arsenal, Carpenter began investigating a breach he had spot-ted in the Sandia system. He had already seen similar attacks a few months earlier on Lockheed Martin's network, the defence giant which ran the labs under a contract, and so he knew that whoever was behind it was particularly good – grabbing what they wanted in a matter of minutes, leaving almost no fingerprints but keeping a backdoor open so they could return later. Carpenter had become adept at using a technique called 'back-hacking' to find out who was behind breaches and what they had taken, pursuing the attackers up into their networks. This was something he had used in the past to retrieve passwords and other sensitive data. It meant following the person who burgled you and perhaps stealing your possessions back off them. Carpenter pursued the attackers online through the places where they stashed their stolen files in Hong Kong, Taiwan and South Korea to what he believed was their source – Guangdong in southern China.[13] Carpenter installed code on the last hop router in Guangdong where the attacks were coming from. Every time the group was active and the router connected, an email would be sent to an account he set up. When he looked at the account after two weeks, there were 23,000 messages. This was much more than one individual. It was a team working all hours.

Carpenter's dogged investigative work turned up a large cache of stolen sensitive documents – many defence-related – hidden on a server in South Korea. Among the documents were hundreds of pages of detailed schematics and project information marked 'Lockheed Martin Proprietary Information Export Controlled' linked to the Mars Reconnaissance Orbiter being built for NASA. Carpenter went to see his superiors. 'I was told it was not my concern,' he recalled a few years later. 'I attempted several times to find a Sandia channel to get the information to the organizations that were impacted.' But Sandia supervisors told him not to. They said they only cared about Sandia computers. Carpenter said that there surely had to be a way of sharing the information about what had happened with federal and military authorities. He was told it was not his job. The reason may have been that Sandia feared talking to others would bring 'unwelcome scrutiny' on Sandia, particularly regarding the methods by which the data had been obtained, including 'back-hacking'. The fact that the data had been lost would also not look good for Lockheed, who managed the lab. Questions would be asked about their security procedures.[14]

It was a case of putting the interests of the corporation before those of the country, Carpenter believed. So he alerted contacts in the US Army Research Laboratory. Because of their rules about working with civilians, he ended up talking to the FBI. Things began to get complicated. They were very interested as they were already investigating a set of breaches, so began using him as a confidential informant as he continued to peer into the computers launching the attacks. Carpenter had found a stunning trail of evidence, but at the same time his work might be questionable under US law because he had hacked foreign computers to find out the information. Carpenter had been using techniques that his office had approved in the past, but this time the rabbit hole he had gone down was much deeper than he had imagined.

In January 2005 Carpenter was summoned to a meeting with Sandia management, who had been informed of his work with the FBI. He was told his investigations were an inappropriate use of confidential information gained from his employment and that he would be stripped of his 'Q' clearance which allowed him to work there. Carpenter recalls that a senior security official, who had previously worked for the CIA, yelled at him: 'You're lucky you have such understanding management – if you worked for me, I would decapitate you! There would at least be blood all over the office.' Shortly after that meeting, Carpenter's boss showed up at his office. He took away his badge and escorted him to the gate. Carpenter went on successfully to sue Sandia for wrongful termination.[15] Carpenter had been exposed to the toxic brew that was Titan Rain – the single most significant cyber espionage campaign in history.

Titan Rain was causing real alarm within the military by 2003 with hundreds of Defense Department systems penetrated, according to a government alert that November.[16] Titan Rain is thought to have stolen 'terabytes' of data from Sandia Labs, NASA and US defence contractors by 2004 (ten to twenty terabytes by 2007).[17] Two years later, not just the US but its allies too were aware that something significant was going on, with the other members of the 'Five Eyes' club of Western powers issuing their own warnings that summer. 'These electronic attacks have been under way for a significant period of time with a recent increase in sophistication,' Britain's National Infrastructure Security Co-ordination Centre warned in June 2005. The centre said that nearly 300 critical businesses and government departments in the UK had been targeted.[18] Titan Rain began by going for the most sensitive industries – those in defence, telecoms and related to national security – as well as government systems.

A leading target was the Joint Strike Fighter (F-35 Lightning II), with terabytes of data stolen, including radar design and engine schematics.[19] The high-tech plane, which relies on 7.5 million lines of code to fly, was built jointly by contractors but with Lockheed Martin leading the project. The company was first targeted in 2003 as part of Titan Rain, reports said. The total investment from design to maintenance of the plane is estimated at a staggering $1 trillion. This was the modern computer-based equivalent of the theft of defence secrets that the KGB had undertaken in the Cold War (and which other countries have also utilised). Britain's BAE had also reportedly been hit as part of the campaign to steal F-35 secrets. This happened in 2007–08, according to reports of what a BAE executive said during a private dinner (the company declined to comment about those reports at the time).[20] One American said that there was not a defence contractor that had not been penetrated. Attacks became more sophisticated as defences improved, going through weak links in the supply chain as the years progressed. The attacks also began to use more innovative techniques – targeting RSA, which provided secure authentication devices for staff at defence contractors to log onto their networks.

Once the Titan Rain codename was revealed, it was changed to Byzantine Hades (and has been changed again since). The more they looked for, the more investigators found. There were at least 500 significant intrusions into the military itself with 600,000 user accounts compromised and an estimated $100 million in costs to assess damage and rebuild networks, according to a once Top Secret document.[21] Not just the F-35 but also other planes, space-based lasers, missile navigation and nuclear submarines had their designs taken. New subsets were identified. Byzantine Candor initially targeted the US Army and later the rest of the military, as well as events like oil deals, with sophisticated emails which when opened would install software that logged keystrokes to allow massive amounts of data to be taken out. Since at least March 2008 these hackers, who were linked to the PLA (People's Liberation Army) Third Department in Shanghai, compromised a US internet service provider so that they could use it to steal a complete list of usernames and passwords from a US government network. A plethora of further 'Byzantine' subsets emerged in the following years with codenames like Raptor, Anchor, Viking, Trace, Prairie and Foothold.[22]

Hacking had moved to a world far away from Cliff Stoll watching Hanover druggies as, from around 2005, cyber espionage went beyond national security-related fields into the wider corporate space. In the days of Cliff Stoll and Rome Labs, it had been about vulnerabilities in the operating systems of computers – the way they were built. They often had open ports to the outside world, no firewalls to control access and poor passwords. Stoll himself had pondered that one of the inhibitors to the hackers then had been that they came across different operating systems against which the techniques they used did not work. In the late 1990s that changed. A single operating system emerged – Microsoft Windows that (almost) everyone used. That had come at the moment e-commerce was just beginning and opened the way for a cascade of criminal hacking that exploited the myriad vulnerabilities in the ocean of code that made up the Microsoft system. By the early 2000s, Microsoft had woken up as hackers voraciously exploited its flaws and sent out viruses like Code Red, latter-day versions of the Morris Worm which crashed websites around the world and gave the company a bad reputation. 'Microsoft didn't even know how to spell security,' says one former British intelligence official of the company at that time. Bill Gates realised the company was standing on a burning platform that hackers were setting alight. This led to a famous January 2002 memo from Gates to all his staff prioritising what was called 'trustworthy computing' over the following ten years.[23] The vulnerabilities in platforms were slowly being addressed and patches were being issued to close them. But as that shell was hardened, the adversaries found a new way in – targeting the humans who worked behind the shell. All they had to do was find one person to let them in so they could roam free. They would conduct campaigns that were stealthy, persistent and advanced, running over long periods. These campaigns became known as 'Advanced Persistent Threats' or APTs – *advanced* because of their sophistication and *persistent* because of their patience.

Countries had spent years perfecting the equivalent of radars to spot anything entering their cyberspace. But APTs were like stealth fighters flying underneath the radar. They did this by persuading a user to click on a link, thereby effectively inviting the enemy into your airspace and making it look like a friendly signal on your system rather than that of a foe. A phishing attack was a generic email sent out in the hope someone would click. The next technique would be spear phishing – a targeted email directed at one person and written in such a way as to make them think it was safe.

Old-school espionage involved breaking into a company in person or perhaps persuading an employee to sneak out some files. Modern cyber espionage has adapted that process for the digital age. A modern espionage heist is a carefully planned operation. The first step is finding someone who works at the office and winning their trust, perhaps by posing as a colleague (drop in a word about an upcoming conference you know they are attending, perhaps). If the first person you try is a bit suspicious, then keep going around the office until you find the one person who is too tired or busy or is just a bit sloppy and clicks on that attachment to the email you send. This allows you to do the equivalent of walking through the security gate behind them or borrowing their pass.

Once they have let you in through the door, you can make your own copy of the key so that you can come and go at will at night without the guards, cameras or gates thinking anything of it. Perhaps once you have this, you might hand the key over to a colleague. You might have had the social skills and the language skills to persuade the employee to let you inside but your colleague is much better at rooting around the office and looking for things without drawing attention to himself while you find your next target. No need for this person to rush. Perhaps spend a few months getting the layout of the building and working out where everything is. Perhaps working out how to get into some of the more secure safes in some people's offices. They will also have a list of what they have been told to look for. Finally, when they are ready they might hand over to a third person. He or she is the muscle. They know how to lift the material fast and move it out. This is the most dangerous moment (for the burglar) since the security cameras may notice someone walking out in the middle of the night with so much stuff. But the package can be deposited at a safe house nearby that your employer rented. This is a staging post so that, just in case you are followed, the trail will not lead back to your organisation. But once the package is there it can be shipped overseas via some unusual route. To a controller in Shanghai. Or Beijing. Or perhaps Moscow, Tel Aviv, Maryland. Or even Cheltenham. Job done. And of course the beauty of this system compared to a normal heist is that you are never even physically in the building itself. It is all done remotely. And you are copying rather than taking the data. If you are lucky, no one will ever even know you were there.

Mandiant, founded by Kevin Mandia, veteran of the US Air Force investigations of the 1990s, saw the first signs of wider corporate interests being targeted from 2004, but activity really picked up a few years later. The company tracked one particularly virulent group dubbed APT1, finding its fingerprints in attacks against 141 companies, largely in the English-speaking world, dating back to at least 2006. Once they got inside a network, the hackers of APT1 stayed there for an average of 356 days; but in one case they were allowed to roam for a remarkable four years and ten months. One organisation alone had 6.5 terabytes of data stolen over ten months. APT1 became the poster boy of Chinese cyber espionage teams but was only one of twenty-five or so groups seen as operating out of the country and achieved notoriety because of its focus on English-speaking targets.[24]

Companies are loath to admit they have been breached. Boards know that an admission may cost in terms of share price (and perhaps in their jobs), and the cost to the business of intellectual property theft may not become apparent for years, if at all. By then, the board members will almost certainly have moved on. What was stolen? Huge amounts of intellectual property, research and innovation, product designs and the like. Western experts started talking about heavily protected research institutes appearing in Chinese cities and the companies linked to them suddenly making huge leaps forward. Intellectual property had always had a more elastic definition in China (think pirated DVDs and computer software), but espionage allowed it to be done to a company's most precious secrets. A new drug or aircraft engine design might have taken years of work and cost millions in research, but all the work could be siphoned off in a few moments. High-tech industries in fields like clean energy, bio-tech, pharmaceuticals and new materials were all top targets – companies whose value increasingly resided in the ones and zeros of what they knew, whether research or customer data – which in turn was vulnerable.

'The greatest transfer of wealth in history' is how the then NSA Director General Keith Alexander described cyber espionage in 2012. Some of the language of Western officials about the damage from intellectual property theft has bordered on the apocalyptic. People have proposed extraordinary figures – hundreds of billions or even a trillion dollars of damage. But is that right? It may well be the greatest transfer of *information* in history. But has that extended to the actual transfer of real *wealth*?

Ask the prophets of doom to cite specific cases where the theft of intellectual property has done material damage and they often pause. Information and intellectual property may have been stolen on a massive scale. This seems indisputable. But how often has that stolen data actually been used to build something that has disadvantaged the original owner? An executive from one very well-known British brand says that even if its highly advanced engineering designs have been stolen by the Chinese, he doubts they have the ability actually to build what is contained in those designs. There may be a lag time between acquiring such data and using it, though. It may take five years, or maybe ten, to be sure what the long-term consequences are. We simply do not know at the moment.

When pressed to be precise about damage, one area Western experts cite is the defence industry. This belongs much more to the ancient tradition of spies targeting defence secrets so as to replicate a weapons design or understand potential vulnerabilities. US experts point to the fact that the Chinese achieved the advanced skill of making a submarine move quietly far faster than the US or Russia. The other example they point to is the J-20 Chinese stealth aircraft. Again, it seems China was able to develop the technology far faster than anyone expected. The new plane arrived around a decade after Chinese hackers (likely as part of the Titan Rain campaign) broke into a US facility called (ironically) China Lake, which is one of the premier research grounds for stealth.[25] However, they may also have got some of the technology from a US aircraft shot down by Serbia in 1999: as is often the case, it is hard to be sure of the exact role of any cyber component.

When Coca-Cola was negotiating the multi-billion-dollar purchase of a Chinese company, the APT1 group is reported to have got hold of their negotiating strategy. The bid failed.[26] Another campaign by a different group called Night Dragon is reported to have targeted oil and gas companies like BP, Shell and Exxon.[27] The thieves appear to have taken highly valuable geological data about where possible gas and oil reserves might lie which may have been worth hundreds of millions of dollars as it could allow a competitor to try to move in. This is the other side of commercial cyber espionage: stealing not intellectual property but business-sensitive information. It might relate to contracts, mergers and acquisitions or bids. China is hungry for natural resources and mining is one industry that is believed to have been hit hard over contract negotiations. Companies trying to buy up a copper mine in the Congo might find themselves losing out but not realising they had been outbid by a fraction based on inside knowledge.

These campaigns are often seen as state-sponsored. Why? It is partly based on what was being stolen. If you are burgled and thieves leave behind the expensive jewellery but take some notebooks, then you know that it was not an ordinary burglary. Modern cyber burglars were not taking information which could be monetised easily – for instance, people's financial information, like credit card details. That suggested another motive. However skilful they are, digital burglars also leave fingerprints – not real ones, but tell-tale signs you can see after they have broken in. How did they do it? How much reconnaissance did they do? How careful were they? How long did they stick around for? What kinds of things were they after – anything they could get their hands on or specific items? And, perhaps most telling: what tools – or types of code – did they use to get in? By examining these over time, you can get a pretty good sense of when you are seeing the same person – or more likely someone from the same team – breaking into different properties. Some of the signs were obvious. Western cyber security analysts noticed the people they were tracking almost always worked office hours (China time) and not national holidays (in China). That of course could be faked. But the evidence was mounting.

In February 2013, Mandiant went public and issued a report saying it believed APT1 was, in fact, Unit 61398 of the Chinese PLA. 'Mandiant had responded to 141 companies,' Kevin Mandia explains. 'Every time the technical and anecdotal evidence took us to Shanghai.'[28] The evidence was enough to trace the attacks to a door in a down-at-heel part of the city that housed Unit 61398, if not quite inside the building itself. The twelve-storey-high facility had a special fibre-optic infrastructure built by China Telecom. The hackers used Chinese-language keyboard settings and their IP addresses were located in Shanghai. Resumés linked to Unit 61398 also matched the skill set of hackers doing this kind of work. Some of the hackers could be identified when they made mistakes and let their guard down. They had hacker names like 'UglyGorilla'.

This was hacking quite literally on an industrial scale, with hundreds, perhaps even thousands of people working in a 130,000-square-foot building with considerable infrastructure and logistics behind it. Cyber espionage had begun with teenagers in their bedrooms in Hanover or in Colindale. Now it was done in specially built factories which even had a children's nursery attached. Downloading material en masse would take for ever on a dial-up modem. Not in the broadband age of fast connections in which data could be downloaded in a moment. This was one of a number of developments that took industrial hacking to an industrial scale. It included a clear division of labour, a factory system like the one Samuel Slater had passed on to America. Foot soldiers would do the grunt work, but if they came up against a tricky problem in getting into a system they might refer upwards to a more senior expert.

Conditions for the workers reflected the huge scale on which electronic hacking was now being done. In some ways they were like those of the British workers at postal and cable censorship offices in days gone by, working laboriously through message after message. One blog posting by a twenty-five-year-old from Unit PLA 61398 described a world of long hours, low pay and boredom. Hours were normally 8 a.m. to 5.30 p.m. but could be longer. He wore a uniform but lived in a dorm and had little time for a real social life other than watching TV and surfing the internet. 'Fate has made me feel imprisoned,' he wrote as he designed viruses to perform specific functions like copying files off a USB device attached to a computer. 'I want to escape.' He worked to targets and if he met them would get an end-of-year bonus but moaned about his boss making the most of expenses while he was told off for claiming the cost of a bus ticket. Life did not sound very glamorous and eventually he seemed to have given up and got out.[29]

Western intelligence analysts spent a long time trying to look for a 'controlling mind' in Beijing behind all these disparate espionage campaigns. But the reality is that there may not be one. Groups may have sponsors higher up the political chain or they may be relatively independent. Some are tightly woven into military intelligence and operating to a clear strategy: some Western analysts can track their work in direct correlation to the latest five-year-plan from China, working to a requirements list just as British spies do. But the same group might still do a bit of freelance work for themselves or for Chinese companies on the side. It can be hard to figure out where the lines between state and purely commercial action lie. That may be because there is no clear line. China is much more fragmented bureaucratically and internally than is often portrayed in the West, where it is often seen as monolithic; its defence industry is even poorly co-ordinated in terms of research, development and spending and riven by a lack of communication and compartmentalisation.[30] Chinese cyber espionage may be far more opportunistic and chaotic than is sometimes portrayed.

Competition means different groups may end up going after the same targets, tripping over each other inside Western companies. One aspect – rarely commented on in the West – is the degree of economic cyber espionage that goes on within China. This involves Chinese hacking other Chinese firms to gain advantage. One cyber expert, who has travelled to China frequently, says that hacking is so endemic that it goes even further. He says that the Shanghai office of a company will even be found hacking the Beijing office of the *same* company as they compete for sales targets and managers seek advantage and promotion. It is not unheard-of either for Western companies to carry out commercial espionage against each other. In the West, boardrooms are regularly swept for bugs and it is not just the Chinese that they are worried about.

Corporate cyber espionage has, like the rest of hacking, become increasingly easy to do. And it has been not just industrialised but commoditised and commercialised too. Hackers, having had to find and exploit vulnerabilities themselves, now sell the tools so that anyone with relatively little training can spy. Or if you still prefer a proper hacker to do it, websites advertise access to corporate networks with fees on a per-hour basis (the burglar acting to order). China is certainly not the only player in town but it is the most brazen. For twenty years Western companies turned the other way, believing that tolerating a bit of spying was a price worth paying for access to the Chinese market. We can innovate faster than they can. steal, was the thinking. As China grew and grew and began to innovate itself more and more, that argument was heard less and less. Corporate America was beginning to have had enough.

Concern over Chinese spying finally reached the highest levels of government when it got personal for senior officials. In June 2007, an email system from the office of the Secretary of Defense Robert Gates (used by his policy advisers) was penetrated. The Pentagon shut down more than 1,500 computers for a week to try and contain it once they were in. Officials at the time said they were highly confident the PLA was responsible. The Pentagon began to push for action. The FBI told both the McCain and Obama presidential election campaigns that their networks had been infiltrated, apparently by the same group. Obama later said that from August to October 2008 someone had got into emails, campaign plans ranging from policy positions to travel plans. 'There are significant compromises in that 2007–09 period,' William Lynn, former US Deputy Secretary of Defense, says of the defence industry. 'China – they are sucking our brains out every day,' is the way a retired American general put the problem. At the tail end of the Bush administration there was a debate as to whether or not to call the Chinese out in public. Hank Paulson, US Secretary of the Treasury, resisted and President Bush trusted his judgement. The concern was partly over the potential impact if China fought back economically, and also the problem that much of the material was classified. Slowly, national security officials began to talk about the issue, not least to try to get executives of companies to take it seriously. The debate as to whether to go public or not continued under President Obama. In 2013, when Mandiant accused Unit 61398 of the PLA of being APT1, it did not make its move without consulting the government. The White House was briefed on the plan. This, according to those involved, helped confirm the facts but also provided the nod to go ahead and publish.[31] The White House had been looking at a range of methods of putting pressure on Beijing, from public shaming to legal charges, and even discussed cutting off the bank accounts of those involved (especially those who had children in school in the US). A big showdown was planned between President Obama and the Chinese leader in the summer of 2013. Mandiant's outing of Unit 61398 was one of the ways to build up pressure in advance. That process was made easier as the US worked through the 'attribution' problem – the issue of trying to prove who was at the other end of the attack. According to former US official James Lewis, in 2007 the US military could attribute about one-third of attacks, but that doubled by 2013, partly by combining other forms of intelligence – like intercepted communications or human spies – with computer-based technical means.

Resolving attribution with total certainty remains hard but, a bit like any detective inquiry, you can increasingly marshal enough evidence to make a convincing case, especially when attackers are sloppy and leave clues. You can also go after them. One of the simplest ways is to switch on the webcam of an attacker's computer so you can physically see them at work – spying on the spies. 'I've actually said to the Chinese at various points, "You know you've got to get your hackers to dress better",' says James Lewis, now of the Center for Strategic and International Studies. 'I saw this one guy, he was wearing a wife-beater T-shirt. I was looking for the pizza boxes in the background.' GCHQ and NSA staff would find themselves wading through data on hackers' computers including pictures of pets and old family photos. But they would also find some more revealing images – such as the presumed hacker in the uniform of the Chinese PLA. One operation (codenamed Arroweclipse) by the NSA's own hackers targeted the systems used by Chinese Byzantine Candor hackers. By gaining access to the billing and customer records it was able to tie attacks to specific user accounts. These came from China's 3PLA.[32]

After the publicity blitz, Unit 61398 went quiet. But only for a time. They abandoned their technical infrastructure but then simply built up a new one. In 2014, the US Department of Justice took the unprecedented step of charging five members of PLA 61398 with hacking into a handful of US companies (who were chosen out of a long list because they were not defence companies). The FBI issued 'Cyber's Most Wanted' pictures of the hackers, who included UglyGorilla. In a deliberately provocative move, two were even pictured in their PLA uniform. It was the most direct link made between corporate espionage and the Chinese state. But will such public pressure work? Not everyone is sure. The reason goes back to the essential strategy guiding Chinese activity. 'There is something almost existential about the Chinese approach to this kind of cyber espionage,' argues Nigel Inkster, a former deputy head of Britain's MI6 and a China-watcher. 'I think the perception in China is that they really, really need to move as quickly as possible across as broad a front as possible to acquire all the capabilities that would enable them to modernise and put themselves more on a par with a country like the United States.'

The US has been trying to pressurise the Chinese state into clamping down on such activity by portraying it as illegitimate and unfair, in both scale and whom it targets. 'I stand back in awe as a professional at the breadth, depth, sophistication and persistence of the Chinese espionage effort against the United States of America. As a professional it's awesome. I don't know how they handle all the data they steal,' says former NSA Director Michael Hayden. 'We steal secrets too, I've already admitted to that – as does GCHQ – but we steal only those things that keep British or American subjects safe and free. We don't steal things to make Americans or in GCHQ's case British subjects rich. The Chinese do. And they do it on a massive scale. That's the difference between what free people do in terms of signals intelligence and what the Chinese are doing. And what makes that so pernicious is that they are a powerful nation state, not attacking a nation state's telecommunications or IT infrastructure but attacking private industries' IT infrastructure. That's an incredibly uneven playing field when the resources of a nation state are massed against even a sophisticated company like Google or RSA.' Some in the US have compared Chinese activity to the piracy hundreds of years ago in which state-sponsored privateers raided the ships of opposing nations to steal their precious cargo. Eventually, that was deemed unacceptable behaviour and states put a stop to it, pushing it to the margins. The hope is that public pressure, along with an internal Chinese trend towards greater central control from the new leader, Xi Jinping, might lead to a reduction in China's espionage activity.

China sees the focus on commercial espionage as hypocritical when the US itself spies so aggressively in cyberspace. 'It is a not completely new thing to obtain the technology secrets for learning from another country. It did exist even before cyberspace,' argues General Xu Guangyu, formerly of the PLA. 'The espionage activities started more than a hundred years ahead of the birth of cyberspace. So we don't need to give special attention to similar things in cyberspace. The key point is that the control power of cyberspace is too concentrated in the hands of the US, which makes cyberspace very unbalanced.'

James Lewis recalls attending a difficult meeting in China to discuss economic espionage. 'In the US economic espionage is a crime and military espionage is heroic,' a Chinese military official told him. 'But in China the line is not so clear.' If Chinese economic espionage is supporting state-owned enterprises and their growth is vital to preserving the country's domestic stability and international position, then does this count as 'national security'?[33]

Samuel Slater, of course, was a hero in America for stealing industrial secrets but a villain in Britain and China today queries America's distinctions on acceptable versus unacceptable espionage. America essentially says that traditional espionage for 'national security' is a free-fire zone in which everyone can do anything. But a state attacking another country's companies is out of order. But in reality, there are no rules in espionage. There might be norms of behaviour that come to be agreed on (for instance, in the Cold War, the KGB and CIA did not normally execute captured members of the other side's spy agency but swapped them). But who gets to decide what the norms are? In Washington and also London, there was another concern about China. As its industries grew, one company in particular sent shudders down the spines of some worried spies.

CHAPTER ELEVEN

IN THE WIRES

When the *Alert* cut German cables in August 1914, Britain was making the most of a significant advantage in what would later be called information warfare and communications intelligence. At that time, it dominated the global infrastructure through which communications flowed, having built an 'All Red' system whose telegraph cables were wrapped around the globe in a tight embrace from Sydney to Sierra Leone, Jamaica to Gibraltar, the nervous system of Empire.[1] This was the result of a deliberate strategy to control the infrastructure, from the companies and ships that laid underwater cables through the supply chain all the way to the Gutta-percha tree of Southeast Asia, whose rubber provided the insulation to protect them. A private company like the Eastern Telegraph Company might run half the world's cables as the twentieth century began, but it worked hand in glove with the British government.

China had been part of the British global communications intercept system. At 4 p.m. on 3 August 1914 a Chinese-speaking British censor and his team had begun work in a building in Hong Kong through which all three telegraph lines passed including those handling Chinese government traffic, according to declassified British files. In other cities the wires of local Chinese telegraph companies were cut to force traffic through loyal partners. A staff of twenty-seven in Hong Kong dealt with about 33,000 messages every week, referring 2–3 per cent of traffic to the censor to inspect in his 'inner room'. It would all be checked against records and blacklists, including a card index of 20,000 names updated daily with 'ELIMINATE' stamped on messages deemed dangerous. America learnt from Britain in the First World War that control of communications was intimately bound up with national power and being a great power

– it allowed you to spy on others but also gave you the security of knowing the favour was not being returned. Decades later America's almost total dominance of the infrastructure of the internet and digital communications also offered power. So what happened when that was challenged?

There are no signs to tell visitors what lies inside the squat, nondescript office in a business park in Banbury in Oxfordshire. It looks like the kind of place a small start-up might house itself in if it could not afford the London rent. But on the third floor there are two thick doors that cost £30,000 each. They are a sign that the office is secure to Britain's 'List X' standard. This means it has been refurbished, down to the air conditioning, to ensure it is cleared to contain classified information.

The first clunking door takes you into a room that is reminiscent of the satirical comedy *The Office* (which is set nearby). There is an everyday reception and behind it a few cubicles where people with laptops tap away quietly. Everyone is expected to store their electronic equipment in a set of lockers before they go through the second door, which requires a swipe and pin card entry system. This inner sanctum is accredited up to Top Secret and no one from China is allowed to enter unescorted unless they have been cleared. Everyone who works there has British security clearance. But this is not a secret British government research laboratory. The whole facility is run and paid for by the Chinese technology company Huawei. The site is called the 'Cyber Security Evaluation Centre' but is often known simply as the 'Cell'. This quiet corner of Oxfordshire is the front line in a global debate over computer security which pits China against America, with Britain in the middle. America has kept Huawei out of the heart of its telecoms infrastructure. The Cell in Banbury is the place where Britain tries to ensure it has not made a mistake.

Britain may have had the cables, but today China – and Huawei specifically – is in the wires. The Chinese company has come from nowhere to being perhaps the largest telecoms equipment company in the world, doing everything from laying fibre-optic cables in the ocean to selling smartphones, providing the infrastructure and the backbone of modern communications.

The fear that haunts a few people – especially in Washington – is that Huawei's influence means it could do in the modern world what Britain did with cables in 1914. It could use its influence over the cables to spy on the world or even hit the kill switch. But compared to a century ago, the impact of such actions would be far greater because so much of modern life flows through telecoms. Telecoms companies used to be about making phone calls. No more. Phone lines and the cables that carry them have increasingly merged with computers and the internet carrying data. As the world has moved online and networked computers have become ubiquitous, almost every industry depends on telecoms and computers to function. Everything from the power grid to the banking industry to the transport system depends on computers being able to talk to each other over commercial fibre-optic lines run by the private sector (the exceptions are separate secure lines used for certain classified government functions, primarily in the military and intelligence world). All of this means that telecoms companies are far more important than the cable operators were just before the First World War. And Huawei's equipment is involved in providing communications to a third of the people in the world in 140 different countries.

Banbury in Oxfordshire is Britain's first – and perhaps last – line of defence against the darkest fears about China. For critics, it is a Maginot Line. For supporters, it is the only way realistically to provide security in a globalised world of computer networks. When you walk through the second of those heavy doors, the inner sanctum of the Cell is larger than the outside office, with a few dozen desks in open plan and smaller rooms ringed around. Here the telecoms kit that Huawei's customers plan to install in Britain's infrastructure is subject to testing. Some of this involves computer-based testing of code. Other checks are less high-tech. One heavily air-conditioned lab has a bench littered with screwdrivers, soldering irons and drills. Here racks of equipment and mobile phone network base stations are taken apart piece by piece and then photographed as well as weighed. That way, it is hoped that any modifications can be spotted.

Even within the secure inner cell, there is one place that is even more sensitive. Here the 'holy of holies' is kept inside a locked steel cage. CCTV cameras monitor every movement in and around it.

But peer in and all you can see is the rather disappointing sight of one computer terminal. It is on this terminal that the source code for Huawei sits – one of the only places where it can be found outside China. This is the company's most precious intellectual property, the magic formula that runs its equipment. A one-way diode means the code can flow into the computer to be examined but not out of the room and is kept carefully encrypted. A two-man rule operates, so a Chinese employee of the company has one half of the password to decrypt the material but he or she can only enter the inner sanctum escorted by a British security-cleared staffer.

The pictures from the security cameras are beamed across the world to Shenzhen in China, home of Huawei's vast mother ship of a headquarters. In a matter of decades Shenzhen has gone from a small border town to a metropolis of 15 million people, larger than London. It has the feel of a classic boom town. Property prices are going through the roof. Everyone is from somewhere else and there to make money. Huawei's astronomical rise is at the vanguard of that change. The company's headquarters is a campus, not an office. You need to be driven round by car down long, wide avenues all entirely devoted to the company, past endless newly built shiny offices and research centres. One houses what is called the 'war room' – a wood-panelled conference room where huge video screens on the wall allow the company to respond to a 'level one' emergency if a country's telecoms network has gone down. It overlooks a network control centre that is like mission control at NASA as dozens of operators watch screens which display the flow of much of the world's communications in real time. Nowhere is the sense more clear that Huawei is everywhere. 'When you walk around the Huawei campus, you are staring into China's future,' wrote one US diplomat.[2] A Western business executive had a different thought about the prospects for his part of the world as he gazed down at the vast army of engineers working in Huawei's offices. 'We're screwed,' he thought.[3]

So where did Huawei come from? The figure at the centre of the story is known by all his staff as 'Mr Ren', a Chinese equivalent of Bill Gates or Steve Jobs. Ren Zhengfei joined the PLA as an engineer. But in 1983, Deng Xiaoping was cutting the size of the army and Ren was laid off. At the same time, Deng was also opening up the economy by creating 'special economic zones', designated areas which allowed, for the first time, companies to operate in something approaching a free market. Ren moved to one of those zones, Shenzhen, which benefited from being just over the border from Hong Kong (then still run by Britain). In 1987, with just a few thousand dollars, he set up a tiny company to supply telecoms equipment imported from Hong Kong. That year there was no mobile phone service and only 3 million landlines in all of China. Two decades later there would be 640 million mobile phone users in the country. The explosive growth would fuel Huawei's rise as it first served the largely ignored countryside, encircling the cities so it could penetrate them by the end of the decade, a mirror of Maoist counter-insurgency strategy in the business world.[4] Although it was a private company and not a state-owned enterprise, its success began to attract attention, especially as its value was recognised in trying to catch up with Western dominance in high-tech industries.

As it grew, Huawei made the transition from importing telecoms equipment to reverse-engineering the equipment and then eventually to innovating and building its own. Western firms like IBM, Motorola, Texas Instruments and Lucent were initially all too happy to join up with this Chinese newcomer. What danger was there in working closely and sharing your know-how with a Chinese company? A joint venture offered a foothold into the vast Chinese market that had executives licking their lips. In the late 1990s, the US had happily signed off on selling high-performance computers to Huawei, following a decision by the Clinton administration in 1994 to remove restrictions on the export of fibre-optic, switching and transmission equipment which had previously been controlled. This was part of the wider shift in that period to emphasise globalisation over traditional national security concerns and was done despite the objection of the NSA, who feared (initially at least) that fibre-optics would reduce their ability to spy.[5]

By the start of the twenty-first century, Huawei was ready to become one of the first Chinese companies to step out onto the global stage. It rose by moving from market to market, offering the same services as Western rivals but at a significantly lower price.

Huawei's critics claim this rise was fuelled by a textbook case of intellectual property theft. Its alleged victim was the American giant Cisco, which had become the internet equivalent of those British cable companies that had connected the world a century before, rising through the 1990s to a dominant position. This time, rather than telegraph cables it supplied routers and switches – specialist computers that directed the packets of data around the global internet with advanced hardware and software. In January 2003, Cisco filed a lawsuit in Texas claiming there was 'overwhelming evidence that Huawei unlawfully gained access to Cisco's source code and copied it as the basis for the operating system for their knock-off routers'.[6]

Cisco's filing claimed Huawei's router used identical commands to theirs, the result they said of 'slavish copying', including even a mistake – the equivalent, the US company said, of a schoolchild copying another's work. Huawei denied the allegations and said if an employee got hold of the Cisco code they were acting without company approval. In the end the two sides settled out of court, with Huawei consenting to withdraw certain equipment and both agreeing not to reveal details of the settlement (a decision Cisco officials now say they find frustrating, not least because the narrative in China about the lawsuit was that it was a big bully trying to crush a newcomer).[7] Cisco – and its supporters in America – seethed as they watched Huawei rise over the next decade.

As Huawei began to push into more and more developing world markets, its products eventually began to be noticed by Western telecoms companies seeking to update their infrastructure and looking to benefit from its cheap prices. But in America they hit a brick wall. The reason was the fear of espionage. In the mid 1980s it was reckoned that 95–98 per cent of US Department of Defense communications were controlled by the private sector – far higher than in Europe, where government monopolies often exercised control. When the AT&T monopoly was broken up, the National Security Telecommunications Advisory Committee (NSTAC) was created to bring together senior executives. This was initially to work on subjects like communications in the event of war, but also to make sure foreigners who entered the supply chain could not intercept sensitive US lines. When the first long-haul fibre-optic cable was being installed between New York and Washington DC in the early 1980s, the award of the contract to Japan's Fujitsu on the lowest bid was overturned on national security grounds. 'It would have been the start of a foreign encroachment into an area of critical importance to the US communications network,' a recently declassified NSA document records.[8] Britain today has secret contingency plans which would involve immediately nationalising the highly sensitive parts of BT if a less trustworthy company made a takeover.

In the late 1990s there was a case that highlighted the fears of espionage, according to the select few former Western security and intelligence officials aware of the details. The case did not involve China. Or even Russia. But Israel. Israel's Mossad is among the most capable intelligence agencies in the world and has a very strong track record in technology (American intelligence officials say that in the 1990s most of their concerns over foreign investment in defence- and security-related technologies were associated with Israel and France). It is also one of the most active countries in trying to spy on the US – particularly in looking for details of US policy towards the Middle East. Washington came to fear that Israel's spy agency had managed to gain access to records for America's phone system. If true, the potential intelligence haul, officials realised, was enormous. It could allow you to see who is talking to whom, up to the highest levels of government: is the President or a Cabinet official in late-night contact with a particular businessman or perhaps a young lady? What leads were law enforcement working on? Which agents were they in contact with? One person with knowledge of the case believes the investigation made people understand the value of this kind of data.

According to one former telecoms security worker, at least one major US company spotted something suspicious. The concerns over Israel extended beyond the US. By the late 1990s, senior British intelligence officials were also briefed and realised the same vulnerability could have been exploited within the UK. Trying to eliminate it from the UK completely would be too costly, so instead the Israelis were confronted by the governments and monitoring was put in place to make sure that information could not be extracted without being detected. One important legacy was a heightened awareness of how other states could work through the telecoms system to carry out espionage. This led to information about vulnerabilities and threats being shared among the Five Eyes countries, and greater sensitivity about the Chinese when they appeared on the scene.

In 2004 a US destroyer was undergoing refitment in Norfolk, Virginia, when someone realised the Cisco routers being installed were fake. It was another case that raised concerns over the potential use of the telecoms supply chain for espionage. Cisco itself spotted the issue as some of its equipment started to break down at a higher rate: an upgrade for a government agency weather communications system failed in 2004.[9] It might be an American company, but the actual equipment was manufactured in China (specifically Shenzhen, where factories churn out goods for Western companies like Cisco and Apple) and then supplied by the company's partners. These routers had been installed in a myriad of government and military systems. 'This caused absolute horror in government circles,' recalls one person briefed at the time. The FBI opened an investigation amid fears that this might be more than just a case of a supplier to Cisco down the chain committing fraud and using cheap parts to cut costs and instead be a way of installing backdoors into American systems. Britain, Canada and Germany were also briefed. No evidence was found to prove it was espionage (and there was no link to Huawei) but it made officials realise that in a global supply chain, even buying from American companies had risks. It was as if the exploitation of hardware used against Russia in the Farewell case of the 1980s was now returning to haunt those who had planned it. Huawei's rise around that time meant it came into the cross-hairs of Washington. Fears of letting the enemy in through the castle walls – like the Trojan horse story of old – meant that Huawei was effectively blocked from taking on a major role in the US communications system. American critics claimed Huawei was a 'national champion' picked by the PLA to leverage its way into foreign markets in the long term, a kind of telecommunications sleeper agent, a claim the company has always vigorously denied and says has been made without offering any proof.

In March 2008, Ren Zhengfei met with the US Consul General in Guangzhou, China. Ren was frustrated that his company's bid to buy a US firm, 3Com (with whom it already had a joint venture), had run into major opposition in Washington. Congressmen were speaking out. A US magazine had published an article entitled 'No Way Huawei' which warned of allowing the Chinese to spy on American conversations. The rhyming couplet of the story's title became the rallying cry for those seeking to block the company's entry into the American market. One Huawei executive asked which country should be more concerned, given that Cisco was all over China. Senior executives including Ren were finding they were only being given single-entry visas to the US, making it hard to travel for business at short notice (they did not know that the State Department had issued guidance that virtually all Huawei employees required 'Mantis clearance', designed to prevent the unauthorised transfer of sensitive technologies). Ren told the US consul they had nothing to worry about and that his firm was not in a position to be a competitor with high-tech firms in the US. In the meeting, Ren denied there were any close ties to the Chinese government or PLA. Ren pointed out that his service in the PLA was unusual given his unfavourable political background. He explained that his parents had both been sent to labour camps during the Cultural Revolution because they were 'intellectuals' (they were schoolteachers) and he had only been allowed to join the PLA as there was a shortage of specialists.[10] However, while Huawei was struggling to get into the US, it had already broken into Europe with a significant contract which gave it a foothold and legitimacy.

In 2003, BT was planning a £10 billion upgrade of the backbone of its network – known as Colossus, in a tribute to Tommy Flowers – to bring it into the digital world. The company was moving towards fibre-optic cables rather than copper wires, with the system controlled over the internet. But the Flowers era, in which a company like BT built all its own equipment, was long gone. No British company could supply all the equipment needed and so foreign vendors had to be involved (Huawei is now one of only two companies, with Sweden's Ericsson, to be able to provide everything needed to set up and run a telecoms network from scratch). As the bidding process evolved, BT mentioned to GCHQ that one of the companies in the mix as a prime vendor was Huawei and asked what it thought. BT itself was desperate to have Huawei involved. One person describes BT as having played a clever poker game to get its way. The company knew that even just having Huawei in the bidding process would mean other companies would drop their prices. And the possible savings for BT in buying Chinese were enormous. Did the British government really want to put BT at a massive commercial disadvantage by stopping it buying much cheaper kit because of its fears? And if the British government really wanted to stop them, would they be willing to compensate BT for the extra cost of going with, say, Cisco equipment? The numbers, one British official involved says, were 'eye-watering'. Not millions of pounds. Not tens of millions of pounds. But hundreds and hundreds of millions of pounds – half a billion in all. Was the taxpayer going to stump up that? And if the state did pay BT the money for not buying Chinese then there was talk that this might fall foul of EU rules by being seen as an unfair state subsidy.

There were a few voices of concern, like the National Infrastructure Security Co-ordination Centre (NISCC), which worked with MI5 on cyber security and particularly on espionage threats. It wrote a report in 2004 on the use of Huawei routers that made fourteen recommendations. The first one was pretty much not to use Huawei kit. But, realising which way the wind was blowing, the other thirteen recommendations outlined what to do if Huawei kit was bought. Some in GCHQ were cautious but their primary goal was to defend government networks, not those of the private sector. Other voices in government seemed keen on doing business with the Chinese. A working group looked into whether it was possible to block a BT-Huawei contract but there were worries about the trade, financial and diplomatic consequences. Britain's Intelligence and Security Committee later pronounced itself 'shocked' that ministers had not been informed of, or consulted about, the deal due to 'complacency' by officials. Officials from the time dispute that and also say that one of the main problems was that it was not clear who was in charge of making a decision. Five different departments were involved, many with no interest or knowledge of the security issues.

GCHQ reported to the Foreign Secretary, MI5 and NISCC to the Home Secretary, but it was the Business Secretary who was at the centre of decision-making.[11] In the end, Huawei was chosen above the venerable but struggling British company Marconi, founded by the radio pioneer Guglielmo Marconi in 1897. If *British* Telecom was going to choose a *Chinese* company over home-grown products, then things did not look rosy for a British supplier. Marconi – already struggling badly – collapsed within months, the decision a final nail in its coffin

Only after the deal was done did people begin to worry. In January 2006, the month after the deal had been signed, Britain's Intelligence and Security Co-ordinator wrote to the Home Secretary asking for an agreement to help BT monitor Huawei and check the equipment which was due to become operational. A special group was set up to track the risks. In early 2008 MI5, charged with preventing espionage in the UK, said that, theoretically, the Chinese state might be able to exploit vulnerabilities in Huawei's equipment in order to gain access to the BT network, providing them with an attractive espionage opportunity.

The Joint Intelligence Committee warned that any exploitation 'would be very difficult to detect or prevent and could enable the Chinese to intercept covertly or disrupt traffic passing through Huawei supplied networks'.[12] These were all couched in terms of 'possibilities'. It did not say that Huawei itself had actively to facilitate the espionage through its equipment (the Chinese could take advantage of it without the company's permission, or demand it) and there was no evidence of it happening. By March 2009, reports emerged in the press that the Joint Intelligence Committee had been briefed about the dangers two months earlier with talk of supplies of power, water and food at risk of being halted by the PLA.[13] The possibility of China switching off the networks remotely and crippling Britain was judged to be a low probability/high impact event – something unlikely to happen, but extremely serious if it did. Was espionage possible? Possible perhaps, even though there was no hard evidence.[14]

GCHQ needed a strategy to minimise the risks. Part of this involved ensuring there was not a single point of failure that could bring down the network – no single kill switch. This is secured by a 'multi-vendor strategy' – making sure the equipment that BT used came from more than one supplier and was distributed around the network geographically. The second stage was making sure all the equipment was checked. It was agreed that a proportion of BT's savings would be invested in employing penetration testers and for other audits. Initially this was also done by BT going to China to look at the equipment and examining with GCHQ everything that came into the UK (some at BT were said to be unhappy that they were given so much responsibility for defending the nation when they were just a private company). Monitoring was put in place so that experts could go back in time over the code and the network to see if anything had been exploited or altered. Other telecoms companies were looking enviously at the huge savings BT were making and the rising quality of the Huawei kit and so began pressing to be allowed to use it too. This would mean that security checks had to be expanded. This led to the creation of the Banbury Cell, with Huawei agreeing to pay the bill.

The 2005 BT deal was a breakthrough for Huawei – a chance to get beyond the idea that China simply provided cheap-cost but cheap-quality products. It announced that Huawei had arrived and was at the heart of modern telecoms in an advanced nation. That meant a lot was riding on making the Banbury Cell work. There were two distinct concerns: one was espionage, the other sabotage. The former is that there is a modern version of the secret censor in Hong Kong picking through the traffic on the cables. British officials say that rules were put in place not to allow Huawei to run sensitive government networks through which classified government information passed. Scooping this up would be too easy.

Telecoms companies in all countries are required to intercept certain people's communications for law enforcement and national security purposes. All telecoms equipment has to provide the capacity to do this. But Huawei was kept out of the direct work of intercept that might allow it to find out which lines had been selected and who the intelligence targets of interest were (which might in some cases be Chinese). Huawei says the company does not provide the monitoring/recording equipment for intercept but only the interface, so that governments can plug in other people's equipment. The company was also kept away from the brains of the network that decide which paths information flows down. This would be too easy to exploit to target an individual's communications. Other areas were open, though. So now, around 10 million homes in Britain have Huawei boxes at the end of their road connecting them to the local exchange. British officials say the consequences for Huawei of being discovered to be spying on behalf of the Chinese state would be catastrophic, destroying trust in the company not just in Britain but also around the world. Chinese intelligence, like other agencies, conducts careful risk-reward calculations about its operations. Former British intelligence officials closely involved with the deal say they believe that if China wanted to spy, it would be more likely to use other techniques – including old-fashioned human intelligence practices like coercion, bribery or blackmail – rather than risk the wrath and reputation of a powerful company by getting it to do its work.

The second fear, of sabotage, is a darker one. Is there (metaphorically if not literally) a man from the PLA in Beijing who has a big red button on his desk saying 'Britain' which – when pressed – will turn off Britain's networks and take down much of the infrastructure on which we depend? This is the equivalent of the *Alert* cutting the German cables. Most British officials say they do not really worry about that. This, they say, is the kind of scenario that comes from reading too many cheap thrillers. Why would China do that? If it had some kind of backdoor then it would be much more likely to use that for espionage than blow it with an attack that might trash the British economy and destroy trust in China and the global economy with it. Only perhaps in the unlikely event that the two countries were at war. And if so, there would be plenty of other things to worry about. But it cannot be discounted entirely. There is a document called the National Risk Register that outlines all the bad things that could happen to Britain and how the government is preparing for them. There is a section called 'transition to war' which specifically relates to the possibility of China shutting Britain down by switching off all the Huawei kit (it would not necessarily

need the connivance of the company itself to do this). British officials believe that if this were to happen it might take down as much
as half of the British network for a number of days, but they could
then bring most of it back up quickly and continue to run it (what
kind of chaos would ensue in the meantime is another question and
officials admit that the system is so complex that no one can be absolutely sure of the impact). Resilience, it is said though, has been
built into the network so that even if something were switched off,
it would be localised and temporary. One reason is that the government would be able to restore what is called the 'last known good'
version of the code running the network that has been stored (a
technique used when Nortel went bankrupt).

The Banbury Cell's job is to make the use of Huawei kit no more
or less dangerous than any other company's equipment. But that
means subjecting Huawei to extra checks. GCHQ has said it is confident that the network has not been at risk at any stage, but officials
say they have to operate on worst-case assumptions – that someone
or some small cell in the 20,000 coders in the company might be
under the control of the Chinese state, and also that they may be
playing a long game or that there is some accidental weakness that
could leave equipment exposed to teenage bedroom hackers. Every
new piece of equipment is tested – sometimes for months – and
then a report is submitted by the Cell to the British government,
Huawei itself and the company planning to use the equipment. But
there are a million lines of code in telecoms equipment and the task
is mammoth. Modern telecoms are not so much a product as an
ongoing service: this might include regular updates or emergency
patches to software using remote access (a kind of backdoor).

So far, officials say, all that has been found are the kinds of programming mistakes and weaknesses you would expect – mainly
sloppy or redundant coding. In one case the lawful intercept interface for a device was supposed to have been switched off but was
left on – apparently by mistake. A review by Britain's top national
security official in 2013 found all the vulnerabilities identified so
far 'could be explained as genuine design weaknesses or errors in
coding practice'.[15] In the early days there were lots of small errors
that created awkward issues for the Cell because they had to tell

Chinese programmers their work was sloppy – something which is more an insult in Chinese culture than Western hacking culture. A few observers in government wondered whether the sloppiness could be deliberate to allow weaknesses to be exploited, but that is something that is impossible to prove – it has its own kind of plausible deniability. These weaknesses were all fixed quickly by throwing people at the problem, even if that meant sleeping at the office. Here is an interesting dynamic, in which Huawei effectively has the quality of its work raised higher and higher thanks to the intervention of the British government.

In January 2011, testers did find something that initially appeared more serious. It looked like an undisclosed functionality – a piece of code that did something other than it was supposed to do (it was a debugging port that had been left open, which meant the software was accessible from the internet). This was the first time that something seemingly significant had been spotted and there was panic in the British government. GCHQ's defensive security department, CESG, went to BT and asked what this could do to the network. What were the implications? The answer was reassuring. It would only affect one small piece of equipment in proximity to another piece of equipment. It was not serious. A team from CESG and BT flew to Shenzhen to talk to the Huawei board about what they had found.[16] At a 3 p.m. meeting the board committed to fix the problem and find out as soon as possible what had happened. By the next morning, they said the problem had been solved. They explained that a piece of debugging code had accidentally been left in and the procedures would be altered so that they were automatically closed in future.

The Banbury Cell had been opened by John Suffolk in December 2010 while he was the UK government's Chief Information Officer. Suffolk had already made clear his desire to leave government and return to the private sector. When the team went out to discuss the undisclosed functionality, Suffolk was part of it. One person recalls a frisson of excitement in the room as it was announced that, unusually, Mr Ren himself would be attending the meeting for the first time. Afterwards, another member of the Huawei board asked Suffolk what he would be doing after he left government. The next

month, Suffolk requested formal permission from the British government to work for Huawei as global head of cyber security.[7]

During a tour of Huawei's internal data centre in Shenzhen, secured by fingerprint checks, Suffolk, who travels the world on behalf
of the company, points out that Huawei is a victim of cyber attacks, coming under a million attacks every week. 'The issue is how
quickly you can spot they have got through and what they're doing,'
he says. A building on the campus known as 'The White House'
(which looks only a little like its namesake) is home to something
akin to torture chambers for Huawei technical equipment. Inside
here, men and women in white coats put Huawei's own kit under
the most extreme pressure to see if it will break. A router box in one
cabinet is subjected to a temperature of 95 degrees centigrade; next
door another is taken to minus 70. Another is subject to thermal
shock – going from heat to cold extremely fast. In a long room, telecoms equipment is blasted with solar radiation. Down the corridor,
it is shaken rapidly back and forth. In another room dropped from
height. One event simulates a sandstorm. The idea is to make sure
the same piece of equipment will operate wherever it is sent. It is a
reminder that this is a company which wants to be everywhere in
the world, whatever the weather.

The boxes are subjected to the same kind of rigour when it
comes to cyber security, Suffolk explains, with the code torn apart
by people with different mindsets who will approach it in different
ways. '[We have] teams of our own penetration testers to break our
own products. Then we say, "let's not believe them". We pass it to
another team,' he explains. The same rigour is also involved at Banbury. 'If they didn't find things I'd be very worried about the quality
of the work that they were doing because every software in the
world has potential for vulnerability,' Suffolk says of Banbury. 'They
have found code that could be exploited depending on the nature of
the configuration [but] . . . there's no deliberately placed code. What
they have found is some poor coding practice, some vulnerabilities
which could be exploited depending on the nature of the configuration and whose other equipment you are deploying.'

Huawei may have hoped that the BT deal and the Banbury Cell
would provide it with the credibility to enter the US market, but

hostility in Washington – and especially Congress rather than the White House – continued to grow. In October 2012, the US House Permanent Select Committee on Intelligence published a report that tore into Huawei and its fellow Chinese company ZTE. ZTE is said to have a closer relationship with the Chinese state than Huawei (which is employee owned) but receives far less attention. Sometimes bundled together, the two companies are locked in a bitter battle, including over alleged IP theft. Ren of Huawei is said to hate ZTE so much he never even utters its name, referring to it as 'that company'. It is Huawei, though, that has become the lightning rod for all the fears over cyber security. Huawei became the bogeyman in Washington, but disentangling real national security concerns from Sinophobia, protectionism and a wider fear of China's rise and America's malaise is not easy.

The US Congressional Report said that US government systems should not use Huawei equipment and that the private sector should consider 'the long-term security risks' associated with doing business with either company. US network providers were 'strongly encouraged' to seek other vendors. It suggested that there was a 'wealth of opportunities' for Chinese intelligence to insert malicious hardware or software implants (which can collect data or carry out other actions) into systems, and even if the company refused to comply with requests, Chinese intelligence could recruit working-level technicians or managers to do their bidding. This could allow 'Beijing to shut down or degrade critical national security systems in times of crisis or war'. The report was a catalogue of the fears that plagued Washington when it came to cyber security, although couched in terms of possibilities rather than realities: 'malicious implants in the components of critical infrastructure such as power grids or financial networks, would also be a tremendous weapon in China's arsenal' was a typical way of phrasing the risks. The report, which differed in tone from a White House review, explicitly rejected the British 'evaluation' model as a means to provide reassurance, saying that it risked creating 'a false sense of security'. Any such testing only provided a snapshot since products and technology were continually evolving and being updated remotely.[18] Finding flaws intentionally inserted by a determined and clever insider, it said, was 'virtually impossible'. The report was tough but lacking in actual evidence of wrongdoing (although there is a classified annexe). Former NSA and CIA Director Michael Hayden claimed that 'at a minimum', Huawei shared with the Chinese state its 'intimate and extensive knowledge' of the foreign telecoms systems it operated within – something different, though, from actually carrying out espionage (more a matter of telling the government how things work so they might be able to do it).[19]

When US telecoms companies like Sprint were pressing to sign contracts with Huawei to modernise their networks, they received heavy pressure from the US government. This went as far as taking executives into the NSA headquarters in Fort Meade. Here they were told in no uncertain terms what the US government's opinion of such a deal was. They may even have wondered if their lucrative contracts with the US government might be at risk if they continued. The deals fell apart, Sprint saying it would exclude Huawei from bidding because of national security concerns in November 2010.[20]

The US Congressional Report raised awkward questions for America's closest ally in the tight Five Eyes club (Australia had kept Huawei out of its national broadband project, some say because of fears of espionage, others as a form of punishment for China after the state, not Huawei, was found spying on Australia's parliament). 'I just think it makes Great Britain more vulnerable,' the then Chairman of the US Committee, Congressman Mike Rogers, says of the British arrangements. 'It gives them a foothold and it allows them to control the pipes basically where that information flows through.'[21] So had the British government missed something? The US report forced Parliament's Intelligence and Security Committee hurriedly to conduct its own inquiry so that it did not look as if it had ignored a problem flagged so vigorously by its American cousins. The British report found no proof of espionage, but raised questions about how the whole Huawei deal had been agreed by British government, criticised some elements of the set-up of the Banbury Cell and recommended a review by Britain's National Security Adviser.[22] His December 2013 report said that, after initial teething problems, the Cell was now working effectively and that Huawei's co-operation had been 'exemplary'. With only a few tweaks, he affirmed the arrangements by which the Cell was paid for by Huawei (originally because the British government did not want the cost transferred through purchasers to consumers). That situation had raised concerns over a company funding a team to check its own kit for backdoors, which was why Chinese staff were not allowed into the inner sanctum and all those working there had to be British and security-cleared. His report also said that having the Cell within Huawei itself was the only way for the company in Shenzhen to trust the group with all of its source code and have the influence to solve problems when they were spotted. The National Security Adviser also pointed out that Huawei was a 'valued investor and employer in the UK'.[23]

Does being a valued investor help? Mr Ren visited the Prime Minister, David Cameron, in Downing Street in September 2012 and announced plans for £650 million investment in the UK and £650 million in procurement. Coincidentally, £650 million was the exact amount that the British government had announced in 2011 it would commit to cyber security and keeping the country safe from threats to computer networks.

The response from Huawei to the accusations from Washington has been 'put up or shut up'– provide evidence of actual espionage or malfeasance or stop making claims about what might happen. The US Congressional Report did claim that it received documentation from a former employee suggesting the company provided special network services to an entity the employee believes was an elite cyber warfare unit of the PLA, although no further details were available. Huawei rebuts the charges. Yes, Ren had a background in the PLA, but how many American firms were started by former military officers or even have former NSA or CIA Directors on their boards?[24] Did the company receive state financial support? Yes, but no more than any other firm. Does the company have a relationship with the state and the PLA? Yes, but does BT have one with GCHQ and do American firms work for the Pentagon? Huawei claims its work for the PLA accounts for less than one-tenth of 1 per cent of its total sales. The question is, not are there ties – but what is the nature of those ties?

Chen Lifang (known as Madame Chen) understands English

well but prefers to answer questions in her own language, reflecting a desire to be precise. In an old railway carriage on the Huawei campus which has been turned into a dining and meeting room, she emphasises that the company is no different from any other. 'There are some people in the United States that say that because Huawei is a Chinese company Huawei cannot be trusted, but whether a company can be trusted or not should be based on facts,' she says. 'I don't know how to respond,' she replies in frustration when she is asked why the opinion persists that the company is a tool of the Chinese state. 'This opinion is imposed on us.' As for Huawei's ties to the state, she denies there is anything unusual. 'The relationship between Huawei and the Chinese government is simply just a relationship between any company with any government.' She argues that globalisation will lead to growing interdependence, requiring more openness. 'Pointing fingers cannot help us to get this issue solved,' she argues.[25]

Huawei has become a surrogate for everything that Washington fears about the rise of China and the erosion of US dominance (much as some Japanese firms became the lightning rod in the 1980s). The company's success terrifies critics of China. Huawei made the transition from bottom-feeding, reverse-engineering, cut-throat-pricing new kid to a truly global, high-tech player in a field the West had dominated. That, as much as the fear of espionage, scares the hell out of the US. What if Chinese companies could rise by (allegedly) using cyber espionage and IP theft but then capitalise on that kick-start to establish themselves as global players before abandoning old practices, going legitimate and developing their own products at lower prices and then driving Western companies out of business?

Huawei's rise could also offer hope: if Huawei is now innovating itself then it has an interest in creating a system in which intellectual property is respected and not stolen. It now files reams of patents. If Huawei is now a global company, then it also has an interest in not being discovered to be hand in glove with the PLA. If it were, its global sales would likely suffer a catastrophic collapse. This is something British officials are trading on. They concede that it is impossible to ensure absolute security. It is all about risk management, they say, and how you balance that with the commercial trade-offs

in a globalised world. Do you want to eliminate any risk to the point
that you ask everyone to pay a lot more for their broadband by ex-
cluding Huawei? Some of them privately talk of encouraging the
reformers in Huawei who want as much distance from the Chinese
state as possible and trying to integrate Huawei and China into the
global economy rather than create a Cold War mentality and dividing line – what you might call an electronic Iron Curtain. Huawei
points to the fact that it is also dependent on the global technology
supply chain, since nearly a third of its parts actually come from
America. This globalisation may pose risks, but its sheer complexity may also offer the best hope of managing them. Equipment is
sourced from so many countries that interfering with the hardware
becomes much harder and risks doing such damage to trust around
the world that it is too risky other than in the most extreme situa-
tions (such as all-out war). A kind of 'mutually assured destruction'
may come to exist in a globalised world of cyber-interdependence.
One former British security official also asks why, if we want to be
rigorous enough to stop Chinese suppliers working in the infrastructure, the policy should not be extended to companies from
other countries with aggressive, highly capable signals intelligence
agencies – for instance Israel, which has a large telecoms and computer industry (and a capable intelligence agency).

The notion of opening yourself up to checks to prove you are not
spying for another country pre-dated Huawei. Microsoft was faced
with the accusation that it was a tool of the NSA as it moved into
a globally dominant position. As a result it instituted its own Government Security Program in which it provided supervised access
to view some of its source code to allay fears when it was installed
in government systems. 'You know if we had put in a backdoor for
the NSA invariably some researcher would find it and our market
cap would go from what it is today to zero almost overnight. I mean
we wouldn't even be able to sell in America,' argues Scott Charney,
head of security. 'I mean it would be complete suicide to go down
that path. But still you have to give people assurance.'[26] The advantage for Huawei of Banbury is that they do not need to give up their
original source code but keep hold of it themselves. These security programmes have the potential to build trust (although some

companies fear governments may also use them to find vulnerabilities they can exploit).

Huawei officials have also argued that the relationship between Huawei and the Chinese government 'is no different than the relationship between Cisco and the American government'.[27] Huawei's supporters suggest that what is really going on is hard-nosed politics masquerading as security. They say this is about Washington protecting an influential American company that has been losing out. But Cisco has itself faced questions about whether it is a tool of the NSA and building backdoors for American intelligence, taking advantage of its dominant position. 'We do not work with any government to weaken our products for exploitation, nor to implement any so-called security "backdoors" in our products,' a senior Cisco executive wrote in response to the accusation.[28] When pictures emerged of NSA staff apparently opening up Cisco boxes so that spyware could be implanted, the chief executive of the company wrote angrily to President Obama. 'If these allegations are true, these actions will undermine confidence in our industry,' John Chambers said. 'We simply cannot operate this way.'[29] What the pictures showed, though, was that a company's products can be used without the company itself knowing. That may be one fear with Huawei. When asked whether Huawei's potential spying for China was any different to what Cisco might possibly be able to do for the NSA, the recently retired head of NSA, General Alexander, said: 'The close ties between the Chinese government and industry are in no way comparable to the relations between the US government and US industry.'[30] But the fact that the question was asked was telling.

When queried as to why he was so worried about Huawei, another former NSA official gave an interesting response. Rather than answer directly, he posed a series of questions. 'Why would the US be worried about Huawei?,' he began. 'Why would it think a Chinese telecoms company might be able to spy?' he continued. 'Why would the US – which until now dominated the world – think it's possible to use this position to spy?' That was as far as he was willing to go. But it does not take a lot to work out where he was heading. The UK's telegraph system provided it with the means to spy a century ago. The internet and its infrastructure provided something similar to America. This does not always take the connivance of a company, but can be built on a deep understanding of how the infrastructure works if your company is the one in the wires. In communications security so much of what people fear will happen to them is based on what they can do to others, and so the question that lurks is: how far was American and British fear of Chinese espionage based on what they themselves were doing rather than hard evidence of Beijing's work?

CHAPTER TWELVE

BRITAIN AND THE CYBER SPIES

There was 'jubilation' as the message winged its way from north-ern India to New York. It was short and simple: 'Hello from Dharamsala' was all it said. It was the early 1990s and the Tibetan community had hooked up to the outside world.[1] The Dalai Lama had fled Tibet in 1959 and set up a government in exile in this remote town, taking refuge in a suburb high up in the hills named after a British colonial governor. From the upper parts of the town, which spread up the hillside, he and his followers campaigned for Tibetan autonomy, making them a constant irritant to the Chinese who have sought to integrate Tibet into China. That also made them a prime target for Chinese spies. Dharamsala – a community surrounded by forests and monkeys – would become the front line in computer espionage.

Thubten Samdup, a Tibetan who spent some of his time in Canada, had seen the advantages that going online offered at the start of the 1990s. Technology in Dharamsala at that time consisted of an old Gutenberg-style movable press down in the basement. A wide diaspora of Tibetans and their supporters scattered around the world were hungry for news. But by the time a message reached them, informing them that a monk had been arrested in China and asking them to pressure their governments, it was often too late. Speeches by the Dalai Lama would have to be faxed individually to offices around the world – perhaps repeatedly when one page came out blurred – and then refaxed on to dozens more offices, a laborious process. So when a friendly professor at the University of Montreal told Samdup that something new called email might solve some of the communication problems, it seemed a blessing. The professor even lent him an account to use for the Tibet Support Group, a body of supporters around the world who tried to get the message out. Speeches and news could reach them in moments.

By the mid 1990s, sending an email from Dharamsala was still time-consuming. It involved going up to the roof of a building and sitting down on a machine to perform what one user called a 'mystical process'.[2] In 1997 Dan Haig, along with a group of other computer experts mainly from the Bay Area of San Francisco, paid their own way to make the long journey to India to expand the Tibetans' internet operations. They carried 165 pounds of cables and routers in backpacks on the forty-hour journey. After an audience with the Dalai Lama they set about wiring up his government by pulling cables through walls to create a local network in the Himalayas. This would help connect the monks to the nascent internet by opening up a presence on the web. The Tibetans had been used to having their post intercepted, but had hoped that computers might be safer terrain on which to communicate and organise. They had no idea they would be subjected to one of the most sustained electronic espionage campaigns the world has ever been seen.

The Tibetans began seeing suspicious activity on their computers by the late 1990s, they say. Machines were mysteriously wiped, resulting in costly bills to have data retrieved. The same also happened to the Falun Gong group whom China saw as a threat to social order. Their websites – including one in the UK – were targeted from 1999. In 2002 the campaign against the Tibetans intensified. In April that year a Chinese Minister of State Security had urged a clampdown on subversion, including over the internet. The websites of dissident groups suddenly became inaccessible to people using Google from mainland China. And the Tibetans were engulfed by hundreds of spoof emails appearing to come from friendly sources but actually carrying Trojan horses, allowing attackers to take control of computers.[3] Tibetans would get emails supposedly from Western campaigners but often written in what is called Chinglish – a slightly mangled version of the language. Sometimes it was so obvious that the recipients would laugh out loud at the mistakes. 'We kindly plead you to find the Invitation Message in your attachment file,' read one. Soon the messages began to get sharper and better, increasing rapidly in sophistication. Activists were taught not to open attachments on emails, but not everything could be stopped.[4] The campaign began to reach activists abroad. It would be the source of the first known cyber breach of the British government, which has been kept secret until now.

In London, an email arrived in the inbox of a Foreign Office diplomat in October 2003. It purported to come from the Tibet Support Group and seemed to be about the Fourth International Tibet Support Group Conference held in Prague the previous month. Foreign ministries and even embassies in Beijing were on the mailing list for the Tibet Support Group set up by Samdup as it was one of the best sources of information as to what was going on in the remote corner of China. The Prague conference, opened with prayers from three monks, had been a major gathering involving 260 participants from forty-seven countries featuring talks and workshops (including one looking at how to disrupt the upcoming 2008 Beijing Olympics). Delegates who attended the conference remember later hearing about a strange incident: lots of paperwork – including confidential material that had the contact email addresses for attendees – had been thrown in the rubbish at the hotel. Word went round that someone had 'liberated' all this material from the rubbish bins. That might explain why, a few weeks after the conference, people on mailing lists around the world started getting emails referring to it. Attached to the email received by the Foreign Office in London was a picture labelled 'Tibetans Prague' which showed some of the attendees. What the British diplomat did not realise was that hidden in the picture was something malicious: a Trojan horse. When he clicked on the email, foreign cyber spies broke into the British government network for the first time. They were then able to roam around the Foreign Office network, gathering information (although not at the highest levels of classification since those were encrypted). This was the first serious Chinese intrusion – or at least the first anyone had spotted. 'How long it had been going on for, we didn't know,' one official serving in a senior position at the time recalls. The Foreign Office itself noticed some kind of suspicious activity after the Tibet email and reported it to the computer security experts at GCHQ.

*

The first British government computer security team had been set up in 1992, called UNIRAS (the Unified Incident Response and Alerting Service). Its early days had a mildly comic element since no one was exactly sure what it was for. In one case, it received a phone call from someone to say they had spilt coffee on their computer keyboard and wondered if the team could help? When a questionnaire was sent out to government departments about attacks, many said they had seen nothing. That was because they had not been looking. The inbox was empty, partly because the British government was not exactly switched on when it came to high technology. It was only in the late 1990s, just as Tony Blair became Prime Minister, that Whitehall officials were finally allowed to access the internet from the office. But despite his talk of modernising the country, the new PM was not computer-literate – a vaguely computer-savvy twenty-something had to show him how to use a mouse to click on an icon soon after he entered Downing Street.[5] But because the UK government came to the internet later than the Americans, it was able to think more about security. The American government's relationship with the internet had emerged organically, leading to a multiplicity of connection points between government networks and the rest of the internet. This made plenty of nice, juicy targets for hackers, as the US had seen through the 1990s. The new British Government Secure Intranet, opened in 1997, offered only one gateway to the outside world. By 2001–02, when the right legal process had been agreed under the Data Protection Act, a box could be attached to the entry point to monitor what was incoming. And as soon as the government team (largely out of GCHQ) started looking, they began to see bad things out in cyberspace. The Tibetan email was the first to get in.

The security experts who investigated the 2003 rogue Tibetan email searched the internet and found details of the conference in something called 'The Tibetan Monthly Bulletin'. There were signs that it was part of a much broader campaign. It emerged that almost everyone who had been named as going to the Tibet Support Group Conference in Prague had been targeted by a series of emails over days, purporting to come from different parts of the Tibetan activist community or government in exile, each one with a form of malicious content attached. This was a large-scale, comprehensive operation. British officials are still far too discreet (of course) to name who they think was responsible, but it is not exactly hard to guess. Apart from the simple issue of motive, the internet protocol address from which these emails were coming was registered in Beijing. Later, experts would find that the campaign was international. The UK government had known about the possibility of this type of compromise before but had never seen it actually happen prior to October 2003. The issue went up the chain to the most senior intelligence officials and the Joint Intelligence Committee. There was a discussion as to how to respond. Some voices asked whether the servers involved in the attack could be taken down. But it was decided that this was too hostile and the attackers could easily move. Others talked about going public. Instead, it seems, almost nothing was done and the issue barely registered with ministers. And so the cyber espionage continued and grew. This was only the start.

The more they looked, the more the cyber defenders found malicious attacks targeting the government in the following years. One person involved at the time describes it as being like a boxer fighting with a paper bag on your head. You could feel yourself getting clobbered but could not see where the punches were coming from. By 2006, intelligence showed that Britain was not immune from the Titan Rain campaign. Even part of the House of Commons computer system was taken offline that year.[6] As in the US, the campaign had begun with government- and defence-related industries. A few businesses with close ties to government had been hit and knew it, including Rolls-Royce and BAE. But slowly the target list began to expand into the broader private sector. One company board saw the attack and went to the government, who offered them help because of their size and importance. When the government looked at what had happened, they could see that the signatures of the malicious code matched what the US had been experiencing. They were textbook Titan Rain. Around this time, another company board asked for help from the government but got nothing back. The government made it clear to that company that it protected 'UK PLC', not an individual PLC. But no normal company (outside the defence sector) had the people or the awareness to be able to spot, let alone deal with, high-end cyber espionage from a foreign state at this time. This meant that for many years British companies presented rich pickings. Although, of course, if they were taken apart they may not have known it (or may not have said that they did).

It is obvious that if one country's military attacks another country, then the task of national defence falls on the government – in fact it is often seen as the government's primary duty. But what if a country targets a company sited in another country over cyberspace to steal information: who is responsible? At first GCHQ seemed intent, outsiders felt, on remaining a secret intelligence agency. Dealing with businesses which were being attacked was not their priority, which was defending the government itself. GCHQ seemed determined to hang on to its precious intelligence data and not share it with the outside world. 'It was the world's biggest self-licking ice cream,' is how one critic describes their stance. Placing cyber security within a top-secret spy agency is problematic in that such agencies like to keep their secrets. In other words, they like to know what is going on but do not necessarily want to tell others what they know, especially if it reveals their capabilities.

At first MI5 and its front organisation the NISCC took the lead, setting up information-sharing exchanges. Chief executives of major infrastructure firms were even brought into MI5 headquarters at Thames House to be briefed by the then Director General, Eliza Manningham-Buller. The NISCC also issued the first public warning in 2005. It said that 'industrial-strength hacking' was taking place on an increasingly sophisticated scale to steal commercial secrets. However the group was forbidden by the Foreign Office from mentioning China for fear of the diplomatic impact and also from using the term espionage, since that would imply a state actor and that would also raise too many difficult questions. The warnings were left vague.

By 2007, the evidence of large-scale espionage was growing. What should businesses be told? Was there a way of confronting and putting pressure on the country thought to be responsible for the attacks? In the end, officials settled on that most British response: they wrote a letter. It came from the head of MI5, Jonathan Evans, and warned the chiefs of 300 British companies of 'electronic attack sponsored by Chinese state organisations' capable of defeating the best protective systems. The intention was partly to warn the companies, but there was also the hope that the Chinese might get the message that they had been rumbled. The notion that this would stop the espionage campaign was fanciful, though. It continued apace.

Some companies were slow to wake up. They thought it was about someone else. 'Going into a major organisation and saying the Dalai Lama got attacked last year, you should really worry about your corporate secrets, didn't really resonate,' says David Garfield of BAE Detica. His company would help one company investigate a breach but see the attacker also stealing from another company. They would then approach that second company's board and hand over to the surprised executives their usernames and passwords in order to get their attention (and their custom).

MI5 had been beginning to get to grips with the issue when its world changed. The bombs that exploded on 7 July 2005 on London's transport system forced the Security Service suddenly to throw almost all its resources against counter-terrorism. Cyber security fell down the agenda. And, soon afterwards, a new Director of GCHQ had taken over who was determined to move cyber security to the centre ground of his organisation's work. The next step for GCHQ was to try to take off the bag and see who was punching you. The way to do this was by plugging into the GCHQ intelligence collection machine and linking offensive spying and defensive monitoring.

GCHQ moved into the Cheltenham building known as the 'Doughnut' in 2004, just as the extent of the cyber espionage campaign was becoming clear. The building serves as a useful metaphor for how GCHQ has operated. From the outside it is designed to be impenetrable, with barbed wire and guards at layered entry points to make sure no one gets in uninvited. But once the security checkpoints are cleared (and any electronic devices confiscated), there is a remarkably open environment. 'The street' – the path that runs right round the ground floor – has coffee shops (with security-cleared baristas) and colourful comfy chairs for break-out meetings. Casually attired staff walk in and out of the offices on either side. This all

creates something of a bubble, a bit like those old books which imagined what life would be like in a colony on the moon in which
everyone lived in an airtight environment. Inherited from Bletchley is a sense of distinct identity. Staff socialise with each other and
quite often marry each other, their offspring going on to work in
the same building; in some cases three generations can trace their
origins back to Bletchley. The sense of apartness is emphasised not
just by the sense of secrecy but also by the fact that London – and
the rest of government – is a long, long way away. Until recently,
GCHQ remained hermetically sealed, only the odd visitor making
it through the air gap, while the occasional staff member traipsed
up to London on the train to try to explain to the rest of Whitehall
what they actually did. That was the old world of SIGINT – signals
intelligence – an esoteric, specialist enterprise running in the background much of the time in the hope of providing early warning of
nuclear Armageddon during the Cold War. The new era of cyber
espionage was going to change all that.

GCHQ's Cyber Defence Operations (CDO) area is on the ground
floor of the 'Doughnut'. A sign on the door reads 'Defending the
UK one bit at a time', a joke for the techies. Its mission is to monitor classified and unclassified sources of information to stay one
step ahead of those attacking the UK over cyberspace. Some of the
casually dressed team look remarkably young. Their day starts with
a morning briefing of events detected overnight before a ticketing
system prioritises the most urgent and important cases. Lines of
green code scroll automatically across the screen of an unattended
monitor. One person has a screensaver of the old arcade game Space
Invaders. They are not quite shooting down rows of aliens here but
they are dealing with wave after wave of attacks.

Iain Lobban took over as Director of GCHQ in 2008. He was
an insider who had risen through the ranks of signals intelligence.
He had joined what seemed like a hierarchical organisation in the
1980s focusing on Cold War targets and spent part of the mid 1990s
on the Middle East desk working closely with MI6. An energetic
character, his years as Director would involve shifting GCHQ's vast
intelligence-gathering machine to point towards the cyber world,
fusing offence and defence. In a 2013 interview for the BBC, Lobban

compared this to the way in which his predecessors worked in the Second World War and the Cold War:

> Bletchley Park built up a worldwide web of listening stations and analytic and cryptanalytic endeavour, which built up a picture of what normal looked like – what was out of the ordinary, what was new, what represented a threat or a vulnerability in terms of the Axis powers. And, as a result, we the Allies were able to exploit faster and to protect better than our adversaries . . . During the Cold War we built up a very rich understanding . . . through SIGINT of Soviet military systems and of what normal activity looked like – what was abnormal . . . Today again we are using our global SIGINT capability – once again with Allies – to understand adversarial intent, capabilities, tactics, developments.

This means monitoring global communications traffic in order to watch and understand your opponent's behaviour and spot the abnormal amid the day-to-day, to find the signal within the noise, to look out for your data being stolen. One person involved in responding to the attack against the American security company RSA in 2011 despaired of the ability to cope with cyber attackers. Every time defences were improved, the attacker found a new way of getting round them. They compared it to the moment in the Second World War when German U-boats were ravaging the Atlantic convoys on which Britain depended to receive war material and support from the US. What was needed was a new defence doctrine – new tactics and technology, he thought. But the crucial component in defeating the U-boats was of course intelligence – specifically that of Bletchley Park – which by breaking the Enigma machine allowed the UK and US to get inside the plans of their attackers and then exploit that knowledge.[7] The modern parallel is that stopping Chinese espionage may require intercepting their traffic and getting inside their networks, which in turn means monitoring global traffic, as Lobban explained:

> We see data moving around the world and if we see data moving between two points which looks like it should not be moving

between two points . . . perhaps if we get into that data we can
deconstruct it and work out that it mentions a company name or
has some particular technology involved – perhaps we can work
out that that might be a British company or . . . a subsidiary of
a British company or the company of an allied or a friendly government – and from that we will if you like trace back up the
chain to where that penetration might have occurred.

This can involve spying on the other side's spies to see how they
are spying on you: Bletchley learnt that if you break the other side's
codes and spy on their traffic, you can understand if they have broken
into your communications. There was a virtuous circle in breaking
enemy codes and strengthening your own. At Bletchley, the two
Allies had often watched each other's backs. In 1942 Bletchley had
read traffic which, in some of the language it used, pointed to the
fact that the American liaison in Cairo had been penetrated by the
Germans. The Americans were informed and dealt with the problem. Lobban argues that the same takes place in cyberspace today as
GCHQ uses what is known as its 'passive SIGINT' capability: 'Today
we are using our SIGINT systems to monitor worldwide communications – we are looking for patterns, signatures, segments which
betray a compromise – the discovery that a hostile actor has penetrated a sensitive network and somehow is exfiltrating data which
had been thought to be secure. Those compromised networks may
be British or may belong to any of the friendly nations we work
with.'[8]

Monitoring an adversary's computer espionage has one added
benefit that Lobban is too discreet to mention. You can watch them
stealing off other countries and organisations and then get hold
of those secrets. It is not quite the same as stealing off a burglar,
since the advantage for the spies in this type of cyber espionage is
that you are copying what they have stolen rather than taking it off
them, and hopefully without anyone knowing. This trick is known
as 'fourth-party collection' and dates back to Cold War communications intercept. In the modern world, if Britain was to watch China
stealing secrets from Russia, it might then copy the data itself (and
so gain Russian secrets and learn about China's capability). 'When a

Chinese hacker succeeds in stealing data from a European company, the data is intercepted in passing by the Americans. Consequently, our secrets head both to China and America,' a French official told a newspaper despairingly. There is even something called 'fifth-party collection' – spying on spies spying on spies. In the hall of mirrors of cyberspace data may also be moved to another country's servers so it can be 'scapegoated' if the theft is discovered. Spies and hackers are often tripping over each other because there are so many of them inside networks.[9]

As GCHQ turned its signals collection machine into cyberspace, what did it see? In 2013 GCHQ saw about seventy sophisticated attacks a month against UK government or industry networks. About 30 per cent were against the government, 15 per cent against the defence sector, 24 per cent against information and communications technology companies, 6 per cent against engineering and 5 per cent against business services (such as law firms and accountants). 'Where you're seeing a state-sponsored activity against individual companies – not against other governments, not against other nation states but against particular companies – this is, if you like, a game changer. This is industrial espionage on an industrial scale,' Lobban said in 2013.

This new world of spying on global traffic to spot a breach has changed the way GCHQ works. In the past the defensive wing of GCHQ, CESG, was something of a 'career graveyard', originally tasked mainly with designing and testing ciphers for the government to use in its communications. Its staff and resources were massively outnumbered by the offensive team of 'collectors'. But it began moving towards the centre to respond to breaches that might be seen. But while the concern over a foreign-focused intelligence agency protecting domestic networks has been less of an issue in the UK than in the US, there have still been tensions due to secrecy. Some businessmen still complain that when they approach spies about a breach in their systems, the spies want to know everything about the details and signatures of the attacks but do not provide much back (one British executive says he gets more from the US than he does from British spies). 'They wanted to have their cake and eat it – to be a secret intelligence agency and also the interface with the outside world on cyber,' says one critic of GCHQ who has worked in both government and the private sector.

GCHQ has found it easier to deal direct with an elite of trusted companies – defence and telecoms primarily. The next level out is described as 'companies of value to the UK'. The definition of companies of value is deliberately broad enough to mean not just UK-owned ones but also foreign companies who might operate in the UK and play an important role in the economy (by paying tax, bringing in research or simply being a staple of pension funds because of their value on the stock market). They are dealt with by the Centre for the Protection of National Infrastructure, which is provided by GCHQ with information on who has been hit. Smaller companies are left to the Department for Business, Innovation and Skills (although foreign spies increasingly targeted big companies through smaller suppliers and the likes of law firms and accountants who serviced the big companies).

MI5 – in its austere, monolithic Thames House headquarters just down the road from Parliament – is Britain's spy-catching service. For decades this meant Soviet spies, its watchers following members of the Soviet Trade Delegation or Embassy from Harrods to the back streets of south London or hunting through the files to find evidence of traitors within, looking for the clues in things like Venona traffic. After the 11 September attacks, it reinvented itself as primarily a counter-terrorism agency (building on its experience in Northern Ireland). But to its frustration it found that it was still having to spend more resources than it wanted on counter-espionage. Some of this was of the old-fashioned kind – dealing with Russian spies who were present in numbers that matched the days of the Cold War. But there was also the new cyber espionage threat. And sometimes the two merged.

'There are now three certainties in life,' MI5's then-head of cyber, who spoke on condition of anonymity, explained in a long conference room overlooking the river in 2013. 'There is death. There are taxes. And there is a foreign intelligence system on your system.'[10] It is a statement designed to shock businesses out of complacency. 'MI5 is here to investigate foreign intelligence service activity against British interests . . . There are hostile foreign states out there who are interested in a company's mergers and acquisitions activity, their joint venture intentions, their strategic direction over the next few years, and that information would be valuable to that country's state-owned enterprises.' In one case, hostile activity amounted to 10 per cent of all activity on one company's network and they knew nothing about it until told by the British government.

All sectors are being targeted, he warns, explaining that foreign intelligence services are carrying out extensive reconnaissance of senior executives. Ultimately, for all its technical aspects, cyber espionage is still about one person trying to target or recruit or subvert another – by first finding someone who has access to the sort of data a foreign intelligence service wants and then working out how to get close to them. Almost all of this can now be done on the internet. This involves understanding their business contacts. 'Which CEO isn't going to open an email from his lawyer?' asks the MI5 officer. It might also involve sifting through social media sites like Facebook, looking at a target's hobbies and crafting an email to maximise the chance of them opening it and clicking on a link. 'Intelligence services are very patient,' he says, arguing this is no different from traditional espionage, where in the past the Russians would have spent many months first researching someone and then cultivating them to establish the best means of persuading them – wittingly or unwittingly – to hand over secrets. 'They do exactly the same thing in cyberspace,' he says. 'But it's also a risk-free way. Instead of having an intelligence officer working on the streets of Birmingham or Manchester trying to gain access to their particular target in order to talk to them and recruit them, they can sit in their home base, in their headquarters tapping away behind a keyboard and trying to do exactly the same thing . . . you don't need to physically meet the people you are working against.' In the Cold War, an agent would have to be subverted – perhaps by offering them money or blackmailing them. Occasionally you might be able to persuade someone to give you secrets without them knowing you were a spy or by making them think you were a spy from a different country (a so-called 'false flag' operation). But in the cyber world, all you need is to persuade someone unwittingly to click on an email and you have the access you require. You no longer need to persuade them to carry out any specific documents for you – just to mistakenly click. And once they have, the old problem of dead drops and brush contacts – leaving pilfered secrets in an envelope behind a tree, or passing them in a shopping bag as you walk down the street – is solved. Once someone has let them in, the foreign spy agency can establish a channel which (until discovered) will always be open to go and retrieve more secrets.

Executives will be carefully watched if they go on business trips to Russia or China as it provides a chance to learn about them. 'The number of Chinese interpreters who happen to be attractive women is extraordinary,' comments one British spy. The same spy recounts talking to a CEO about his trip to a sensitive country. The CEO said it was all fine: 'The first thing I do is lock my laptop and phone in a safe.' To which the spy said, 'That's the first place our spies go.' The businessman, he says, looked shocked. The types of precautions only spies and diplomats took are now required for businessmen who do not want their secrets stolen. And it does not just concern companies. In April 2013, university chancellors were specifically warned: 7 per cent of the attacks GCHQ saw in 2013 were against their institutions. Universities, focused on sharing information, are usually easier to penetrate yet are often working with industry and conducting cutting-edge, innovative research and development, for instance on quantum computing, aerospace engineering or new high-tech materials like grapheme.[11] 'It's an easy way for a foreign intelligence service to get that sort of data,' says the MI5 officer.

The list of those who have been penetrated is long but largely anonymous if you listen to speeches by intelligence chiefs and ministers. A large international manufacturer is said to have been hit during a period of negotiation with a foreign government, leading the hackers to access the email accounts of the entire leadership team. A defence contractor received a file posing as a report on the Trident nuclear missile programme from someone apparently from another contractor but containing malware. A 'well-protected international company' was also breached through a foreign subsidiary to be robbed of 100 gigabytes of intellectual property, equivalent to 20 million pages of A4.

The head of MI5 said in 2012 that his service had worked with one major London listed company which estimated that it had incurred revenue losses of some £800 million as a result of a hostile state cyber attack: this was not intellectual property theft but an estimate of the value lost from a contract which the company would have received if it had not been hacked. Journalists sought to identify this company at the time. It was, a number of individuals believe now, the mining giant Rio Tinto, though the company declines to comment. The extractive industries have been a major target for Chinese spies since the country has a vast appetite for raw materials to fuel its growth – including iron ore for the steel to build its skyscrapers, cars and for use in its factories. Any attack may have been linked to a major contractual negotiation with China about the price of iron ore. The contracts with suppliers at the time involved fixing a price over long periods for vast sums of money. Renegotiation of these contracts is a game played with high stakes, especially as differentials can open up between the market and fixed price. China was pushing hard for lower prices at a time when relations with Rio Tinto were already strained. In 2007–08, when BHP Billiton and Rio Tinto were discussing merging, a deal with huge consequences for China, both companies' networks were reported to have been penetrated. China was also angry when a Rio Tinto partnership with a Chinese company fell through in 2009. A number of Rio Tinto staff have been arrested, including in 2009 when China accused some of bribery and espionage (there were said to have been 200 attempts to hack into the networks of the legal defence team).[12]

In the case cited by MI5, it appears that China may have used cyber espionage to understand and manipulate negotiations. However, individuals with knowledge of the industry say they do not recognise the £800 million originally cited by the MI5 chief (which may be due to price fluctuations in the following years altering the original estimate).[13] China may have achieved its goal through a mixture of more traditional means – such as arresting or pressuring staff – alongside cyber espionage at the crucial moment in negotiations in order to ensure a shift to the most favourable pricing deal. When asked to comment, the company will only say that this is 'speculation'. But it is indicative of a wider experience that many companies are unwilling to discuss, partly out of embarrassment but also out of concern for further inflaming relations with a country that remains a crucial market. In another case, someone looking into the sale of a high-profile British company to a Chinese firm wondered why there were suspicious (but inconclusive) trails in the data and why the Chinese purchasers seemed to know what the British negotiators were thinking at every step along the way.

British government departments have also been hit. One group targeted 200 email accounts at thirty of Britain's forty-seven departments to try to gain sensitive information. In December 2010, a spoof email from the White House was sent to the UK government. The next month Foreign Office staff were sent an email apparently from a British colleague about a forthcoming visit which was actually from a 'hostile state intelligence agency'.[14] The office of the Director of GCHQ has been attacked. The Treasury said it was seeing attempts on average once every day by 'hostile intelligence agencies'.[15] In 2010, an email relating to a G20 meeting went to the Treasury. A few minutes later it appeared to have been resent. The second email had swapped a legitimate attachment for malware. 'Before international government conferences we will very often see ministers, their senior officials, their special advisers, targeted by socially engineered emails which purport to come from somebody that they know, may well come from the address of somebody in the same department and which may be instructions for next week's conference.' These attacks, Lobban said, were directed by other countries' intelligence agencies. But when asked if Britain did the same thing, he gestured in a way that made it clear an answer would not be forthcoming.

Britain's classified systems have been breached. 'The number of serious incidents is quite small, but it is there,' Major-General Jonathan Shaw, the head of cyber for the Ministry of Defence, said in 2012. 'And those are the ones we know about. The likelihood is there are problems in there we don't know about.'[16] Shaw said it still surprised him that the Ministry of Defence headquarters 'is the only building, main defence security establishment, where you don't leave your mobile phone and iPad in a box outside your office . . . people's personal behaviours are not good enough.'

Which 'hostile foreign intelligence agency' was behind these attacks? Ask officials who are normally vocal about the threat and they suddenly go all shy. 'This is being sponsored by some other state,' Iain Lobban replies. 'We know who it is . . . we're sure we know who it is.' But when pressed to name the country, he declined. 'I can't say who it is to you,' he replied. When it was suggested that the finger was often pointed at China, all he would say was 'So I've heard.'

A 2010 secret report (later leaked) made it clear that, privately, GCHQ saw Beijing as responsible. 'China has a capable and very wide-ranging cyber programme targeting the full spectrum of governmental, military and commercial targets. The Chinese mount a large number of relatively unsophisticated attacks, often using publicly known vulnerabilities and have successfully compromised networks globally. This assessment is based only on the attacks that have been detected, and does not preclude more sophisticated and targeted attacks from China.'[17]

The desire not to accuse China in public is the result of an argument over how tough to be which goes to the very highest circles of government. Those who worried about the state of the economy, rather than just national security, argued that it was easier for America – still the world's number-one economy – to call China out over cyber espionage. Britain was not in quite so strong a position. The British Cabinet in 2013 was reported to have been split on how tough to talk with Beijing, with the then Foreign Secretary advocating a harder line that was resisted by the Chancellor and Prime Minister, who feared the impact on trade.[18] British politicians have been frequent visitors to China, soliciting investment in major British infrastructure projects including the nuclear industry, as well as bringing along British business executives desperate to sell into the growing Chinese market. One former senior British intelligence official believes that ignoring economic cyber espionage and intellectual property acquisition for the sake of trade is a dangerously short-term mindset. 'The next time [Britain] tries to sell something to China, we will find they do not need to buy it,' he says exasperatedly.

Other European countries have also been hit by cyber attacks from China. Germany began seeing activity from 2007, the Chancellery being targeted by hackers in China just before a visit to Beijing by Angela Merkel. From 29 September to 2 October 2008, Germany's BfV held a meeting on the Chinese cyber espionage threat. Their conclusions mirrored those of the US intelligence community about the threat from Chinese socially engineered emails. Five hundred such email operations were conducted against German military, diplomatic, high-level government systems from October 2006 to October 2007. German intelligence detected increased activity immediately preceding times when the German government or commercial interests were negotiating with the Chinese.[19] Companies based in Germany like the Munich office of European defence manufacturer EADS and ThyssenKrupp were also targeted by the Chinese campaigns. For Germany this is proving particularly thorny since the country's economy rests on high-value exports and has been working to maximise access to the Chinese market. Most of the attacks on German companies were traced to three Chinese cities: Beijing, Shanghai and Guangzhou, including Unit 61398.[20] Australia has even had the blueprints of its new intelligence headquarters stolen, it is believed by the Chinese.[21]

The Chinese get the attention in the West because they are the busiest and the noisiest. But they are far from the only actors in commercial computer espionage. In 2013, the Chairman of the US security company FireEye said it was tracing 'callbacks' to 190 countries involved in such activity.[22] After the Chinese, British officials cite the Russians as the most serious danger. Russia is said to have built a global espionage infrastructure of computers around the world which can be used when needed to move data with few obvious links back to Moscow. These are like Cold War deep-cover 'illegal' agents waiting for a call to action. A leaked GCHQ document of 2010 says Russia 'operates a sophisticated, mature and successful cyber programme, using an extensive global internet-based infrastructure' which 'poses a significant threat to UK networks'. It says: 'Targeting of UK government departments is assessed to be a priority for Russia, and is likely to be ongoing. Governments, industry and academic institutions across a range of sectors have been targeted.'[23]

American officials also say the Russians may be a distant second after China in terms of volume, but are more skilled at what they do. China has many more hackers and works on much higher volumes (back to the thousand grains of sand point), while Russia is more selective and also more clandestine. China's programmers are said to employ well-known tools; Russians are more likely to write their own code. China's hacking is often sloppier and easier to spot (hence all the attention), while Russia's hackers are more expert and operate below the radar, as James Lewis explains: 'I was talking to some Japanese officials once and they were telling me about the number of intrusions they saw from China and I said "That's consistent with what everyone else sees. But how about the Russians, do you see Russians on your network?" And they said, "No, we have never seen the Russians on our network." And I thought to myself, "That's a true statement, just not the way you meant it."'[24] Another former official puts it in a different way. The Chinese use big nets to trawl for their catch. The Russians pick their spot on the side of the shore and choose their bait carefully to catch exactly the fish they want. These Russian operations are highly targeted. One is said to have involved researching an executive, leading to the conclusion that he was gay but not out of the closet. The hackers then sent him an email from a gay rights organisation which they suspected he would open since it looked as if it was sent to him, but in fact held malware. They then counted on the fact that, even if the executive did suspect it was malware, he would not be willing to go to his company's IT department or security team for fear it would reveal his sexuality. This is classic, high-level, targeted Russian espionage.

In the Cold War, a spy would have to get instructions from his or her handler. This might be done by a classified ad in a newspaper which provided instructions but was hidden from everyone else by using a secret code, or it might be done through a random stream of letters sent across radio that needed to be listened into and decoded. Now, that entire process has been automated. A compromised computer will receive a command to visit a particular website every second – say, for instance, that of a news organisation. But a page of that website has been hijacked and a photo replaced by one which looks identical but has commands embedded in code. So when the photo is loaded by the specific, compromised computer it provides instructions. Because it is checking the site every second rather than waiting for the classified ads every day, the controller far away can effectively issue second-by-second instructions telling the agent computer exactly what to do.[25] This is semi-automated, computer-controlled espionage.

The Russian economy is heavily dependent on exporting energy, and the country's intelligence services appear to have prioritised this field. One major energy firm was told by the British government that it had something on its system and that it needed to look itself because it could not be shown everything (suggesting the information might have been intelligence from spying on Russian traffic). Within a day, the company found that one PC was beaconing out after a spear-phishing attack in which a small number of individuals had been sent carefully researched emails. 'Beaconing' is when malware is trying to communicate with a pre-established command and control server computer, effectively saying it is ready for a connection to be made so that the attacker can begin issuing instructions (this is the moment when espionage is often easiest to spot, like watching an agent contact his handler). Forensics on the machine showed the malware had been in place for nine months. The company analysed the email, the target and the timing (correlating it with business transactions going on at the time). That left them 99 per cent sure that it was a state-sponsored attack by the Russians on behalf of their energy industry. The company could not be sure, but believed the attackers might have gathered what they wanted but left an access point open for the future. They appear to have been looking at pricing and discussions about nuclear plans.

Russian spies are also said to be interested in plans for shale gas drilling (or fracking), since this threatens dependence on Russian energy supplies. What would happen, one energy insider speculates, if the Russians could hack into companies planning to frack and then selectively release secret reports so as to try to undermine public support for the process (perhaps by focusing on flaws in design or seismic issues)? This could be put into the public domain by passing them to a third party who might not know they were being used by the Russians. There were also reports that Russian hackers might have been behind the selective leak of thousands of emails from a climate change research project at a British university. One UN official compared it to Watergate. 'This is not climategate, it's hackergate,' they said. 'Let's not forget the word "gate" refers to a place [the Watergate building] where data was stolen by people who were paid to do so.'[26] Hacking a climate change project and leaking information or doing the same with the witness list in an international investigation which the Russians wanted to hamper is very much classic KGB work – what was known as 'active influence operations'. During the Cold War, this involved trying to shape public opinion and views within other nations, often using propaganda which might be spread by third parties. The Russians have long been masters of this. Russian spies are doing what they always did, but now in cyberspace.

For all the noise about China, the US also found the Russians to remain the most adept adversary. In 2008 America had to respond to a serious espionage attack codenamed Buckshot Yankee. This aroused high-level attention because it penetrated classified systems that were supposed to be 'air-gapped' from the internet. It had jumped the gap by deploying a virus carried on a USB stick from an infected machine which was inserted into a military laptop at a Middle Eastern base, then spread from device to device looking for classified information (not that dissimilar from the malware on floppy disks the Russians had distributed at the conference a decade earlier). It took four months to be discovered by an analyst from the NSA in October 2008, who saw something beaconing out from the US military's Central Command. Russia was believed to be responsible, and it seemed to be another example of how the old Cold War foe remained the opponent working at the highest level of sophistication against the US. 'We thought they were extremely well protected and we turned out to be wrong,' says Bill Lynn, who was a Deputy Secretary of Defense.[27] 'We don't think we suffered significant damage because we caught them early enough but the fact was a shock. We did not think we were vulnerable in that way.' This attack was important in raising awareness of the issue at the highest levels of the Pentagon and the administration. This event, as with the Eligible Receiver exercise a decade earlier, was also cannily used by the cyber hawks to push their case that more power was needed in their hands in order to respond. It is not clear if the virus actually did that much damage or was that sophisticated, but it was the fact that it reached classified systems that made it a useful recruiting sergeant for those sounding the alarm. Buckshot Yankee led directly to the creation of the US Cyber Command which would be run by the tough-talking General Keith Alexander, who already ran the NSA. Power was being increased and concentrated. In response to the attack, the number of connection points between classified networks and the internet was reduced (from a peak of 10,000 to just under twenty) and they were monitored intensively. This led to a wider campaign by Alexander and the NSA to safeguard not just the military itself, but also the defence industry. After that, Alexander would push to stand guard at all the internet gateways through which traffic was coming into the US.

The NSA, like GCHQ, also married defensive cyber work with global intercept and signals collection. In 2009, for instance, intelligence suggested an attack was being developed by the Chinese. This meant that, by the time highly targeted emails arrived in the inbox of the Chairman of the Joint Chiefs of Staff and three other senior military figures the following October, a counter-measure had been developed to prevent their systems being penetrated. The NSA developed a global system called Tutelage, whose sensors monitored threats and provided the ability to alert, intercept, redirect, block or substitute malicious code it could see coming through the network, even working back to spot when that code was designed and before it was deployed. All of this meant that an attacker could be spotted and attacked back automatically so that their machine was itself compromised (including by implanting malware). This made it possible to gather intelligence or stop them. The merging of missions was highlighted in a July 2009 story when the NSA's Threat Operations Center worked off a tip from the defensive team to target a command and control node use by China's Byzantine Raptor group. They then collected all the traffic passing through this system. It included not just data being stolen from the US (which could then be better protected), but also documents stolen from the UN (so-called fourth-party collection). The NSA's UN team was then able to issue reports about 'high-interest' events based on intelligence originally stolen by the Chinese (a similar trick was used to piggyback on South Korea spying on North Korea). 'Steal their tools, tradecraft, targets and take,' as one NSA document boasted of this kind of work.[28]

The push to intercept and monitor at bulk for cyber espionage was becoming the new trend. Traditional defences were too easy to get around. The only way to find the other side's spies was to monitor dynamically what is normal on a network to spot the anomalous. Companies also began to monitor the traffic on their corporate networks. This can raise privacy issues. When do employees normally log on and download data? Is someone installing encryption software and working in the middle of the night? Do they normally do this because it's part of the job? Establishing what is normal and abnormal requires gathering as much data as you can for as long as you can, including the innocent behaviour on a network. This might all be gathered by a machine, a person only being called in when something anomalous is spotted; but it does mean those machines are watching everything employees are doing online. This could be used to find a foreign hacker stealing data or an insider leaking secrets, but also someone just acting against company policy in the websites they look at. Or it could be used to check on their productivity. You might want to correlate network traffic with the records of employees swiping into work. That might lead you to ask why Mr Corera is downloading files from a remote login when he is registered as being in the building. But it is only a small step to asking how much time he spends in the building compared to his colleagues. Some companies are open about it: for instance, financial companies often tell employees that all their phones and emails are monitored for compliance. But in other cases it may be surreptitious or just not talked about. This is a company defending itself and its own network against threats from within and without. But what happens when the state wants to do it on a vast scale?

This goes back to one of the central tensions for both GCHQ and the NSA in the cyber age. They are the repositories of expertise on cyber security – they know how to hack systems and they can see through their own intelligence systems how others do it. But they are both historically 'foreign' intelligence agencies built and tasked to spy on foreign countries. How far, then, should they operate in the domestic sphere to protect companies at home from attacks that originate abroad? In the UK, GCHQ has stayed out of monitoring domestic traffic, including defensively monitoring company networks on their behalf, preferring instead to focus on watching international data flows as part of its global intelligence mission. In the US the NSA, charged with monitoring domestic classified networks, also has relationships with defence companies and ran trials of automated defence on military and contractor networks using Tutelage. But the NSA also repeatedly pushed for a greater role in defensively monitoring traffic across a broader range of private companies and the telecoms carriers themselves. At one meeting, NSA chief General Keith Alexander is reported as saying: 'I can't defend the country until I'm into all the networks.' But having the NSA inside 'all the networks' meant monitoring all the traffic at the gateway points, and that came up against resistance from the Department of Homeland Security, as well as companies and privacy groups who argued that a military spy agency was the wrong body to head domestic protection.

That resistance would be fortified mightily by the revelations in 2013 of what NSA had been up to secretly in terms of collecting domestic data for counter-terrorism, which seemed to stymie any hopes it had of expanding its role. The Department of Homeland Security – with its Einstein programme – remained the middleman when it came to broader defensive scanning with the private sector; but the NSA loomed over its shoulder since it had all the expertise and, especially, the most complete database of the signatures of cyber attacks that the system was trying to spot. Einstein monitors traffic to seek out attacks and try to automatically disable them before they hit government systems, with sensors installed at the points where those systems connect to Tier One internet service providers. The importance of monitoring the data pipes that carry traffic is another reason why Chinese telecoms companies may be so unwelcome in the US.

The vast signals intelligence machine built by Britain and America had been turned towards defensive monitoring as well as offensive intelligence-gathering. Defensive monitoring means watching global data flows, looking for signatures of cyber attacks by the Chinese or other countries coming into your country, and perhaps for large, unauthorised flows of data going out, and acting to stop attacks before they reach their target. Officials say they do not necessarily have to engage in intrusive inspection of the content of traffic to spot the signatures, although the full extent of the intrusion is often unclear.[29] But the technology behind this kind of monitoring – watching the data flow – has its risks. The tools for cyber security are, in a technical way, closely related to those for domestic intelligence-gathering. The only difference is how and where they are deployed. A government watching out for cyber spies can also look for other things, at home and abroad.

There is a major difference between the days of Enigma and the cyber era, other than sheer speed. In the Second World War the German Enigma and Tunny machines were discrete systems used by the German military or High Command alone. The same applied to the Soviet systems that GCHQ monitored in the Cold War. But in the modern age, the traffic which is being intercepted to find compromising evidence is part of the great global tide that makes up the internet. Foreign spies will be working through the same networks as ordinary people and businesses. Private individual and commercial information is interwoven with the foreign intelligence you are after. The tension between offence and defence grew. Now everyone is using the same communications system – one that needs to be defended when your friends are using it, even while your colleagues down the corridor might be trying to break into it to steal secrets.

CHAPTER THIRTEEN

DISSENT

The compound that housed the Dalai Lama and the leadership of his Tibetan exile in Dharamsala was protected by three rings of security. Furthest out were Indian police, next were members of a paramilitary border force and closest in were a group of elite commandos. In the summer of 2008, a pair of computer experts passed through these layers and then an airport-style security check before finally reaching the inner sanctum. These were tense days. Beijing was about to host the Olympics and in March of that year some of the worst violence seen in years had flared up in Tibet itself with riots on the streets. At the same time, the cyber espionage campaign against the Tibetans was intensifying. Because they were outsiders, it took a few days for the two visitors, Greg Walton and Shishir Nagaraja, to win the trust of the Tibetans. But by the third day they were allowed to hook up their monitoring equipment to the Tibetan computer system. 'We turned over a stone and found a whole bunch of worms,' says Nagaraja. For all the physical security around the Tibetan government in exile, their electronic systems had been penetrated. There was no need to sneak past the commandos and infiltrate a spy into Dharamsala – the Chinese were already inside the walls.

In the Second World War, Britain had made sure that every time it used intelligence based on breaking Enigma there was another reason for the successful action that would be plausible to the Germans – a spotter plane seeing a ship rather than the traffic being read. But the Chinese had not quite learnt that lesson. They had shown their hand by being a bit too sloppy in the way they used the intelligence they were gleaning. The Dalai Lama's office had noticed that meetings it privately arranged with leaders of other countries were almost immediately being cancelled. The Chinese have long run campaigns against foreign officials who meet the Tibetan spiritual leader (including British), but around 2008 these meetings were being cancelled almost as soon as the invitations were mailed out and long before they were public. The diplomatic traffic was one of two sensitive areas where a fear of compromise had led to the visit of the two computer experts, Walton and Nagaraja. The other was a database of refugees who had fled Tibet and who might be wanted by the Chinese authorities (and whose families back in Tibet might be at risk). Nagaraja was shocked by the implications of that database being compromised. 'I have never actually come across data which, if leaked, could lead to the death of a human. I've never honestly come across that before or since.'[1]

Thubten Samdup, who had first helped bring the internet to Dharamsala, had gone on to set up a project in which teams of activists went into chat rooms to communicate directly with young Chinese people in order to try to present a different picture of Tibet. Websites about Tibet were blocked in China, but online chat rooms offered a way to reach ordinary people. But when one young woman who was part of Samdup's team returned to her family village in Chinese Tibet, she was detained as she crossed the border and held for two months incommunicado by Chinese security personnel. She denied being involved in political activity and said she had been away simply to study. But then a Chinese intelligence officer pulled out a dossier on her activities. This included transcripts of her internet chats as part of Samdup's team going back years. The Chinese had clearly been spying on the conversations. 'That scared everyone in my office,' says Samdup.[2] Social media had facilitated dissent but also its surveillance.

During their week-long visit, Walton and Nagaraja put in place real-time monitoring of Tibetan networks and email service. As they did this, they could immediately see confidential documents on the move. 'You have all this security,' Nagaraja says of the rings of commandos and troops on the outside, 'but the Chinese never actually needed to breach any of them. They came through the wires.'[3] It appeared the Chinese had been all over the private office and email systems back to 2005. Walton recalls watching documents being infiltrated off the system in front of him and the Dalai Lama's aides. 'It was having their worst fears confirmed,' he says. The penetration was global. The team eventually located 1,295 computers that had been infected in 103 countries with a Trojan horse that provided real-time control of a targeted computer for an attacker, allowing files to be extracted. The microphones and cameras increasingly available on a computer could also be turned on remotely. This allowed audio and images to be sent back in the way a bug or covert video camera might have done in the past but without all the effort of breaking into a premises and installing them. A person's computer could, unbeknownst to them, be turned into a spy.[4] Emails had come to monks, seemingly from other monks, but containing malware. Messages were intercepted in transit and legitimate attachments replaced with malicious payloads, giving attackers full access.[5] One monk said he saw the ghostly sight of his own email system opening up and then sending an infected attachment to other people without him touching the keyboard. The investigators found something else: a website used by the attacks for command and control. And the attackers had failed to protect it with a password. That allowed a group called Citizen Lab in Canada, which supports activists and a free internet against state surveillance, to run a classic counter-espionage operation, spying on the enemy spies. 'The attackers had meticulously compiled an inventory of all of the organisations that they'd managed to infiltrate,' says Ron Deibert, head of Citizen Lab.[6] As well as Tibetan machines, Citizen Lab saw hundreds of other computers compromised around the world including diplomatic traffic from Iran, Bangladesh, Latvia, Indonesia, India, South Korea, Germany and Pakistan as well as international organisations like the UN and even the mail server of Associated Press in Hong Kong. 'It was a mind-blowing experience' watching it all, says Deibert. Researchers set up a 'honeypot' to watch what happened to their own computers and saw that the attackers seemed to be connecting from Hainan Island in China – home to a signals intelligence facility of the PLA. China's use of electronic espionage to seek out dissent would wrap round the globe. And in doing so it led to a collision between one of the most powerful tech companies in the world and the Chinese state, a battle that drew in a young student.

*

Twenty-year-old Tenzin Seldon received an unusual phone call in January 2010. She was a student at Stanford and one of her resident fellows was on the line. He sounded worried. 'Something very important came up,' he said. 'It's not life-threatening but I need you to be here in your dorm right now.' Seldon headed over to meet him. He explained that he had just received a phone call from the President of Stanford University, John Hennessy. Hennessy was a well-connected man who sat on the board of Google. Hennessy said he had a warning to be passed on to the young student. Her computer had been compromised, her account breached and the Chinese government were likely spying on her. She needed to call a senior Google official. Looking back on the scene, the funniest thing for Seldon was the way the resident fellow conveyed this to her, as if this were some dramatic revelation in a spy film. Her attitude was 'Okay. And?' For Seldon, the spying was anything but a surprise. 'For me, it was just every day.'[7]

State espionage seemed mundane to Tenzin Seldon because she was an activist and organiser in the Free Tibet movement. Her parents had fled Tibet on foot for the Himalayan mountains in India, where she had been born. She came to the US as a teenager and became involved in the Tibetan cause at high school. She was used to the idea that the Chinese government might be spying on her. Her first experience dated from 2008. Seldon had been organising protests against China when the Olympic torch relay for the Beijing Games had been going through her hometown, San Francisco. It was a hugely symbolic moment for China and Seldon and her friends were determined to use it to draw attention to the plight of the Tibetans living under Chinese rule. That meant organising protests. But Seldon and her fellow activists noticed they were getting emails from each other that they had not sent. They would say things like, 'I need you to open this attachment, it's related to our media strategy.' Other members of the Tibetan community were getting emails saying, 'Hi I'm Tenzin. I'm doing well here and here's an attachment that I really wanted to forward you regarding what's going on in San Francisco.' She knew she had not sent this message. Someone was manipulating email accounts to try to gain access to the group and work out their plans. The route for the torch seemed to change at one point to avoid a protest they had been organising secretly over email.

Seldon was also taking part in a US film production about Tibet. From 2008 that team found themselves under cyber attack: one member of the crew realised that her laptop was occasionally capturing images of what was on the screen, the cursor would move round by itself and the machine would switch itself off. The team found that when they arrived in China, cyber espionage was combined with traditional physical surveillance as they were followed by eight to ten vehicles and videotaped by security personnel. Hotel rooms were searched. Back in the US and Europe, producers and editors involved in the project or linked to the team found their email accounts accessed and other machines failing.[8]

That history was why in January 2010 she was almost nonplussed by the revelation that she was again a target. Once more emails were being sent round the Tibetan community from her account – but not by her – and her account was mysteriously suspended. But this time she was caught up in a bigger game. A few days after the initial warning, a senior director of Google called up and asked to inspect her laptop. Why? she asked. 'We want to see what kind of techniques the Chinese government used,' he explained. Curious, and so that she could learn how to protect herself better and find out what she had done wrong to let them in, she agreed he could come and pick it up from her dorm. A few days later he returned it to her with a message. 'Google will be making a very important decision today,' he said. 'And you should watch out for it in the newspaper.' The Google official was cautious in what he revealed about their investigation. But it seemed as if, at the same time as she was accessing her account from Stanford, two or three other people were accessing it from other places in the world. What about the laptop, Seldon asked, did they find anything? 'The strangest part in all of this, Tenzin, is that we did not find anything,' the Google man said. That was strange, since often an account is compromised by spyware being placed in the laptop and its software. The attackers had got in another way – not through the laptop and Seldon but through Google itself. That day, Google did something unusual. It went public about being attacked. On its own blog the company announced it had been the victim of a sophisticated cyber attack. That decision reflected outrage at the highest levels of the tech giant.

Google – like many companies – had not thought much about state-sponsored espionage until it was too late. In the company's early years, the main threat was people taking the site offline. Then it was criminals. But this time the hackers showed no interest in things like the credit card database for customers but were after email accounts and the like. 'Suddenly we were looking at a different kind of adversary – one where there wasn't necessarily a monetary component to the crime or the event,' recalls Heather Adkins, who had worked on the company's security back to 2002. 'This was a broad and deep attack.'[9] It first became aware of the attack in mid December after technical checks the company had put in place (which it is reluctant to discuss in detail, but are thought to involve large-scale data monitoring) spotted something anomalous occurring within the corporate network. Google seemed to have been penetrated initially by a carefully targeted chat message containing a link to a photo-sharing website. This used a previously undisclosed vulnerability to give the attackers a foothold from which to explore Google's internal system.[10]

The security team quickly realised the sophistication was a step up from anything they had seen before. Suspicion fell quickly on the Chinese government, although it was impossible to prove definitively. 'You have to do a mind-shift,' says Adkins of the realisation that the team was up against a state with all its resources and not just a lone hacker or two. Concern escalated to the top of the company, with some of the most senior executives, like Sergey Brin, taking a personal interest in the battle against the intruders.[11] The attack was codenamed Aurora because the term appeared in the malicious code. The impact on Google as a whole was colossal – like a tornado moving through the company in terms of momentum and velocity, insiders say.

The security team worked long hours over Christmas – some to the point of exhaustion – trying to pin down what was going on and rooting out the parasite which had got inside. The team also saw other companies being hit by the same campaign – often high-tech companies – and so began notifying them. The target at Google seemed to be people. 'This was certainly something we believed was aimed at our users ultimately,' says one person. Those users included Tibetan activists like Tenzin Seldon. But the more it looked, the more the security team found to worry about.

When it went public, Google focused on the espionage against human rights groups by getting into the Google network. This was only one part of the attack. The attackers had also been targeting the source code of Google's applications. If you have the source code you can understand how a system is made and where the vulnerabilities are. Google has remained quiet about what exact source code was stolen. Some accounts said it involved the password system that controlled access to all services, known as Gaia.[12] Going after the source code was a common feature of the wider campaign known as Aurora. The attackers also seem to have targeted the database and portals that contained requests from government and law enforcement for wiretaps or access to people's accounts. This, along with other attempts to get similar information from Microsoft around the time, suggests a counter-intelligence function. By finding out who US law enforcement was asking for data on, a foreign intelligence agency could learn which of its agents the American authorities might have discovered. Attackers also looked for other sensitive information like the certificates that validate a piece of software before it is downloaded (although they did not capture this). This suggested that Google was both a target in itself and because of its widescale use, a means by which to get to other companies and people.

Google began to reach out to cyber experts who worked specifically on espionage against activists to see if they could help, Brin calling them late at night. But when some of those people picked up their messages the next day and called back, they found that Google no longer seemed as keen for their help. There was a reason. Google had approached not just them but America's NSA (originally the FBI, who pointed them to Fort Meade). And once the NSA had signed a secret agreement with Google, it did not want any outsiders involved as its teams began to analyse the way in which the attackers had got in. The NSA had worked with parts of the defence industry in the past but Google had not come knocking, thinking perhaps that as it was an American company the American government would automatically protect it against other states. According to one account in *Vanity Fair* magazine, Google called the NSA and said, "'You were supposed to protect us from this!" The NSA guys just fell out of their chairs. They could not believe how naïve the Google guys had been.'[13]

When it went public in mid January 2010, Google stated that the 'primary goal of the attackers was accessing the Gmail accounts of Chinese human rights activists' and said it believed they had failed, only getting into two accounts in a limited fashion. But they also saw that, independent of the direct attack on Google, the accounts of dozens of US-, China- and Europe-based Gmail users who are advocates of human rights in China had been 'routinely accessed' thanks to other vulnerabilities. Google announced on its blog that the attacks were forcing it to reconsider its business relationship with China. Reports said the attacks could be linked to universities in the country, which might be acting in concert with the state.[14] The company said it was no longer willing to continue censoring results on its Chinese website google.cn and would begin discussions to see whether that meant shutting down the site along with its offices in China (it moved to Hong Kong, where the restrictions were looser).[15] This was a direct challenge to Beijing – American corporate power against the most populous state on the planet.

This salvo from Google to Beijing was about more than just the Aurora attack. It was a shot across Beijing's bows and the latest skirmish in a battle going back years. When Google entered the Chinese market in 2006, there was a big question. How could a company that believed in the free flow of information operate with a government that sought to control the free flow of information? The answer initially was an accommodation. After discussions in the US, Google said it would censor results according to Chinese law but would put up a disclosure notice when this happened and would provide an uncensored, US-hosted site which was subject to US law. It also said it would not disclose to the government any personal information about its users or their search habits. For some activists, like Tenzin Seldon at the time, this was still a betrayal of Google's famous motto: 'Don't be evil'. 'You are kowtowing to Chinese law,' she had argued, believing the companies were in a stronger position than they thought to dictate terms and not compromise their principles. 'It really sickened me.' Critics thought the company was selling out to get into the massive Chinese market. But it did not work out that way.

The requests for censorship began to come in to Google from the Chinese authorities. Most in the first few years related to pornography and illegal activities, much the same as elsewhere. But there were also requests to remove political information about sensitive subjects like Tibet and Tiananmen Square.[16] In all, about 1 per cent of search results were blocked. But Google still found its website periodically blocked over the next three years by the Chinese government, allegedly for failing to block pornography. Events took a turn for the worse from the run-up to the Beijing 2008 Olympics onwards when the government was trying to clamp down on dissent. Many of the requests to take down content were thought by Google to be frivolous – particularly when dealing with official corruption or embarrassing stories about officials. The company tried to resist, leading to anger from the authorities.

Chinese officials made it increasingly clear that they wanted a link removed that allowed people to move to the uncensored google.com site from the censored google.cn. By the spring of 2009, the request became even firmer. This apparently came about after one Politburo Standing Committee member in charge of propaganda discovered that if he entered his own name, a raft of critical results turned up. Google resisted but the Chinese started to put commercial pressure on them, telling telecoms companies to stop doing business with Google. Google decided at this point it did not want to go public on the pressure so asked the US government for help in contacting Chinese officials in support of the company.[17] But the Aurora attack pushed Google over the edge into direct confrontation. Some of the staff inside Google had themselves been deeply uncomfortable with the accommodation with Beijing, which in turn heightened their anger when the hacking was discovered: 'we did a deal with them and they still hacked us,' is how one outraged person who worked at the company at the time puts it (ironically, they would say exactly the same thing about the US government in a few years' time).

In Beijing, a 'well-placed contact' of the US Embassy told its diplomats that the Aurora intrusions were directed from the very top. The contact said the Politburo Standing Committee had been behind it. Another contact suggested that one top official was also working with the leading Chinese competitor search engine Baidu against Google.[18] There was a perception in Beijing that the US government and Google were working hand in hand to drive a vision of internet freedom into China which the government did not want.

Hillary Clinton, the US Secretary of State, waded in. On 21 January 2010 she made a major speech calling for internet freedom, but Chinese internet censors were, somewhat predictably, deployed in force to block coverage of and commentary on it within the country. The reaction from the Chinese Foreign Ministry and newspapers was deeply critical, talking of a 'Cold War' mentality. And even some of those supportive of her views in China were dismayed. In conversations over the next few days, they told the US Embassy that it was turning the issue of internet freedom into an 'us versus them' debate between the US and China which would make it harder for them to make their voices heard without being seen as pawns of the United States. They privately warned US diplomats that Chinese officials regarded US efforts to promote internet freedom as an 'attack', and that members of the Politburo saw the insistence on the issue as an attempt to undermine social and political stability and instigate unrest. The speech, some warned bitterly, would now give more power to those trying to establish control.[19]

To the Chinese, the timing looked suspicious. Hillary Clinton was making a major speech just after the Google announcement. She had also had dinner with Google's CEO Eric Schmidt earlier in the month. For the Chinese, this confirmed their impression that Google was an arm of American power. They were two sides of the same coin. Western tech interests were aligned with Western foreign policy interests. 'The West's so-called "internet freedom" actually is a type of cyber-hegemony,' wrote a PLA lieutenant-general involved in foreign relations and intelligence in January 2013.[20] Battle was now well and truly joined in a simmering conflict over cyber espionage and who was in charge of the internet.

Evangelists for the internet often say that technology knows no borders. And when the Clinton administration liberalised technology in the 1990s, it undertook the move believing that the internet was an unstoppable force for globalisation. It certainly has connected people around the world in ways no one could have predicted. It also made American companies rich along the way. States were at first caught in the headlights of this seemingly all-powerful force which threatened to challenge their authority and legitimacy, and even their tax revenue. But then they began to fight back. It turned out that the internet could have borders for good and for ill. States might need to find new ways to exercise their power and sovereignty in cyberspace, and in some cases their power would be constricted but in others it might even be enhanced. This was true for every state – from authoritarian to democratic – but China, home of the world's largest population of internet users, is where that has become most apparent. In the Chinese state, computer espionage was a double-edged sword. On the international front it offered the opportunity for the Chinese to spy abroad, but also for other, more powerful players to spy on China. Domestically it provided the means for dissidents and others to organise and potentially subvert the rule of the Communist Party. But could it also provide the means to spy on them? Maintaining internal stability is the number-one priority for the Chinese state, and so while cyber espionage on intellectual property may have attracted the most attention, the domestic use of computers for internal control is far more significant for the regime.

The first email from China went over an academic network in 1987. 'Across the Great Wall we can reach every corner of the world,' it announced.[21] Two years later students occupied the vast space of Tiananmen Square in Beijing, demanding the opening-up of their country and greater democracy. From Hong Kong, then still a British outpost on the edge of China, GCHQ intercepted PLA military communications as a tense stand-off developed. As they listened in, the British analysts realised that rumours of splits in the army were false. The tanks were sent in and crushed the protests, leading to a still-unknown number of deaths.[22] The Communist regime had survived its greatest scare but now it was determined to retain control. By the mid 1990s it was becoming clear that the internet could be used for more than professors talking to each other: activists and hackers were emerging online and the Chinese state was beginning to think seriously about how to exert control. In 1997 it passed a Public Security Bureau Regulation which required domestic internet providers to monitor and report anti-regime sentiment, a concept cast in broad terms.[23]

The internet is not some ethereal creation which exists in a cloud. It depends on a physical infrastructure of cables and routers, like the British telegraph system. This may be built and operated by either the state or the private sector in the form of national telecoms companies. Whatever the case, it offers the chance for states to exert control over the system – to exercise the modern version of cable censorship. In China the desire to exert control meant building what has become known colloquially as the 'Great Firewall of China'. This is modern border control. Rather than a man in a uniform, a technical border post operates at the entry points where the cables and routers bring the internet into China. They will let most data through and into the country, but keep undesirables out. This will include a spectrum of content ranging from pornography to political dissent. Sit in a hotel room in China and search for certain sensitive websites you know are out there and you hit a blank, an error message as if the page does not exist. You have just reached the Great Firewall. It began by blocking reports from certain foreign websites – primarily news providers – sometimes completely, sometimes only certain articles which were deemed subversive. As social media emerged, sites like Twitter and YouTube were excluded entirely.

As time has gone on, the system has become subtler in seeking out what is deemed subversive material, blocking certain pages and also peering into the packets of data travelling past. This process is known as 'Deep Packet Inspection'. This is the equivalent of opening up all the mail at the sorting and censorship offices that Britain ran in the First World War. Except it can now be automated so that, rather than having a censor sitting and reading everything, a machine can look for suspicious words or items or traffic coming from a particular internet address. This technology and technique, like so many in the cyber world, can be used defensively or offensively – the boundaries are often very hazy. It can be used to look for a malicious piece of code that you fear is being sent into your country (the equivalent of a letter from a foreign state to a spy in your country) or for something sensitive being sent out (details of troop movements to a foreign power), but could also be used to filter or monitor a broader category of 'subversive' material. Of course, the crucial question is who gets to decide what is subversive. In China's case this might mean blocking material coming in from a foreign newspaper if it mentions Tibet or Falun Gong, for instance (or discusses the personal wealth of senior members of the Chinese Politburo). It is possible to jump the Great Firewall using technology, but this takes effort and risks drawing attention to yourself.

China's fear is that the free internet is a Western Trojan horse designed to introduce subversion by undermining social stability and promoting political change. Western companies are viewed both as ideologically subversive and also potentially as the actual vectors for espionage. The Chinese have been extremely conscious of the Western domination of the internet, from hardware through to commercial internet companies. They became determined to challenge that dominance. Blocking certain foreign websites may have been motivated by internal security, but it also had the useful side effect of acting as a form of protectionism which ensured the development of a powerful, indigenous Chinese software industry – with Chinese versions of Google and Twitter growing to be giants and not, of course, having the potential to spread Western ideas (a process aided by China's language). 'The Chinese are doing to the internet what they did to Buddhism a thousand years ago – which was to comprehensively Sinicise it and turn it into something very different from what it started out being,' argues Nigel Inkster, a former deputy head of MI6 and a long-time China-watcher. 'The Chinese internet has become a discrete phenomenon.'[24]

This is the next significant development in the history of computers, spies and the internet. The systems can be not just a tool for espionage – for stealing secrets from other countries and states – but a tool for domestic surveillance and monitoring – spying on your own population. Tibetan activists are not the only ones targeted.

Falun Gong, democracy activists, people trying to mark the anniversary of Tiananmen Square are also subject to cyber espionage and attack. And China, of course, is not the only country to do this.

China has aggressively demanded information about subversive activity from internet providers – for instance, insisting on information from Yahoo about dissidents that led to their imprisonment. The government is also believed to have ensured backdoors were installed in certain products, which meant that conversations people thought were secure and encrypted were in fact being recorded by the state when they used designated keywords like Tiananmen and democracy.[25] Telecoms and internet companies say that when they operate in a country they have to comply with that country's laws, providing lawful intercept or access to data for criminal investigation and national security. If they really do not want to do so, then the answer would be not to do business in that country. One telecoms insider says a fascinating difference in Chinese equipment is the higher capacity for lawful intercept. In the UK equipment typically has the capacity to collect perhaps 1 per cent of calls; but some Chinese kit can collect 14 per cent, indicating the difference in appetite between the two states. This is all about following the law and the demands of national security, companies say. But the problem comes with how different states define these needs, and in particular how more authoritarian states do so.

China began adding a system for domestic internal monitoring rather than just controlling what came into the country under its broader 'Golden Shield' programme. This is done at a variety of different levels – local, regional and national – and tracks public opinion and dissent. It involves an army of thousands of individual censors patrolling websites to remove inappropriate content and post pro-government messages on social media. The system of surveillance also creates a form of self-censorship by Chinese citizens online, who are careful about what they say. That is not to say that the internet in China has not become vibrant. It is a powerful and also disruptive force. Chinese 'netizens' are highly active and vocal, often holding power and authorities to account and forcing the pace on issues like corruption at a very local level. This can be a useful outlet for public anger, the authorities seem to believe, but only if it

is kept channelled within certain paths and away from others. This
also applies to the work of its hackers.

In April 2001, a US spy plane was flying near Hainan Island. The
island was home to a major Chinese PLA intelligence facility – it
was the place where the intrusions into Tibetan computers would
be traced to a few years later. The US military plane was collecting signals for the NSA – part of that global collection mission that
had been built for the Cold War but persisted beyond its end. The
Chinese, unsurprisingly, disliked the Americans flying so close and
regularly sent their own fighter jets up to keep watch. On one occasion a Chinese pilot got so close to the Americans that he held up a
piece of paper with his email address on it for them to read.[26] But
that day in April something went wrong and there was a collision.
The same Chinese pilot who had showed off his email died and the
American plane crash-landed on the island. Anger erupted in China.
The capture of intelligence-gathering technology on board was a
nightmare for the NSA. But another problem was the way in which
'patriotic hackers' in China took to computers, defacing American
websites. On the Chinese side this was led by a group known as the
Honker Union of China (it was not one-way traffic though; American 'patriotic hackers' also targeted Chinese websites). There was
speculation at the time that the damaging Code Red Worm released
onto the internet in July of that year might be the work of the Chinese and that hackers probing utilities in California might also be
connected – although there was no proof for either.[27] This was the
second major incident in clashes between 'patriotic hackers' on both
sides – the first having been in 1999, when the US bombed the Chinese Embassy in Belgrade during the Kosovo conflict, leading to a
spontaneous outburst of fury in China on the streets and online.
The hacking of US websites was tolerated or even encouraged. But
as the authorities began to fear things might get out of control and
lead to serious unrest on the streets, they began to clamp down.

'Patriotic hackers' may be just that – independent hackers motivated
by nationalism – but they can also be highly plausible proxies for
state intelligence activities seeking deniability. When the Honker
Union or Red Hackers go after those critical of China, such as sites

in Japan, are they acting independently or on the explicit orders of the state? Or is the state deliberately turning a blind eye to activity it could otherwise stop? This might offer some plausible deniability for governments carrying out a clandestine action. However, the more authoritarian a regime is – and especially the more control it exerts over domestic actors in cyberspace – the more likely it seems that it is involved in such activity.

In China, hacking groups, which emerged around the turn of the millennium, seemed to have shifted to more advanced cyber espionage work. This may be out of patriotism, but also perhaps because they were recruited by the state. 'Patriotic hackers' sent out messages suggesting more organised activity was a far better use of their time. One message from the Honkcr Union of China in 2010 asked: 'What benefit can hacking a Web page bring our country and the people? It is only a form of emotional catharsis, please do not launch any pointless attacks, the real attack is to fatally damage their network or gain access to their sensitive information.'[28]

There are huge hacker forums in China where tips and tricks are shared, but the fact that there are more hackers in China than any other country is simply a reflection of the fact that there are more internet users in China than any other country. In these forums, it's thought, the government can watch them – pressuring those who are going down the wrong path and potentially recruiting those who are particularly talented. Western hackers are also active against China, for instance targeting those responsible for building the Great Firewall because they view it as an affront to internet freedom. They may also publicise weaknesses in Chinese control systems and ways of evading censorship. The US may maintain that hackers within its shores who do this are entirely independent, but the Chinese see things differently.

In 2004, China got a fright that confirmed the country's fears regarding its reliance on foreign computer technology. Microsoft was trying to clamp down on pirated versions of its operating system around the world and came up with a clever idea. Anyone not operating a properly licensed version of Windows would see his or her screen slowly turn to black and a message appear. The problem for China was that, thanks to its liberal interpretation of intellectual property, pretty much every copy in the country was pirated – including those of government systems. And so computer screens all over the country suddenly went dark. What was known as the 'black screen of death' produced a terrifying realisation. An American company had just remotely switched off their computers. Maybe only for a short period. And maybe only to make a commercial point about privacy. But, seen from China, someone in faraway America had just hit the kill switch. There had been fears of just such a possibility from the start. And so began a huge campaign to protect hardware and software across the country and especially in government, amid fears of the type of sabotage the CIA had conducted in the Soviet Union in the Farewell case.

China introduced procedures to check that there were no backdoors hidden in imported Western technology. This involved companies disclosing technical details and having them tested in labs. For instance, for Microsoft to sell its products to China it had to agree that the government could look at some of the source code for the Windows operating system, something it had once jealously guarded. Under its government security programme, the company provided a number of countries with managed access to some of the source code to reassure them that they were not installing backdoors on behalf of the NSA.[29] By getting access to the source code of foreign technology, China could also learn about vulnerabilities to exploit it; but this is similar to what Huawei has to do in Britain and may be the only way of building trust in a globalised world. Increasingly, though, rather than import Western products and check them, China has moved on to build and develop its own software and hardware. The Chinese state and many of its netizens continue to believe that the US has a kill switch it can use to cut off China from the internet at will, and knowledge of an armful of vulnerabilities that could be deployed to the same effect. They may even be right.

China thinks of 'information security' in a different way from the West. It is not just about protecting information and keeping it secure. Rather it is about the fact that information itself can be potentially subversive. Control of hardware, software and information is therefore vital in order to guard against the risk of foreign subversion. This is the underlying reason behind the struggle with Google and for sovereignty in cyberspace.

The Chinese Foreign Ministry is a grand building in central Beijing whose guards click their heels and stand to attention as visitors approach. In an ante-room of its lobby, Dr Huang Huikang, a legal adviser to the Foreign Ministry and one of the most senior negotiators on cyberspace, outlined the Chinese position in 2013. He agrees there is an ideological battle over cyberspace but argues that the kind of state control China exercises is the type many countries use to ensure social order. 'Some people think this is a control of cyber flow of information and it is a violation of human rights but we don't think so. It is necessary for all the countries to establish a good order and make a balance between the inflow of information and the public security.' He compares the internet to a traffic highway – one that requires rules of the road agreed by states. He also – without naming the US – makes it clear that he believes certain countries have been driving dangerously. The diplomat repeats the oft-heard refrain that China is the victim not the perpetrator when it comes to hacking. Such activity is illegal and punishable by law, he explains, adding that 'China is one of those countries suffering most from hacker attacks'. In 2013 Chinese officials claimed that more than 10 million Chinese computers were maliciously controlled from overseas, 30 per cent by computers in the US.[30]

The realisation that the state could, after all, exercise sovereignty over the internet has set the stage for a significant global struggle. Britain, the US and others talk of fighting for a 'multi-stakeholder' internet in which governments, companies and civil societies jointly set the rules for cyberspace and in which free speech is prioritised. But China and Russia lead the way in arguing that states should be free to exercise sovereignty within their borders in the way they traditionally have in the physical world and then negotiate with each other about international rules through traditional state-on-state diplomacy. Russian and Chinese officials believe the Western talk of 'multi-stakeholder' means (largely) American companies being free to do what they want at the cost of their own control over what happens in their country. Internet freedom, they think, is simply the freedom to be exploited and spied on by the US and its companies. One problem for Western countries is that many other states agree with the Chinese-Russian position. To them the internet looks like a rather large, scary creation in the face of which they feel help-less. Whatever their populations might want, the leaders, perhaps unsurprisingly, prefer a model in which they retain control to one whereby they lose it.

Social stability, subversion and dissent are all in the eye of the beholder. States beyond China, initially fearful of the net, have in-creasingly learnt to exert control through the infrastructure – often through telecoms companies (sometimes because they are state-owned, but even if not, regulations or pressure can be used). Some have created 'walled gardens' where people are forced to stay within certain boundaries (or, more correctly, it is made harder or more expensive to leave), a trend that analysts call 'network authoritari-anism'. The number of national firewalls has been growing as more and more countries institute internet border controls. Turkey for in-stance says it does that to make sure it is 'family-safe'. But it does not explain what exactly that means. Social media has been restricted in the country as part of broader regulation, with Twitter in particular targeted for spreading messages.

When a coup is under way or a government fears protests will bring it down, its tanks now head to the internet and telecoms op-erators rather than the TV stations for control. The extreme act of hitting the kill switch and closing down communications is one that governments have also attempted. China took this step when there was unrest in Xinjiang province in 2009 (a technique it had used as early as 1996, when computer bulletin boards were shut down in some universities to prevent anti-Japanese demonstrations being or-ganised).[31] Countries in the Middle East also hit the kill switch to try to prevent news of social protests and the organising of those protests spreading during the 'Arab Spring'. Social media were a way of escaping the grip of the state and pliant media. In Egypt, the authorities at one point took down Facebook because they feared its power to organise, and demanded that government-dictated text messages be sent by phone operators to all customers. Officials also went to operators at gunpoint, demanding that networks be taken down – leaving the companies in little doubt about the consequences. When one country places a black box at its gateway to monitor all the traffic, it is legitimate law enforcement activity and vital for social stability. When someone else does it, it is often described as mass surveillance and censorship.

Russia and China also point out that even ardent supporters of a free internet place controls and filter or monitor. Sometimes this is to keep out pornography, but can be for political purposes – for instance, neo-Nazi content is banned from online social media in Germany. Pressure to remove content or exert controls also grew in the West amid fears that social media was being used by groups like Islamic State to reach out and radicalise. But there were even debates over internet freedom when it came to less dramatic threats.

In the summer of 2011, riots broke out on the streets of Britain, shops were vandalised and burnt to the ground by mobs of young people after a pair of trainers or simply to vent their anger. For a few days, the social fabric seemed to fray. The Prime Minister, David Cameron, focused on the way in which those involved were organising and communicating over the internet. He said authorities were looking at 'whether it would be right to stop people communicating via these websites and services when we know they are plotting violence, disorder and criminality'. Such talk about blocking online activity for the sake of social stability might make the Chinese smile.

When Colonel Gaddafi's brutal regime fell in Libya, visitors to ransacked offices of his secret police found evidence of how his grip on power had been maintained. There was the transcript of a sixteen-minute online chat in which a man flirted with a woman but also confessed that he feared he was a target of the regime. Another message revealed a plea from an activist to a Human Rights Watch staffer for help. Just as revealing was the evidence pointing to Western companies having sold the monitoring technology.[32] This apparently included the most advanced equipment which can inspect the content of packets of data. Similar evidence was found in Egypt after the fall of the Mubarak regime. And it was Western technology companies – those proponents of internet freedom

– who had initially helped build China's monitoring apparatus, supplying both the software and hardware for the Great Firewall.

The tools of cyber espionage have now become a commodity that can be bought off the shelf, another product sold by the West to others to make money. Western companies are now among the most active in supplying monitoring software to authoritarian companies around the world, their brochures (not normally available online) boasting about the ability to provide surveillance and intercept. These are practices all governments undertake but which assume a darker hue depending on how repressive a regime is. A brochure by a group called 'The Hacking Team' talks about being able to 'monitor a hundred thousand targets' through a single 'easy to use interface'. It talks of offering the ability to 'attack your target; while they are browsing the internet, opening a document file, receiving a text message' or 'crossing the borders with his laptop'. This is the kind of technology China developed against Tibetans, but it is now sold to Middle Eastern states going after human rights activists and dissidents whom they want to pursue in cyberspace. These companies maintain they only sell their products to governments for lawful interception, to catch criminals and terrorists. The problem, of course, is how different governments define terrorism and criminality. Many countries are now monitoring all their traffic and finding out that there may be less need to invest in training a spy service when you can purchase a monitoring capability off the shelf. This is a powerful tool for the state against forms of dissent. The ability of the internet to support dissent may increasingly be matched or exceeded by its ability to spy on dissenters.

The issue of state surveillance of dissidents raised complicated questions for governments in the US and UK and made them frequently look schizophrenic. The crypto wars tensions had gone global. On the one hand they wanted to support the use of the internet by activists abroad to challenge authoritarian regimes. To do this, those people needed the anonymity that the web provided with programmes like TOR (originally developed with US government funding) and forms of encryption. But other parts of those same Western governments disliked the anonymity the internet offered and the power of encryption when it was used by criminals, enemy spies and terrorists who challenged them. One part of the US government (the State Department) might back such tools while another (the NSA) would at the same time be trying to find ways to hack into TOR and deprive people of their anonymity. Different countries want to talk about different things when it comes to cyberspace. Some focus on intellectual property theft, others crime, others militarisation, others threats to social stability. Even within Western governments, different parts of the bureaucracy want to talk about different things – defence, democracy, security or exports, to pick just a few. The issue cuts across too many departments and is often too complex for any single person to get a handle on. That has meant the dominant voices have been of those who know the most and wield the most power – spies and the military.

CHAPTER FOURTEEN

SABOTAGE

In early 1943, a young Norwegian named Joachim Rønneberg was summoned to the offices of Britain's Special Operations Executive on Baker Street in London. He was asked to find a team of six whom he could take back to his homeland, now occupied by the Nazis. It was a secret mission and Rønneberg thought it was probably a one-way trip given that the British kindly supplied a cyanide pill in case of capture. Rønneberg and his team went in by parachute in February. 'We jumped out at midnight and the landscape was covered with snow,' he recalls.[1] The team landed miles away from the planned drop site but eventually made their way to their target – a factory in Vermork, a remote part of Norway. Rønneberg used wire cutters first to get through the perimeter before crawling through an access tunnel to lay his satchel bombs. Once out, he waited for the bang. The explosion was faintly disappointing but his escape was not lacking in excitement. Rønneberg had to flee 200 miles on skis with an entire German division chasing him. With a wry smile, he looks back on it as 'the very best skiing weekend I ever had'. Rønneberg and his team had succeeded in putting the factory out of action. Why such a risky mission though? The factory produced heavy water – a key component for an atomic bomb. Britain feared the Nazis could use what was made there to change the direction of the war.

Sixty years after Rønneberg skied across Norway, spy chiefs again sat down and wondered how to stop a nuclear programme they saw as a threat. This time the country was Iran. The decision was again made to undertake an undercover operation. The effect would be comparable to Rønneberg's Operation Gunnerside – setting back, but not destroying, a programme. But this time, rather than parachute men in to lay their charges and then escape, the damage would be done by the click of a mouse, through computer code.

Underground in a vast, cavernous hall in the mountainous region of Natanz, Iran was spinning thousands of centrifuges. A centrifuge is a slender marvel of engineering. Inside the cylinder is a rotor which spins so fast it can separate out the heavier parts of uranium gas from the lighter – a process called enrichment. If you enrich uranium enough you can use it as fuel for a nuclear reactor. Keep enriching it and you can use it for a nuclear bomb. But spinning so fast is an intensely demanding technical challenge which few countries have been able to master. The materials have to be strong enough to withstand huge stress but also perfectly balanced. The electric current that spins the rotor has to be maintained at precisely the right level. A centrifuge has to rotate, but vibrations are its mortal enemy – the slightest imbalance in the system and a rotor can spin out of control, crashing around inside the cylinder with such force that everything disintegrates. Centrifuges are arranged in cascades so that after one centrifuge has enriched the uranium a little it passes it on to the next one to continue the process. If one centrifuge breaks down – for instance by spinning too fast – it does not just clunk out but can take out a whole cascade.

The first thing the Iranian engineers heard was a screeching sound. That is the machine skidding round inside its case as it loses control. By this point, if your control panel has not warned you of a problem, it is already too late. You will hear one machine taking out the next and the next like dominos. There was no explosion, just a clatter as the delicate, precious machines destroyed each other. The Iranians had already been seeing smaller problems. Machines were failing, parts breaking down. It was not always clear why. Was it poor engineering standards? Bad parts or designs? No sooner would one problem be fixed than more centrifuges would go awry, forcing them to be stopped and checked. What they did not know was that a hidden hand was remotely manipulating the controls to take advantage of the delicate nature of the devices. The code that struck Natanz was a work of engineering bravado every inch as much as the centrifuges it was designed to destroy. And it worked by stealth over years rather than with a single bang.

One attack targeted the valves that transfer the gas from one machine to the next, including the isolation values that protect each centrifuge from a faulty neighbour. Another targeted the controls that dictated the speed of the centrifuges – so-called frequency converters – sending an instruction to spin faster and then slower. These attacks were undertaken carefully, introducing stress on the materials so they would break down over time but without ever attracting too much attention. 'The attackers were in a position where they could have broken the victim's neck,' says Ralph Langer, who has studied the attack, 'but they chose continuous periodic choking instead.'[2] Deviously, the attacks first recorded what normal operations looked like and then fed back that data when the attack was under way so no one would spot anything until it was too late.

One visitor to Natanz remembers seeing the Iranians desperately trying to understand what went wrong – running tests on the motors to find out why the speed was changing. Without knowing the cause, engineers were left wondering what they had done wrong and what would happen if they started up a cascade again. This had the potential to sow confusion and even paranoia within the programme. Was one of their number a traitor sabotaging the machines? This had been the intention of the covert Farewell programmes in the 1980s that sent sabotaged equipment to the Soviet Union. Make your opponent no longer trust technology. By 2010, the destruction became more dramatic. In all, at least 1,000 machines are thought to have been damaged. But this was also the time when the secret escaped. Perhaps because of a programming error, the virus began to spread around the world, infecting tens of thousands of computers from the UK to Azerbaijan. That meant it got noticed. It was supposed to be covert and deniable but, as Robert Morris had learnt a quarter of a century earlier, things have a habit of spreading further than you think on the internet. As experts around the world analysed the code, it acquired the name Stuxnet. It became clear that Iran had the highest number of infections and the virus did not appear to inflict damage anywhere else other than Natanz.[3]

A group of his top virus-hunters walked into the office of Eugene Kaspersky, the flamboyant Russian founder of an eponymously named anti-virus company. 'You know we have been waiting for something like this,' they said to him. 'Well, it has happened,' Kaspersky recalls. 'That was the first time we had the cyber missile in our hands. That was really a scare.'⁴ Kaspersky had graduated from a state-backed specialist institute for cryptography and computing in Soviet Russia. He had begun his work on computers in the days when viruses and worms were practical jokes by what he calls 'hooligans and vandals'. He had watched in the 1990s as the internet emerged and then seen criminals and hacktivists move into the space. He had begun to worry about attacks on infrastructure from around 2002, but says he decided not to speak out in case it gave attackers ideas – that was until he realised the cat was out of the bag when he saw the film *Die Hard 4*, in which Bruce Willis battles cyber terrorists ('Thank you, Hollywood,' says Kaspersky wryly). Now he was watching state actors launching attacks for real. 'That was the most sophisticated malware we ever had in our hands,' he says of Stuxnet, estimating that it cost millions of dollars to develop. Both the missile – the delivery mechanism to get inside a system – and the payload – the code that did the damage once in – were like nothing seen before. Normally, attackers build on existing tools and code. But Stuxnet was different. The final version employed no less than four previously unknown vulnerabilities called Zero Days. Once a vulnerability in a system is spotted, the 'hole' in the defence is patched in a number of days. A Zero Day gets its name because the vulnerability has not yet been spotted. This means its signature will not be detected – it is a surprise attack. Using four for one attack was unprecedented. Once they saw the reports, Iranian officials at Natanz called in the Ministries of Intelligence and Communications and began examining the code. 'Our first measure was that we transferred the virus to the lab. In the lab, we attempted to completely identify the virus's behaviour,' an Iranian expert later said.⁵

An emergency meeting of top US officials was called in the White House situation room once it was clear Stuxnet was out in the wild. The rest of the world was now able to dissect the worm and would be asking which state – because it had to be a state to do something so complex – was behind this. That was a problem for these officials because they knew the answer lay with them. The operation – codenamed Olympic Games – was started under the

Bush administration, according to US accounts, and President Bush personally recommended his successor keep it running. President Obama is reported to have taken a keen interest, studying maps of Natanz as he ordered its work to be accelerated (likely leading to the more advanced attack in 2010).[6] Senior British intelligence officials at the time of the operation say they were 'not surprised' when it took place, indicating they were at the very least aware of the plans by their close ally.

Stuxnet was stunning in its ambition and its highly targeted precision. 'You need to know a lot about the centrifuges to do that,' says one person who has visited Natanz and seen the machines.[7] It was not something that a bedroom hacker could manage. The virus was looking for a specific model of a Programmable Logic Controller made by Siemens, and even then only became interested when it was sitting in a particular configuration indicating it had found Natanz. If those conditions were not met, it would do nothing. In other words, the aim was to avoid 'collateral damage' to other systems (it was also timed to self-destruct in mid 2012). Veterans of the US government say they can imagine the endless inter-agency meetings over whether to go ahead or not, with a final agreement to do so only if it could be guaranteed not to hurt other industrial systems. 'It just says lawyers all over it,' former US cyber tsar Richard Clarke has remarked.[8]

The attack required an intimate knowledge, not just of the Siemens controllers that ran Iranian centrifuges but also the specific configuration at Natanz. That needed inside knowledge. It would also have required extended testing to see how the manipulation of the controls would affect centrifuges, including your own cascade to conduct dummy runs. The US already had a set of similar centrifuges which Libya had handed over when it gave up its nascent nuclear weapons programme, which – like the Iranian programme – had been assisted by AQ Khan of Pakistan.[9] The CIA had penetrated the AQ Khan network by turning some of the businessmen who supplied materials. Along with MI6, this allowed them to watch the supply of centrifuges to Libya and Iran and introduce tracking devices into the parts being delivered. The Iranian programme was also sabotaged with faulty electrical converters and the like even

before Stuxnet was deployed (again echoing the Farewell operation). Stuxnet took sabotage to a new level – into the cyber world.
It is believed to have been a joint operation between the US and
Israel. Working closely with Israel was important because it had
good intelligence on the plant, but also because the operation was
a way of showing the US was willing to do something and trying to
forestall Israel conducting a military strike. Unlike an airstrike, this
operation was designed to be hidden. The attackers even set up fake
football websites to act as command and control servers for Stuxnet. So when it needed to report back from Iran, the network traffic
would look like an employee who was checking out sports results.[10]
Stuxnet was designed to be stealthy and to work over an extended
period without being spotted, sowing confusion rather than being
a single strike like an act of war. That is why it is far more within
the tradition of intelligence operations and covert action than overt
military action and cyber war.

A covert attack by an intelligence agency first requires reconnaissance – gathering intelligence on the target. Because the language
of Stuxnet was so distinctive and unique, analysts at Kaspersky Labs
could work back and see the other variants from the same family
dating as early as 2007 (perhaps even 2005), which appeared to have
been designed by the same team (the Labs have an automated tool
called a 'similarity engine'). Two were codenamed Flame and Duqu.
Duqu was twenty times larger than Stuxnet but was an espionage
rather than sabotage tool, targeting only a few machines from 2006.
Flame was double the size again, and even more advanced, some
analysts claiming it involved breakthroughs that could only have
been achieved by 'world-class cryptographers'.[11] Flame was made
to look like a Microsoft update and worked secretly for many years,
with the capability to activate microphones and webcams on computers to gather information and send data by long-range Bluetooth
connection. This was reported to have been another joint US-Israeli
programme collecting information on Iranian computer networks.
'This is about preparing the battlefield for another type of covert
action,' one former high-ranking US intelligence official told the
Washington Post.[12] Both Duqu and Flame may have been the initial
reconnaissance spies for Stuxnet to follow. Curiously, one leaked

document talks of a major surge by both GCHQ and the NSA to respond to the Iranian discovery of Flame but provides no mode details: this seems to suggest Britain may also have had a hand in Flame.[13] There were other attacks on Iran, including one in spring 2012, perhaps by Israel, which wiped Iranian oil and gas ministry computers; another is said to have led to the song 'Thunderstruck' by AC/DC being blared out at full volume on computers in the middle of the night.[14]

How did the Stuxnet code get onto the centrifuge control system when it was not connected to the internet? The original Stuxnet virus had to jump the air gap. This required original old-fashioned human espionage – a person. It is believed that lists were drawn up of companies and engineers who had access and who might be able to carry the virus in a USB stick that could be plugged into a computer (most likely it was done by an engineer who did so unwittingly, with some Iranian reports suggesting it was a foreign expert who visited the sites).[15] This part of the operation may have been facilitated by the CIA's team that specialises in working at the junction between technical and human intelligence, although the Israelis possibly had better access. The later version of the virus may have been able to replicate itself without the need for the same kind of human intervention, hence the problem of it spreading more uncontrollably.

Israel and the US avoided confirming their role in the attack. 'It would be irresponsible for someone of my background to even speculate,' argues former NSA and CIA Director Michael Hayden. 'But it's not speculation to know that someone just used a cyber weapon to effect damage, not in the cyber domain, but in the physical domain. That's the first significant crossover that we've seen. Now look, I tell audiences that crashing a thousand centrifuges at a time is almost an unalloyed good, but when you describe what just happened there in a slightly different way – someone just used a cyber weapon during a time of peace to affect physical destruction in what another nation would only describe as critical infrastructure – you've got to realise that, although that was a good deal, it was also a really big deal and it does have second- and third-order effects.' He acknowledges the moment's significance, saying it has the 'whiff of August 1945' and the first use of the atomic bomb. 'A

new class of weapons has been used,' he explains. 'Go deeper into
history and say somebody's crossed the Rubicon. We've got a legion
on the different side of the river now.'[16] An American legion. The
US has claimed China broke with norms with its economic cyber
espionage. But critics say the US broke with norms by carrying out
the first destructive attack. When the US accuses China of exploiting the internet to spy, the Chinese ask who went first in militarising
cyberspace – whether by massively expanding its military cyber
command or deploying weapons in the form of Stuxnet.

Those who support the deployment of Stuxnet argue that its
precedent-setting nature has to be matched against the other al-
ternatives. These included Iran getting a nuclear bomb – or Israel
attacking Iran and the US being drawn into a bloody confrontation
in which many might have died. Crossing a line in the cyber world,
they say, was a big deal but a smaller deal than going to war and
leaving the Middle East in flames. Those kinds of calculations are
the type leaders make: is dropping a nuclear bomb on Hiroshima a
valid way of shortening a war or not?

Stuxnet may have been a rational choice, but it has consequences.
Unlike the atomic bomb, it was supposed to be stealthier and more
deniable than an overt use of military force. But now the secret is
out. Stuxnet was incredibly hard to develop and few nations could
yet manage something so complex. But, as was the case with the
atom bomb, the use of the weapon by the US is almost certain to
act as a spur for others to try to develop the same capability as fast
as they can. And Western countries may be most vulnerable to
weapons like Stuxnet because they are most connected.

How much did the operation set Iran back, though? The most
bullish commentators talked of years. But others were sceptical,
saying it was only a matter of months. Perhaps three. Perhaps six.
'We may have incurred some slight damages here and there, but generally speaking we have been able to manage the issue very well,' an
Iranian expert said in November 2012, a year and a half after Stuxnet
was first exposed.[17]

Whatever Stuxnet did in terms of damage to Iran, it certainly had
an impact on the country's thinking about cyber security – offensively as well as defensively. 'We need to have a better cyber army

to be able to stand up to cyber attacks aimed at infiltrating various information systems and companies in our country,' said one Iranian expert in its wake. Iran accelerated development of its own cyber militias, or Basij, to build up capability and moved to work more closely with 'patriotic hacking' groups. In November 2010, a Revolutionary Guards commander was reported as saying that 'the Basij Cyber Council has trained 1,500 cyber warriors who have assumed their duties and will in future carry out many operations.'[18] In March 2012, various universities were reported to have opened up cyber training academies as part of the country's attempts to catch up with the West. There appears to have been a similar transition in China, where 'patriotic hackers' moved from website defacements to more sophisticated attacks, probably with state support.

Iran had already been working towards building a 'national internet' in which it exercised greater control over what information entered the country from abroad (and greater surveillance of what happened inside the perimeter), creating a walled garden similar to China's. This was a reaction to the Green Movement in 2009–10, when people went onto the streets to protest at what was seen as a fraudulent election. Their use of social media to organise scared the regime into action and greater control of communications infrastructure (reportedly with help from China). Iran blocks millions of web pages including social networks, although many Iranians manage to circumvent controls on Facebook and the like in a cat and mouse game with the authorities. Government-linked hacker groups then found ways of installing malware on the anti-censorship software, allowing those trying to evade state control to be identified and spied upon. But in the wake of Stuxnet, an Iranian hand would also be seen in striking back abroad.

In the space of a few minutes on 15 August 2012, the computer network of the Saudi oil giant Aramco was crippled. Thirty thousand computers became as useful as bricks, rendered useless by something called a 'wiper'. The code had not quite been executed properly, but hidden within it was an image designed to send a message – a burning American flag.

The company's computers, not just in Saudi Arabia but around the world – including Europe and the US – were taken down for eight days. Aramco's exploring and engineering centre, which was responsible for upstream oil and gas technology development, lost valuable production and drilling data.[19] Some of this had not been backed up at the central database, possibly because Ramadan was starting as the code hit. As a show of brute force, it lacked the subtlety and sophistication of Stuxnet, but the message was all too clear.

A group calling itself the 'Cutting Sword of Justice' said it had carried out the attack in retaliation against Saudi support for those carrying out 'atrocities' in the region. On 27 August, RasGas in Qatar was also hit. Aramco initially tried to keep the details as secret as possible but it was a cyber shot that would echo around the world, especially in Western corporations, as word filtered out. They had seen cyber espionage for years and had – to varying degrees – turned a blind eye. But now they were witnessing the actual physical destruction of machines. This was something that could affect their bottom line, their reputation and their share price in the immediate term – not ten years down the line. 'That focused minds,' MI5's head of cyber explained in 2013. '[It] makes chief executives realise the power of a destructive cyber attack.'

Experts were flown into Saudi Arabia in a hurry. The Saudis were nervous though. They would not let one team see all the code, and so different teams were assigned overlapping tasks to try to piece together what had happened, which made their job harder. One of those involved said they had 'never seen single-minded destruction on this scale before . . . the tools were blunt, but effective . . . A breach is bad – scorched earth where all your computers used to be – is a whole different game.'[20]

In December, the company finally began to talk about what had happened, holding a press conference at Aramco's Dhahran headquarters. They revealed that the aim of the attack was even more ambitious. 'The ultimate aim was to stop the flow of oil and gas to domestic and international markets,' according to Abdallah Al-Sa'adan, Vice-President of Aramco and head of the company's own investigative team.[21] He said the hackers had tried for a month to bring down the system before finding a weak point. 'Not a single drop of oil was lost during the crisis,' he said to reassure the world.

Aramco says the attack never moved from its corporate network to industrial controllers. That was fortunate. If it had, the shock-waves would have reached almost every person around the world as energy prices would have rocketed.

Rumours had swept Saudi Arabia that the attack had been an 'inside job'.[22] Reports said the code – not nearly as advanced as Stuxnet, and possibly written by a single individual building on commercially available software – had been deployed through a USB stick inserted by an employee into the company's internal net-work. A specific employee who was logged on at the time was even identified in one report.[23] That was an idea officials were keen to dispel. 'The attack originated from foreign soil,' said Major-General Mansur Al-Turki, spokesperson for the Interior Ministry at the press conference. He said it was the result of spear phishing from a team based across four continents. Not everyone was sure of the origins. A few experts believe that the attack was the work of a radicalised, fundamentalist employee with high-level privileges who had come to hate Saudi support for the US, pointing to language used in the code. But the consensus has been to attribute the attack to Iran. Iran had motives: revenge for Stuxnet, showing its capability, hitting a major regional rival and hurting the US where it counted by dam-aging oil production at a time when the Iranian oil industry was being placed under US sanctions. None of this was proof. But talk at the time had been of how Iran could close the Straits of Hormuz to block oil supplies. That would have led to war with the US, which had made it clear that it would use its navy to keep the Straits open. But, just as Stuxnet seemed to offer a way of carrying out an act in the physical world without the level of violence or the consequences of a traditional strike, so the Aramco attack may have carried the same aspirations for Iran (although not quite with the same degree of success). And for all the shock in the West at the attack, what happened was not that dissimilar from a wiper attack in March of the same year which had deleted information on Iran's own oil and gas ministry computers. Again, the question was: who had crossed the Rubicon first?

One US financial company said that on a normal day they had about 15,000 hits per minute on their website. But a few weeks after the Aramco attack they were receiving 3 million. US companies like Bank of America found their websites subject to some of the largest denial of service attacks that had been seen. In these, websites are flooded with requests so that they are overwhelmed and shut down. Some banks were briefly taken offline as they struggled to put in place measures to cope. US officials, without offering definitive proof, suggested they were sure that Iran was again behind this campaign, with hackers in the country acting on behalf of the state and using a 'false flag' by pretending to be a group of Sunni Muslims called the Izz ad-Din al-Qassam Cyber Fighters. Iran denied the claim. Just as the attack on Aramco may have been linked to the Saudi oil industry making the most of sanctions on Iran, so the attack on these institutions may have been motivated by their role in financial sanctions, again providing a means of retaliation for Stuxnet but in a way that did not involve violence and risk further escalation. It was Iran saying that it too had the ability to strike in cyberspace, an attempt to establish some sort of deterrence.

'I think they have come back and sent us a message: we can do destruction of networks as we did to Aramco, we can do denial of service attacks on US banks,' says Richard Clarke. 'The implicit message to America from Tehran, I think, is: "What if we did the wipe-out attack on American banks? We could do real damage to the American financial institutions." And therefore: "America stop hacking into our networks because we can do it too." I think the Iranians have sent a very sophisticated message and I think the American government has heard it.'[24] Some wondered why the Iranians would strike the US and not Israel in cyberspace, given that Israel was also behind Stuxnet. The answer may have been that deterrence works. Iran may have realised that the Israelis would have felt less restrained than America in retaliating.

It was the belief that Iran – not previously seen as being in the premier league of cyber powers – was able to carry out these attacks that surprised and worried many officials in the West. It appeared that the capability to carry out cyber attacks was proliferating far faster than people had expected. Experts talk about Iran now having some of the most advanced attack capacity of any country. It may also have spread some of its lower-level hacking skills to Syria and to its 'patriotic hacking' group the Syrian Electronic Army, which attacked Western news media aggressively as the conflict in Syria worsened, accusing them of bias and seeking to embarrass them by hijacking Twitter feeds. Established cyber powers were finding their advantage was narrower than they thought. Developing cyber attack capability was a lot easier than building nuclear weapons and a lot easier to use, making it an attractive option for weaker states.

It is hard to imagine anything less digital than a stone statue. But it was a statue that opened the way for a significant cyber campaign in Europe. When Estonia announced in spring 2007 that it was going to move a statue of a Red Army soldier from the centre of the capital to the outskirts, it touched a nerve. For many Russians the statue symbolised the sacrifice of the Red Army in the Second World War fighting the Nazis. But for many Estonians it reminded them of the decades of Soviet occupation that followed the war until their independence. The Estonian plan to move the statue sparked outrage from the large minority Russian population of Estonia and from Russia next door. Protests and riots erupted on the street.

President Toomas Hendrik Ilves of Estonia knew something was wrong when he tried to click on newspaper sites to get the latest reports on the crisis and nothing happened. Someone then called him to say that government sites were down as well. Within a few hours, the head of his Computer Emergency Response Team told him they had a bigger problem.[25] They were being subjected to something which one former American general calls 'digital carpet bombing'.

There was something rather fitting about Ilves being in the hot seat of one of the first sustained cyber attacks on a country since he is one of the few heads of state who can claim to know how to write computer code. His interest in technology, dating back from time spent in the US, had been an ideal fit for a tiny Baltic state which saw that the best way of escaping its Soviet past and catching up with the modern world was to become a wired and connected nation. Alongside the physical violence over the statue came an intense three-week cyber barrage against Estonia, with hackers launching a sustained attack on its electronic infrastructure, attempting to take as much offline as possible through denial of service attacks.

Inside the Estonian Computer Emergency Response Team in the first few hours of 27 April, the screens which showed traffic coming at the sites fell over and went offline because the sensors were overwhelmed by the volume. The team – which normally consisted of three people – struggled to cope with the scale and range of what they would see in the coming days. On 28 April the Estonian Defence Ministry said: 'We are under cyber-attack.'[26] The emergency phone number – 112 – also came under fire for a while.

The first Estonia's banks knew was when they saw the number of customers reaching their systems online dropping off. Ninety-five per cent of bank transactions were done electronically in the country. Banks went offline for an hour or so on a number of occasions during the attacks. With riots on the streets, people were desperate for news, but their search for information online only added to the loads that hackers had already pushed onto the media websites, leading them also to crash. But both banks and media were only taken down for short periods. For those working on cyber defence it was an intense period, but for ordinary Estonians the experience of cyber attack was more one of curiosity and annoyance than catastrophe. It barely affected real life. Estonia is a member of NATO, but the Alliance decided that the cyber attack did not classify as a 'real attack' which would trigger the Alliance's agreement on collective self-defence. Few would really argue that taking down some websites justified a fighter jet dropping a bomb (although many scholars, including some in NATO, say the Stuxnet attack would). And anyway, in the Estonian case whom would they have sent their tanks and warplanes to attack? How could you work out who was behind it all? The attacks were clearly highly co-ordinated: one that targeted a website started on the dot of midnight GMT and ended precisely twenty-four hours later.

'How do you explain that?' President Ilves asked an official.

'Well, the money ran out,' the official replied.

'What do you mean?' asked the President.

'Well, these are done by botnets . . . owned by criminal gangs,' he explained.

In other words, someone had hired hackers who in turn controlled networks of hijacked computers (known as botnets), but the arrangement had been highly transactional by nature. 'Criminal groups don't generally get involved in politics. They're paid,' says Ilves. Those botnets would have to have been taken over ready for an attack well in advance. He is reluctant to say directly that there was Russian state sponsorship, but that is clearly what he believes. 'Who else would bother?' he asks.

Estonian officials suggest that this was an attack by the Russian state, organised by its domestic security service the FSB (the successor to the KGB), but it remains hard to prove. The attack seems to have been carried out by Russian 'patriotic hacking' groups along with criminals. Instructions on how to attack were widely distributed on hacker forums. Some claim the youth movement Nashi was involved, others point to criminal gangs or a mixture. At the very least, the Russian state is likely to have tolerated the activities, perhaps even encouraged them.

The Russian-Georgian conflict the following year saw a further raft of cyber attacks on government and key national websites alongside the bullets and bombs of a real shooting war. Georgia was a less wired country than Estonia, but still the government and banking systems were taken offline. And this time there was more sophistication, both in the technology that was used and in terms of what was being attempted. The attackers targeted Georgian news media websites, not just to take them down but as part of a wider campaign to shape international perceptions of the conflict – a form of information war – in which fake websites were set up and polls on websites rigged. Images of the Georgian leader were juxtaposed with those of Adolf Hitler.

Russia's campaign was an updated version of the *Alert*'s cutting of cables. The choice of targets was calibrated to supplement Russia's military activities by sowing confusion, making it hard for Georgians to understand what was happening as tanks and soldiers began to move and by reducing the ability to communicate. There were no destructive cyber attacks on infrastructure of the Stuxnet type – a type the Russian military could no doubt carry out if it wanted to. In 2011 a further espionage campaign against Georgia was discovered. News websites were hacked so that anyone visiting specific pages (for instance, about a NATO delegation visiting Georgia) would be infected with software that would search for specific 'sensitive words' in files held on the computer hard drive – words like NATO, USA, NGO, FSB and CIA, which were then uploaded to a server. The attacker could also capture video and audio from a computer's microphone and camera.[27]

Investigators found little sign of direct state involvement, yet the hackers seemed to have advance knowledge of the parallel military campaign, giving them time both to marshal their forces and carry out reconnaissance ready for action. This suggests a level of co-ordination with the state even if the attack was carried out by criminals or other hackers.[28] Western analysts have long pointed to the overlapping power centres between Russian intelligence and organised crime, and it is highly likely that this extends into cyberspace. As in other countries, cyber intelligence activity tends to mirror the structures and values of regular intelligence activity. Russia began battling in cyberspace in the late 1990s after Chechen hackers replaced Russian news sites with propaganda, after which organised groups of hackers tied to the FSB reportedly went after the Chechens.[29] Large criminal networks of hackers operate in Russia, most famously a group called the Russian Business Network which was linked to the Georgian attack in 2008 and 2011. Western experts believe that Russian hackers are left to their own devices by the state on two conditions: firstly, that they do not attack within Russia, and secondly, if the state asks them to do something, they agree. It has even been claimed that Russian hackers who are convicted are offered the chance to work for the intelligence services rather than go to jail. All of this would provide a significant but also largely deniable capability for the Russian state, wielded in conjunction with intelligence services.[30]

The campaigns against Estonia, Georgia and, from 2014, against neighbouring Ukraine after it sought greater independence from Russia do not fit with what we might think of as 'spying' – stealing information. As a result, some place this activity in the category of warfare; but that categorisation may be a mistake, and partly relates to Western stereotypes of intelligence work. As is the case with China's use of economic espionage, Russia has long used its intelligence services for activity it called 'active measures' in the

Cold War – spreading negative propaganda about opponents, de-stabilising them and influencing public opinion and politics (the CIA also engaged in similar activity, known as 'political warfare', in the early stages of the Cold War). This is not intelligence in the sense of gathering information, but it is done by intelligence agencies as it is designed to be covert. It is a tool for intimidation and subversion that Russia can employ to keep neighbouring countries from align-ing themselves too closely to the West. Russia sees itself as coming under information attack from the US as it tries to spread American values, and so it portrays its own actions as defensive – an attempt to prevent Western covert subversion through support for civil society groups. The online cyber component of its push-back is simply the subset of a larger struggle, also fought by funding non-governmental organisations and even international TV channels like Russia Today.

Other conflicts and crises also began to witness hacking take place alongside regular violence and propaganda – for instance between Israel and Palestinian hackers and Hezbollah, which has always been adept online. Cyber has become integrated with wider information warfare and is increasingly the dominant strand of that activity. As states recognise the value of hackers, many are creating so-called 'cyber militias' which can be a means of formalising the alliance between 'patriotic hackers' and the state. In some countries like Russia this may be between criminal and underground groups. China meanwhile has organised 'cyber militias' drawing on experts in private-sector companies and at universities to support PLA activities. These have caused alarm in the West, although they do not sound that different in principle from what the UK is doing in creating Joint Cyber Reserve, composed of people from industry working with the armed forces.[31] Russia, as seen in Ukraine, has de-veloped a form of irregular or special warfare which is below the level of full military action but uses intelligence and (semi) clan-destine military activity. The tools of cyber espionage and attack dovetail with this kind of work. For Russia, computer espionage is merely a new way of doing old things.

CHAPTER FIFTEEN

THE LIGHTS GO OFF

A phone call at 4.45 a.m. woke Oliver Hoare, the head of cyber security for the London Olympics, in July 2012. An early wake-up call was especially unwelcome when it was the day of the opening ceremony and the call was from GCHQ. 'There was a suggestion that there was a credible attack on the electricity infrastructure supporting the Games,' Hoare recalls.[1] Attack tools had been found in the possession of a hacker, along with what were thought to be schematics of the systems underpinning the Olympics. If the lights had gone off that evening and the Queen plunged into the dark on global TV, the reputational damage to the UK would have been enormous. Emergency meetings were held in the Cabinet Office to try to bottom out the threat and work out how to respond. 'We effectively switched to manual – or had the facility to switch to manual,' Hoare says, explaining how technicians had to be stationed at various points to keep the power flowing in case it was switched off remotely. An hour before the opening ceremony he was reassured that if the lights went down they would be back up within thirty seconds. But thirty seconds of dark during the Olympic opening ceremony with billions around the world watching would still have been a disaster. In the end, the feared attack turned out to be a false alarm: it would eventually emerge that the plans the hackers possessed were similar to but not the same as those of the Olympic systems. However, as when the lights went out in the US Super Bowl in January 2013 and everyone wondered why, the incident revealed how jittery officials have now become about the dangers of cyber threat to infrastructure. 'It is just too serious a matter to ignore,' says Hoare.

One of the reasons officials are so worried is because they understand how vulnerable infrastructure is and they have seen what

Stuxnet can do. Many industrial control systems (known as SCADA) are decades old and often have minimal security measures. In the past this did not matter, as an engineer needed to be physically present to manage them and they were not accessible from the outside. But companies have increasingly hooked them up to the internet for convenience: for instance, a manager may want to monitor and manage the flow through a gas pipeline remotely or know what reserves there are in order to quickly buy extra capacity on the market. That might all be done from the same laptop on which the manager sends his or her emails. It makes life easy, but also dangerous. By putting public-facing front-end computers on top of old insecure systems you immediately have a major problem. Replacing or updating them would be expensive, since they are embedded within large industrial plants. Now, if hackers can get into your system, they can also get into the controls which are 'sitting ducks'. Researchers have been able to find half a million SCADA systems accessible over the internet.

The 'human factor' of insiders was most apparent in one of the earliest attacks on infrastructure in February 2000 when 800,000 litres of raw sewage were released into parks and rivers in Queensland, Australia. This turned out to have been the work of someone who had failed to get a job with the company and issued commands to the computers controlling the sewage over an unsecured network. A more serious wake-up call came with a test by the US Department of Energy's Idaho lab in 2007 which showed that remote hacking into the operating cycle of a power generator could send it out of control to the point where it effectively blew up.

These infrastructure systems are often in private hands, so whose responsibility is it to defend them? Government or industry? Industry has often proved itself either incapable or unwilling to spend the money. It has also fought against ideas to impose security standards, fearing the cost will make companies uncompetitive globally. Government is reluctant to get into the business of protecting anything but the most core national assets in the private sector because the job is so vast. Infrastructure is so complex and interconnected now that no one really understands the points of connection or the vulnerabilities or what is actually critical.[2] The private and public sectors are interlinked, often across national borders, with foreign companies running parts of a country's infrastructure. A dense mesh of cyberspace is emerging which is vital to the functioning of our world but also poorly understood. As with financial systems, the danger is that there is no one who fully understands the vulnerabilities and the way actions can ripple out and cause a crash.

When natural gas pipeline operators are targeted by Unit 61398 of the PLA, the spies have not been stealing corporate data but seeking information on how controllers that run the systems operate.[3] The fear is that this could open the way for Stuxnet-type attacks. Stuxnet showed just how much work is required to carry out an effective act of sabotage, but there are plenty of signs that the kind of reconnaissance needed is being done.[4] Cyber reconnaissance of infrastructure even infiltrated a speech in President Obama's March 2013 State of the Union address. 'Our enemies are also seeking the ability to sabotage our power grid, our financial institutions, our traffic controls systems,' he said. China, as ever, gets most of the attention but it is not the only actor.

A Russian spy named Oleg Lyalin was arrested in 1971 careering down Tottenham Court Road in his car, drunk and with a blonde at his side. Ostensibly a knitwear representative for the Soviet Trade delegation, Lyalin was in fact an expert in hand-to-hand combat and part of the ultra-secret Department V of the KGB. This dealt with sabotage in the event of war, the latest incarnation of the 'stay behind' networks of the Second World War and the early Cold War whose job was to activate when a conflict started and do as much damage as possible. As a defector, Lyalin revealed plans to land teams of Spetsnaz Special Forces in Britain, flood the London Underground and blow up Fylingdales radar station. This was a classic aspect of Russian espionage that again takes spying beyond the narrow field of gathering information: it is also preparing for and carrying out covert action. Just as they invested heavily in this kind of spying during the Cold War, so the Russians are also believed to have become masters of similar activity in cyberspace, expertly probing infrastructure for weaknesses which can be targeted if the order is given. It should surprise no one that American and British spies have also hacked into the infrastructure of Russia and China as well. This may be partly for deterrence – to send a message to your opponent that you can do to them what they can do to you – to create a form of mutually assured destruction. But it can also be to prepare for war.

The penetration of these systems is a form of intelligence-gathering much as states have carried out in the past when preparing themselves for conflict. For hundreds of years they did this by making maps of a potential adversary's key facilities, perhaps after despatching spies or interviewing people who returned from far-off lands. In the Cold War it would be done through more technical means, such as satellite reconnaissance or signals intelligence to try to identify enemy military units and associated infrastructure. Now this is done in cyberspace. And, crucially, it is done against the private sector and not just government and military networks, since that is often where national power resides. But does the act of reconnaissance actually constitute an 'attack'? It may involve pene-trating networks and even leaving behind implants and backdoors to allow a future attack. But it is not the same as actually pulling the trigger. In that sense, this kind of activity is closer to traditional military intelligence and reconnaissance. It only becomes sabotage when deployed covertly (as with Stuxnet, and therefore still a tra-ditional clandestine intelligence activity) and only cyber war when used overtly as an open act of aggression. This is something states have yet to do. And why would China do this to America or Britain do this to China? Only if the two countries were at war or about to go to war. In which case cyber attack would be the least of people's worries, given the presence of far more lethal weapons. How likely are the US and China to go to war when their economies are closely connected – and far more interwoven, for instance, than that of the US and USSR in the Cold War? Seen in this way, cyber war is merely a new route that warfare will take in the unlikely event of an actual conflict.

There is one problem, though. Cyber reconnaissance is hard to distinguish from warfare. The act of getting into a network and leav-ing a backdoor to be able to carry out an offensive action in the future is 99 per cent of the work required to take a network down or switch off the power – all that may be missing is a command.

That makes it different from traditional intelligence-gathering and much harder to distinguish from attacking. 'You've got to know about an adversary's network before you want to work your will on it,' says Michael Hayden. 'But in a very interesting way the reccie [reconnaissance] in the cyber domain is actually the higher-order action. It's actually operationally and technically more challenging to penetrate someone's network, live on it undetected and extract large volumes of information from it – far more difficult – than it is to do something once you're inside that network. And so when you see someone in a SCADA network, one that controls industrial processes . . . power grids or banking systems, what's really scary is that "foreign" – whatever that means – presence in that network tells you that that agent already has the ability to do harm because they've penetrated the network and have lived on it undetected. That's what makes "foreign" – read Chinese – presence on these industrial networks quite scary. That already indicates the ability to do harm. It's not like in the physical domain where okay, I get it, they're conducting espionage, they're learning about targets. In this case they've already mastered the target.'[5] This type of espionage may therefore create a sense of vulnerability and fear that is itself destabilising.

What if a country could use even non-classified knowledge – gathered by cyber espionage – of what supplies are being ordered, whether food or oil, not just to work out where military units might be moved but also, in times of crisis, to disrupt those supplies in order to prevent troops or ships being deployed? What if penetrating defence companies allowed you not just to steal designs but also implant vulnerabilities which could be turned on during time of war? 'My nightmare scenario is that the United States tries to use force or is contemplating using force in a region of the world and when it trots out its military nothing works because there are Trojan horses inside the software in the American military arsenal,' says Richard Clarke. 'If you look at something like the F-35 fighter plane, there are tens of thousands of computer chips in it and very few of them made in the United States, very few of them made under secure conditions. And the software that we rely on is also filled with errors that can be exploited, so the supply chain for American weapons is very vulnerable.' Kill switches hidden in the hardware of guided missiles are the 'ultimate sleeper cell', others fear.

This means that cyber reconnaissance is not just drawing up maps of your opponent's terrain; it is more like sneaking in and leaving a few satchel bombs hidden in air ducts and underneath the floorboards ready to be triggered remotely if you ever need to. The act itself involves interfering with a network and can be misinterpreted as hostile, even if the purpose is only reconnaissance. In this way there is a greater danger of escalation in cyberspace, both because intrusion is so easy and also because it can be misread. Cyber reconnaissance exists in a new place, sitting uneasily and dangerously between traditional espionage and real warfare.

Fort Meade, the long-time home of the NSA, sprawls across a chunk of Maryland. An old signals intelligence collection aircraft sits near the museum that houses America's cryptologic history, a reminder of the past. Through the gates and into the ominous black building the sense is clear, not least from the number of uniformed personnel, that visitors are entering an institution which is firmly part of the military (unlike Britain's GCHQ, which is civilian). The sign outside also tells a story. As well as National Security Agency, it reads US Cyber Command. The US has created an almost (but not quite) seamless join between espionage in cyberspace and military action. This is reflected institutionally in the fact that Cyber Command, whose job is to carry out military attacks, is joined at the hip with the NSA, whose job is to carry out intelligence missions, with the same military man running both. The NSA has always been close to the military, growing in size to support it in Vietnam; but the computer age added a new factor – that of actual offensive work rather than intelligence support. This came partly because the deep understanding of computer networks resided in the NSA and the idea of replicating that level of capability in a separate organisation was seen as making little sense. But it also reflects a US view that the two activities are closely intertwined: the same skills needed to penetrate a network to gather intelligence are required for the reconnaissance and execution of a military attack. Chris Inglis, former NSA Deputy Director, puts it this way: 'What is needed is finding, fixing, holding in your mind's eye the thing that you would either defend, or exploit or attack. And then and only then do you make the final choice about what you are going to do with that.'[6]

The closely bound nature of military and espionage work in the NSA is reflected in the way the military men who have led it have thought about cyberspace as simply another domain in which to wield power. This was a way of thinking the air force popularised in the 1990s and which was explained to Air Force General Michael Hayden when he was briefed about the NSA on his arrival as its head in 1999.

> They introduced me to this thought of a domain – land, sea, air, space, cyber. Once you are in that place a man of my background begins to understand that, just as in the other domains – land, sea, air and space – the United States wants to be able to freely use that domain and to deny its use to others who would will us harm . . . the language we use to describe what we want to do in cyberspace, it feels an awful lot like air force doctrine of air superiority and air dominance . . . We want to control the space and then, after we control the space, we will work our will there. Now look, that sounds very aggressive. What I'm talking about is in a wartime situation . . . now unfortunately American law, American congressional oversight, divides what you want to do in cyberspace into attack, and defence and exploitation, the espionage thing. But those of us who work in that space know they are all the same thing. They are all about controlling the space.[7]

From the 1990s there had been a division in the US military. On the one hand were those who saw information warfare and then cyber as a revolutionary new form of warfare – a game-changer akin to nuclear weapons – and who talked about cyber Pearl Harbors. On the other hand there were those who saw it just as a way of exploiting vulnerabilities in systems during conflict, much like the old techniques of electronic warfare. So far, it has only rarely been deployed. In the late 1990s there was discussion about shutting down the Serbian banking system during the conflict over Kosovo. But

the US decided against, fearing it would set a precedent and open
the way for hackers to take their revenge on the much larger American system. Similar fears expressed by the US Treasury Secretary
stopped the manipulation of Saddam Hussein's bank accounts in
2003 as well.

Apart from the targeted Stuxnet attack, there seem to have been
mainly smaller-scale attacks against machines. One leaked document claimed that the CIA and NSA carried out 231 offensive cyber
operations in 2011 – different from espionage because they might
involve shutting down someone's network or scrambling the data
on a machine. Nearly three-quarters were reported to be against
top-priority targets like Iran, Russia, China and North Korea.[8] A
2012 Presidential Directive noted that cyber attack 'can offer unique
and unconventional capabilities to advance US national objectives around the world with little or no warning to the adversary
or target and with potential effects ranging from subtle to severely
damaging'.[9]

Cyber Command has grown rapidly – to at least 6,000 personnel,
almost all military. It consists of more than a hundred teams, some
assigned to support each regular combat command, others to focus
on defending – or attacking – particular sectors or countries (like
China and Iran).[10] This involves exploiting NSA spying skills. Insiders say there are differences in culture. Cyber Command is, as said
above, almost entirely military and the chain of command is rigid.
The NSA is less than 50 per cent military and, insiders say, people
will ignore seniority to defer to the smartest person in the room.

The rapid advance of Cyber Command looks scary to the outside
world. That perhaps is the point. 'No offence to my friends in Cheltenham, the greatest concentration of cyber power on the planet is
at the intersection of the Baltimore-Washington Parkway at Maryland Route 32,' Michael Hayden says. The fact that he refers to a
place rather than the NSA and Cyber Command is important. Also
located at the intersection are a vast array of defence contractors
feeding off the growing trough of money associated with the buzz-
word 'cyber'. In recent years this has included private companies
being contracted to carry out offensive hacking. Sometimes they are
asked to race to see who can get inside a target system first or find

a vulnerability, with the pot of prize money going to the winner. These private contractors provide the US with its version of plausible deniability (as well as profit and pay packets for the people who move back and forth with government). Russia may engage criminal and underground gangs; the US uses companies. When he left office, President Eisenhower warned of the scale and power of a military-industrial complex. Today there is a cyber-industrial complex.

Contractors are also involved in buying up computer vulnerabilities that can be exploited for attack – creating a market, critics say, in which private hackers sell exploits to contractors and middlemen rather than tell the software companies so they can be patched up (who can easily be outbid if required, with some vulnerabilities going for over $100,000 a time). These arms brokers (a bit like shady intelligence brokers operating in places like Istanbul and Brussels a century ago) now hand out business cards at hacker conferences trying to recruit new staff who are expert at finding vulnerabilities. This is another step in the industrialisation and commercialisation of hacking towards even the intelligence space. Intelligence agency recruiters are now looking for the successors to the Hanover crew whom Cliff Stoll found working for the KGB for money in the late 1980s. They were the first, but now this is business. The balance between the old defensive mission of protecting computers and the offensive one of breaking into them has been tilted, critics fear, far too much towards offence by the increasing desire of the US military to 'stockpile' vulnerabilities so that it has an arsenal bigger than anyone else.

The UK is also developing cyber weapons for use in the case of war. However, their exact utility is not always clear. One person who attended meetings on the subject remembers that discussions were reminiscent of a particular scene in an Austin Powers movie. In the scene, the villain Dr Evil explains that he is going to dispose of the captured British superspy Austin Powers by coming up with an overly elaborate and exotic scheme which will lead to his death. At which the villain's son, Scott Evil, asks the obvious question everyone who has seen a James Bond movie always wants to know: 'Why don't you just shoot him?' He offers to get a gun but Dr Evil then threatens to ground him. The British official had exactly the same thought as Scott during the meetings on cyber weapons. In other words, it might well be a lot easier to drop a real bomb to do the job than go for some elaborate and destructive cyber weapon which may or may not work (and, as in the Austin Powers and Bond movies, give your opponent the chance to get away). The only time you might prefer a cyber attack to a computer network is – as with Stuxnet – when you want it to be an act of sabotage that is covert and not immediately traceable back to you.

A row took place behind closed doors between GCHQ and the military. The military wanted to wrest more control of cyber weapons from GCHQ, the generals and their officials arguing that cyber weapons are increasingly a core part of fighting wars. They may have feared becoming irrelevant if they lost control of the one part of the budget that seemed to be growing while the rest of their empire was shrinking. GCHQ argued that cyber attack was one end of a spectrum of capabilities (ranging from espionage), rather than something that could be isolated and separated off. They also argued that the number of times it would be used overtly by the military would be low: once this happened you would blow your capability by showing your hand. The military tend to work more by having overt capabilities that can be used to deter opponents, but in cyber, the spooks argued, this did not apply. As soon as you reveal what you can do your hand is blown, and your opponent will patch up the vulnerabilities or Zero Days that you had exploited. Far better to keep the capability as intelligence-led and therefore clandestine, they said, with some in GCHQ even suggesting that if the military wanted their own overt military capacity, then they should build it in parallel.

In the Second World War, penetrating the opponent's systems was most valuable in deception; senior intelligence officials say that this may still prove to be the most valuable aspect of computer network espionage in the future. It might be electronic warfare to take out certain enemy systems, but it could more fruitfully involve inserting false information to confuse the enemy – like Operation Fortitude, when dummy communication networks made the Germans think that the main thrust of D-Day would be in Calais and not Normandy. Even if your opponent knows you are inside their network, that in itself can lead them to not trust their own communications and sensors and undermine their ability to act (turning the red dots on screens blue so you attack your own people or don't know who is friend and who is foe, as one retired American general puts it).

Cyber attack may well become integrated with regular warfare to the point where they are indistinguishable, just as cyber espionage becomes entirely interwoven with regular espionage. The comparisons often made to nuclear weapons are misleading: everyone knew what a nuclear weapon could do – they had seen them used in Japan. The truth is that no one knows what other countries can do when it comes to cyber weapons. People who have worked at the highest levels of the British effort concede they do not really know what the US is capable of doing. Military thinkers are struggling to define what constitutes an attack (as opposed to espionage) and therefore what a proportional response would be. Should a cyber attack be countered by shutting down the computer responsible, wherever it is? What if you strike back and your enemy diverts your attack to shut down a hospital's computers and then blames you? Can you – as the Pentagon suggests – return fire from a cyber attack with a real-world missile? A missile comes with a return address in a way a cyber attack does not; in a cyber attack the problem is knowing who is attacking you, and whether it is even a state.

In January 2008 a CIA analyst surprised a gathering of infrastructure protection engineers from the US and Europe with a candid statement. 'We have information, from multiple regions outside the United States, of cyber intrusions into utilities, followed by extortion demands. We suspect, but cannot confirm, that some of these attackers had the benefit of inside knowledge. We have information that cyber attacks have been used to disrupt power equipment in several regions outside the United States. In at least one case, the disruption caused a power outage affecting multiple cities.'[11] He made it clear that there had been a debate about whether to disclose this, but the agreement was that it was better that the experts assembled at the conference understood what was really going on. The ability to carry out cyber reconnaissance and attack is not restricted to states. For criminals, the threat of destructive cyber attacks offers a route to extortion by holding to ransom companies that rely on websites – threatening, for instance, to take gambling sites offline on the day of the Grand National. The first reports date as far back as the mid 1990s, when Britain's Department of Trade and Industry said that it was investigating reports that firms in the City of London had been extorted to pay millions to avoid their computer systems being wiped. However, it said it had not seen hard evidence.[12]

The dog that has not yet barked is destructive cyber attack by terrorist groups. A staple of Hollywood thrillers and alarmist briefing papers from the 1990s onwards, there has been relatively little evidence so far. The fears of attacks on infrastructure grew from the 1990s and after the Oklahoma City bombing. Intrusions into utilities in California in the summer of 2001 were traced to Asia, with some wondering if it was Al Qaeda, others thinking it was the Chinese. Laptops found in Afghanistan after the fall of the Taliban in 2001 showed that Al Qaeda may have carried out reconnaissance over the internet, but only in the sense of searching for the schematics and engineering designs of things like nuclear power plants and water systems on the web rather than actually planning to attack them over computer networks. Al Qaeda and related groups have not managed to use the internet to carry out a destructive cyber attack. What no one is sure about is whether this is a calm before the storm. It may be because of a lack of capability (as Stuxnet showed, it takes real effort and work), but it may also be because it is a group that prizes real death over online disruption. That may change, and other groups may develop more effective capabilities as cyber attack techniques proliferate faster than anyone had expected. 'It's got to be a worry, and speaking personally I think it's only a matter of time,' MI5's head of cyber explained in 2013. 'The intent is already there, the capability can only follow in a few years' time.'

The internet may be more fertile territory for those wishing to spread fear and confusion rather than cause mass casualties. But the border between hacktivists, states and 'terrorists' is often in the eye of the beholder (or the accuser). 'In the morning a person could be a hacktivist, but at the end of the day he needs money. So it is very difficult to draw the line and I'm afraid that criminals and hacktivists will be employed by terrorists,' argues Eugene Kaspersky. Michael Hayden sees three threat actors: states, criminals and a third group with an agenda. 'I haven't developed a good word for them yet, but "hacktivists", "anarchists", "nihilists" – people living in their mums' basements who haven't talked to the opposite sex in five years,' he says. 'Now you've got this third group, blessedly the least capable. But I don't know what motivates them and I certainly don't know what deters them. I don't know what kind of demands they'll make in the future and I'm not so sure they care much about collateral damage.'[13] The term hacktivist, though, can easily be applied broadly to those using cyber tools to dissent – definitions are rarely simple and often contested. And separating hacktivists from states is also getting harder.

Spear-phishing emails arrived in the inboxes of employees at the Sony film studio in September 2014. Once they were in, hackers began exploring the network carefully. Finally, in late November they were ready to act. When employees logged on they saw a message from a hacking group calling itself 'Guardians of Peace' and found their systems not responding – a similar experience to that of Saudi Aramco. Next came a data dump. Masses of personal information and corporate emails exposed on the internet. Movies the film studio was working on were published, and some of the emails, in which studio executives talked in none-too-flattering terms about celebrities like Angelina Jolie, were there for everyone to see. The attack was linked to the release of a film called *The Interview*, a comedy which featured a CIA plot to kill the North Korean leader. 'They came in the house, stole everything, then burnt down the house,' Michael Lynton, the movie studio's CEO, told Associated Press. 'They destroyed servers, computers, wiped them clean of all the data and took all the data.' Staff had to be paid with paper cheques and dig out old phones to communicate. Lynton admitted he had 'no playbook' to deal with the crisis.

The exposure of corporate data was embarrassing, but it took a threat of real physical violence to escalate the crisis. A message suggesting that cinemas might face some kind of terrorist attack if they showed the film was enough for them to back out, leading to criticism from the White House. Was it just a group of hackers? The US authorities were confident in attributing the attack to North Korea. Some computer security experts questioned that, and wondered if an insider was involved. But the North Koreans had made mistakes, and the US administration could also be confident because they could use their wider intelligence machine. The US had been spying on North Korean activity for a few years, implanting malware in its computers from at least 2010. It knew what they were up to.[14] That focus had intensified after 2013, when it became clear that North Korean hackers were capable of destructive attacks after they targeted South Korean media and banks.

Look at a global map of internet activity and the northern part of the Korean peninsula looks almost entirely dark – in stark contrast to the south. But while the country's citizens may be almost entirely cut off from the global World Wide Web and instead relegated to a domestic walled garden, a select few hackers working for the state have honed their skills. From the early 1990s, North Korea seems to have wised up to the idea of using the internet to gather intelligence from its enemies: like many other countries at the time, it saw its value as an equaliser against the technologically advanced West, especially in the wake of the US operation against Iraq in 1991. According to one report, this realisation came after some of the country's computer experts visited China and saw that it was already undertaking intelligence collection. A team of fifteen were reportedly soon sent over the border to a military academy in Beijing to learn the tricks of the trade. A steady stream of hackers were sent to China and Russia over the years who were envied for their experience of the outside world and the luxuries they were allowed, defectors said. South Korea now reckons there are 6,000 North Korean hackers working for military and intelligence agencies, some using infrastructure over the border in China. North Korea, like other states, was coming to value computers as not just a means of espionage but also of 'asymmetric warfare' – levelling the playing field with the US and others.

US spying on North Korea had enabled it to allocate the blame for the attacks, although some asked why, if the intelligence was so good, the US had not warned Sony. But that goes back to the wider question of how far the NSA is there to protect corporate America and how far it wants the secrets of its work exposed. The case was one of the first, though, where one state directly accused another state of attacking a corporate network. A few days after that happened in December 2014, the entire internet and mobile phone data network for North Korea went down – only for a short period, but it was perhaps a signal of capability. The North Koreans blamed the Americans. America certainly has the capacity to attack, but it also knows how vulnerable it is.

The film *Dr Strangelove* satirised the Cold War desperation to ensure that you could obliterate your enemy before they obliterated you (including even building a doomsday device run by a network of computers to destroy the earth as a last resort). The attacker has a huge advantage in striking first and knocking out an enemy's systems so they cannot respond – something contemplated in the early, dangerous days of the Cold War with nuclear weapons. The fear of not being able to respond fast enough is driving research in computers today as it did with SAGE in the nuclear past. And defending countries in cyberspace is becoming increasingly automated. Taking humans out of the loop, defence officials claim, is the only way of stopping an attack at network speed (before perhaps it confuses you or knocks out your systems). The only way of blocking malicious cyber attacks, they say, is by monitoring all the traffic and analysing its patterns to understand what looks dangerous and then stopping that coming in – a kind of automated defensive monitoring. The next question, though, is whether you also want to return fire automatically. A US system, almost comically named as MonsterMind, was reported as being considered to do this.[15] Even during the dark days of the Cold War, the President was expected to have a few minutes to decide whether or not to retaliate to a suspected Soviet missile strike. In the cyber world this could be down to milliseconds to judge whether or not to shut down a machine that is sending malicious code to you wherever it may be. But in an online world in which so much can be obscured and confused (hosting an attack from a third country's computers, for instance) and in which deception (or at least anonymity) is a fundamental tenet of the internet and the work of spies, could we really be sure we were striking back at the right computers in the right country? In a sense this returns us to the era of nuclear-tipped missiles ready to be launched automatically against Soviet bombers, and Roger Schell's question to the US Air Force back in the 1960s: do we really trust computers with decisions of life and death?

CHAPTER SIXTEEN

REBIRTH – CABLES

In January 2002 a storm battered Land's End, the furthest tip of Cornwall where England stretches out to touch the Atlantic Ocean. Not long before the end of the road, a narrow lane leads to the village of Porthcurno. In the aftermath of the storm, strips of thick, black wire were exposed on the beach like fossils, dredged up by the wind and rain. These were fragments of disused telegraph cable, remnants of the days when the beach had been home to the largest telegraph station in the world, the dots and dashes of Morse code messages flying across the Atlantic. But the storm exposed more than just the past. As well as the skeletons of their predecessors, two modern cables pulsing with life were also unearthed by the wind and rain. These were fibre-optic cables bearing beams of light which carry the ones and zeros that connect together the modern world. The modern cables that land in Cornwall surface in nondescript huts and connect into a pulsing global telecommunications infrastructure.

The first international fibre-optic cable landed in Cornwall in 1988. It made landfall in a hut originally built to withstand a nuclear blast with its own five-ton blast doors and air filtration system. As the internet took off, this cable and the many that joined it began to carry the world's emails and web-browsing before these were joined by every other type of data, from downloaded movies to corporate video conferences to personal webcams, as our lives moved online and the physical and the digital worlds converged.[1] In 2015, more than 200 million emails traverse the world every minute. A modern submarine cable can carry up to 60 terabits per second across up to six fibre pairs. A terabit is a trillion bits of information. In 1870 it would have taken a patient Morse operator thirty-seven days to send the novel *War and Peace* over a telegraph. Today, a fibre-optic cable could send 300,000 copies of the book every second, according to a rough estimate.[2]

In 2015 fibre-optic cables carry more than 90 per cent of the world's data and follow many of the routes of the British imperial telegraph system of 100 years ago. And, just as in the First World War, Britain would learn how to exploit the cables to gather intelligence from across the world, this time as part of the alliance forged at Bletchley Park. But, as it first gathered pace in the 1990s, this digital revolution marked the most profound challenge for the spies who had first exploited signals at Bletchley. The data tsunami, many inside British and American spy agencies believed, would engulf them. The internet and the digital world threatened to disrupt their business models, like those of other traditional institutions, and relegate them to the past. When you are approaching a mountain, you do not think you will be able to climb it, a former senior British official recalls. But, he adds, when you do and you get to the other side, the view can be breathtaking. Instead of being overwhelmed, the spies would learn to ride the digital wave and to master the internet, leading to what some inside that world have called the 'second golden age' of signals intelligence. But it was not a foregone conclusion.

The fall of the Berlin Wall and the end of the Cold War was a time of triumph for Western spies, but it was one that quickly led to questions as to whether they were really needed any more. GCHQ and the NSA struggled to adapt. They were static organisations because the targets they had been watching were static. One stove-pipe in GCHQ would work on Soviet radio, another on Chinese satellites. An analyst might have joined in the early days of the Cold War and been assigned to monitor the signals of a particular Soviet missile battery. They might have spent the next forty years learning every nuance of the way that that unit communicated, all so that they could listen out for the 'launch order,' or spot something out of the ordinary that might provide advance warning of an attack. The same analyst might retire and draw their pension never having seen such a signal. A huge intelligence-gathering machine had been built – men clamped with headphones in far-flung corners of the planet, satellites scouring from space, dishes on earth picking up signals – all to try to provide a few, precious moments of warning. 'We lived on their network,' is how one former British analyst described the way the US and UK enmeshed themselves inside Soviet communications. But with the end of the Cold War, this vast bureaucracy seemed redundant. And a further problem emerged at the same moment.

The end of the Cold War coincided with the digital revolution. Over the 1990s, the number of internet users grew from 4 million to over 360 million. This new technology was going to be used by everyone, including the new targets that intelligence agencies were now looking at – those involved in spreading biological, chemical and nuclear weapons or terrorist groups or drug barons. But finding them online would require a totally new way of working. Previously, most communications followed a fixed path that could be intercepted. The internet broke up the message into packets and then distributed them separately around the network, taking the least congested route, to be reassembled at their destination. That required a different collection model. The internet was helpful on one level, as there was now only one communications system and it was transparent – there was no need to do the equivalent of stealing an Enigma machine to work out how it functioned. But the problem was that it was an open system that was moving at a bewildering pace, driven by commercial innovation.[3]

This confluence of events – the loss of the old enemy and the emergence of new communications – created an existential crisis for hidebound, inward-looking, highly secretive organisations. Were they dinosaurs about to become extinct? We are 'hanging on by our fingernails', a Deputy Director told NSA staff in 1995.[4] When a Senator said he could learn more from CNN than from an NSA briefing, an analyst chased him down the corridor to try to disabuse him of that notion. 'It will be fatal, in an age of austerity, to be seen as an expensive alternative to *The New York Times*,' said an official.[5] Budget cuts began to bite on both sides of the Atlantic.

The special relationship woven at Bletchley seemed to be fraying. As the members of the Five Eyes club shifted to looking at issues like drug-trafficking, they found they were not viewing the same parts of the world in the same way as each other. The sense of a common mission was fading. Intelligence capabilities – eager for something to do – also turned more towards spying on other countries to gain national advantage. America and Europe got into a flap over a US-UK system called Echelon. This was actually the codename for collecting microwave satellite communications, but European capitals were convinced it was a programme to spy on them and their companies. Former CIA Director James Woolsey penned an article in the *Wall Street Journal* in 2000 that started by saying that 'most European technology just isn't worth our stealing,' before then going on to admit that Europe was indeed being spied on (mainly by the NSA, although he did not say so). But, Woolsey said, the spying was needed to catch European companies bribing foreign governments to secure contracts in places like Saudi Arabia and Brazil against American competition. 'That's right, my Continental friends, we have spied on you because you bribe.'[6]

The volume of communications was exploding like a supernova, and encryption was beginning to spread as systems like PGP took it beyond places like banks and into the hands of the more technically adept and security conscious members of the public. Reports started to reach the media that there was a problem in the secret world. The new digital world was shifting to fibre-optic cables, and it turned out that these could not be intercepted as easily as microwave satellite communications, where you just had to point a dish in the right direction to catch a signal which was broadcast like radio. The talk was of the NSA and GCHQ 'going dark' and losing their ability to listen in. 'The dirty little secret is that fibre-optics and encryption are kicking Fort Meade in the nuts,' a recently retired senior CIA officer told *The New Yorker* magazine in 1999. 'It's over. Everywhere I went in the Third World, I wanted to have someone named Ahmed, a backhoe driver, on the payroll. And I wanted to know where the fibre-optic cable was hidden. In a crisis, I wanted Ahmed to go and break up the cable, and force them [the communications] up in the air.' Forcing the communications of a country into the air and onto radio was what the *Alert* had done to German cables at the outbreak of the First World War. And now again it seemed it was the only hope the NSA might have of listening in (as had been attempted to Iraq in the 1991 Gulf War).[7]

When General Michael Hayden took over NSA in 1999, it appeared he was inheriting a sclerotic, bureaucratic institution struggling with the modern world, at once bloated and yet one in which old-timers were unable to cope with a cut of one-third of its budget since the end of the Cold War and the emergence of new technology. The failings extended to one of the new missions: finding terrorists. The NSA had been slow to take up the problem when it emerged in the 1970s. The central tension was already clear: the Soviets had their own distinct communications systems and the NSA organised itself around targeting these, but terrorists used the same systems as the general public. Finding terrorist communications amid the vast public tide created unique problems, both of sheer scale (the needle in a haystack analogy) and culture and ethics (spying on systems used by ordinary people rather than on the unique systems of a state). It was much more like the domestic world of spy-hunting – finding someone hiding amid the public at home – than the traditional foreign intelligence work of the spy agencies, studying communications of militaries and governments. There had been the odd success, such as detecting a plot to assassinate Henry Kissinger during a visit to Syria, and a plot to bomb Israeli targets in New York, but counter-terrorism had been marginal.[8]

By the late 1990s the counter-terrorist mission had moved to the forefront – especially after the 1998 bombings of US embassies in Africa by Al Qaeda. A satellite phone belonging to Osama bin Laden was being intercepted. Or at least it was until word leaked out and he switched it off (a leak sometimes blamed on journalists, although the NSA seem hardly to have been discreet, playing visitors' recordings of bin Laden talking to his mother). What intelligence was collected often sat on the shelf, waiting to be translated by the few Arabic speakers. That included a message talking about 'Zero Hour' approaching. It was only translated after 11 September 2001 when Al Qaeda wreaked devastation on America, killing 3,000 people in a single morning, sending the Twin Towers crashing to the ground and even breaching that symbol of American military might, the Pentagon.

The airspace was closed, but one plane was allowed to fly over the Atlantic the next day. An old freight DC-10 had taken off from a British military airfield. The plane was largely empty save for the

most senior figures from British intelligence, including the head of
GCHQ. As they flew over the eastern seaboard of the United States,
the spies could still see the smoke emerging from the vacant space
where the Twin Towers had once dominated the Manhattan skyline.
From Edwards Air Force Base they were taken to CIA headquarters
at Langley. Unlike the Americans travelling the other way during the
Second World War, they bore no tangible gifts other than an offer
of support. When they arrived, the Britons immediately sensed that
everything had changed for their American counterparts – they inhabited a different world now. No one at the time had any sense of
where that would lead.

The sprawling US intelligence community, born after the Second
World War, had one overriding mission: to prevent another surprise
attack like Pearl Harbor. On the morning of 11 September 2001, those
inside the three-letter agencies felt they had failed. Accompanying
the pain was a sense of responsibility – a feeling that if only they
had done more then perhaps they could have stopped the loss of
thousands of innocent lives. The CIA and NSA had often swung
between poles of excess and restraint. There had been plots to
assassinate foreign leaders and spying on domestic anti-war activists
for a while in the 1960s and 1970s; then came revelation, condem-
nation and new laws saying what you could and could not do, and
senior officials desperate to avoid being hauled before Congress.
After 9/11 the pendulum swung again, this time more violently.
That was partly because of the utter shock at what had happened,
but also because of the presence in the White House of political
leaders who were determined to unleash the spy agencies. Those
agencies are a tool of the executive. It issues directions and requirements but also sets the tone for their work. And after 9/11 the
Vice-President, Dick Cheney, made it clear he wanted the limits to
be pushed. 'What more can you do?' he asked the spy chiefs, indicating his own lawyers would find a way to help. Cheney would say
publicly that the country would have to take a walk on the 'dark
side'. The US was deemed to be at war with Al Qaeda, which put
new powers on the table. For the CIA, this would mean secret pris-
ons emerging, dotted across the world, into whose dark confines
people would disappear. For the NSA, it meant boxes appearing in

the darker corners of buildings dotted around America into which data would disappear.

Hayden's ambition was to 'live on the network' of his target's communications as the NSA and GCHQ had done with the Soviets. The difference was that the terrorists' network was the internet, which was also everyone else's network. Even before 9/11, Hayden had been pushing to make the NSA 'a powerful, permanent presence on a global telecommunications network that will host the "protected" communications of Americans as well as the targeted communications of adversaries'. He had been arguing for bulk access to communications before 9/11 (one CEO was allegedly punished by not getting contracts after he refused to co-operate in February of that year). After 9/11, the NSA was now able to move into this new space.[9]

The mantra after 9/11 was that US intelligence had failed to 'connect the dots'. It was said that the leads were there in the system but had never been collated and followed up to understand that known Al Qaeda members had entered the United States, one hijacker making a call from San Diego to a known Al Qaeda safe house in Yemen. This perceived failure became the justification for a wide expansion of intelligence powers to monitor communications – even though critics would argue that the real failure was not to do with a lack of powers or connecting dots but due to a failure to share intelligence because of bureaucratic infighting. Before the Second World War, army-navy competition in code-breaking had hampered the chances of spotting Japanese intentions over Pearl Harbor. Before 9/11, the CIA had not shared information with the FBI to identify the operatives, and the NSA had likewise not talked to the FBI to get a warrant on the San Diego call. A sharp line had existed to keep foreign intelligence-collectors like the NSA and CIA out of the domestic spying game, but an inability to share had made that a problem.[10] However, in the aftermath of the attack, the agencies and the White House chose to focus on new powers.

Terrorism creates a demand for pre-emptive intelligence – finding information before a bomb goes off rather than investigating after the fact. And it is intelligence about people not states, and those people are sometimes living within your own country. In the rush to find anyone else in the US planning an attack, the previous requirement for a warrant for intercepting communications in which one party might be in the US was abandoned if there was reasonable suspicion that the person was communicating with Al Qaeda overseas. This was part of a new set of powers known formally as 'The President's Surveillance Program'. In the past there had been a clear distinction: the NSA spies abroad but not at home. But now the powerful capability it had built up to target others overseas was also being turned inwards to look at the point where foreign met domestic. The fear of more attacks carried out from within was breaking down the hard divide of the past.

It was now easier to monitor targets – that was one part of the new powers. But how do you find them? When Hayden had taken over the NSA, he had found it overwhelmed by volume. Rather than fight volume, he was going to swim with it and turn it to his advantage. That meant bulk access and analysis. Get hold of as much data as you can and then use computers to work on it. A new programme involved collecting domestic phone and email records in bulk – the 'who was contacting whom' information, rather than what they were saying. This was supposed to seek out the unknown terrorist operatives – to find the proverbial needle in the haystack you first needed to acquire a haystack.

On the weekend of 6 and 7 October 2001, a small number of NSA staff received a call at home. They were asked to report to work on Monday – Columbus Day, a public holiday – for a highly classified mission. Hayden personally briefed the assembled analysts, mathematicians and programmers. He explained that the President had signed a special authorisation. The aim was to provide early warning of impending terrorist attacks. It would involve a major expansion of powers into the domestic space. Hayden made it clear that lawyers had concluded this was all legal. He told the staff they would do what the authorisation allowed and 'not one electron or photon more'.[11] The programme was first called Starburst and then StellarWind. The President's authorisation was kept in a safe in the NSA Director's own office, with only a tiny handful of people allowed to see it. This was an 'aggressive' move, an internal report found, and had been based on a view that additional attacks on US soil were 'imminent'. But what was initially a temporary response would evolve into something more permanent.

By early October, a twenty-four-hour, seven-day-a-week watch operation, called the Metadata Analysis Center, had been formally established. Fifty computers were ordered to help with storing and processing all the data. Many of the ninety staff were veterans of traffic analysis on Russia from the Cold War who were experts in 'call-chaining' or 'contact-chaining'. This was a particular technique that required access to bulk data about phone calls (metadata), but not the content itself. You could take the number of a known terrorist and then look at the whole data set to find whom they had communicated with directly (one hop), and then who has communicated with those people (a second hop), and even go on to a third hop. This could potentially lead you to a huge amount of numbers (up to a million in some cases), and this was where the need for skilled analysts came in because, if used correctly, it allows you to work out which connections are significant and hopefully identify networks and unknown operatives: perhaps you might find that person C was in contact with suspects A and B and therefore might be a missing link. It was a technique used in foreign intelligence but was now being applied to domestic US phone calls. Many would later wonder why it was that the NSA – a foreign intelligence agency – rather than the FBI was doing this? The reason seems to be that it was the only organisation that had experience with chaining, and the only one that had the technology (the FBI was sclerotic when it came to computers). In 2012 the NSA queried the domestic telephone metadata set with 288 phone numbers believed to be linked to foreign terrorist activity, leading to twelve tips for the FBI to investigate.

Supporters of bulk collection argue that it is the only way of carrying out certain searches to find 'unknowns'. If you do not have a full set of data then you cannot carry out contact-chaining to find unknown links. Nor can you find an 'alternate identifier' when a known target changes phone or email suddenly to avoid being tracked. It is no good, they say, looking for a needle in a haystack unless you are sure the needle is in there, and that can only happen if you know you have all the hay. They say that a system that just filtered and extracted what you wanted from traffic as it went past would not have the same advantage as a full set of stored data. This is because you do not always know what will be of value until later. If a bomb goes off, they say, you want to then be able to go back and access the phone and email contacts of a suspect to see who else he was communicating with in the run-up to the event to find out if they are dangerous. It is no good just tapping his phone now. Obtaining the metadata in bulk had a crucial advantage: it meant that you could go back in time. But that requires keeping everyone's data, which inevitably means that of innocent people as well, even if, as the spies maintain, this will be not be looked at.

The point of the new systems was not to look at known conspirators but to find unknowns. There was a jitteriness about America, a sense of vulnerability and a sense that, after the shock of that September morning, suddenly anything was possible. And the proponents of new programmes were convinced there were unknowns out there. Overall, StellarWind ended up targeting around 37,000 phone numbers and addresses, about 3,000 of which belonged to Americans. Terrorism was changing the equation in terms of computers and spies, increasing the pressure to see what could be done and pushing the boundaries of what it was permissible to do.

Where were these vast amounts of phone, email and internet data to come from? (The email metadata programme was later stopped.) The answer was simple: the raw data would come from companies. In some cases court orders would require them to hand over phone metadata in bulk. But they would also provide access to the communications choke points inside the US. For decades back to the days of telegrams, these companies had provided access to international traffic. The NSA had classified relationships with over a hundred companies to assist with 'essential foreign intelligence-gathering'. Two companies in particular had been compelled to provide access to large volumes of foreign-to-foreign communications transiting the US through fibre-optic cables, gateway switches and data networks. Losing their help, NSA directors told an internal secret report, would mean the US signals intelligence system would be 'irrevocably damaged because NSA would have sacrificed America's home-field advantage as the primary hub for worldwide communications'. This so-called 'home-field' advantage in communications and data was the crux of America's intelligence capabilities.

America had created the internet and also controlled the pipes through which much of it flowed. In 2002, global worldwide data bandwidth was slightly more than 290 gigabytes per second (gbps). Of that, less than 2.5 gbps was between two regions that did not include the US.[12] In other words, at that time more than 90 per cent of the world's data touched the United States. That made it potentially accessible.

How much of an advantage was American dominance? Officials, even retired ones, talk in careful terms. 'I don't know if I could answer that,' says Richard Clarke when asked what it allowed, admitting only that the fact that much of the internet traffic flowed through the US would have provided 'opportunity', without saying if anything was made of that. 'I don't want to comment on operational matters,' says Michael Hayden when asked a similar question. 'But I will share with you the thought that the fact that it was American ingenuity that created much of the internet and the fact that information was simply readily available in the American technical community, of course that helps.'[13] In other words, America understood the internet and how it worked better than anyone else – how the routers worked and what their vulnerabilities might be, for instance – which would also provide an advantage in knowing how to exploit the system for intelligence. One former British official says that it used to be that up to 80 per cent of the internet went through the US and so was accessible to Fort Meade (making a general point rather than using precise figures). And, they add with a wry smile, the other 20 per cent went through the UK 'and so was accessible to Fort Meade'. Access to global communications at the level of the piping and an understanding of the infrastructure itself was one of the twin pillars of America's home-field advantage. This was the twenty-first-century equivalent of Britain's dominance of the global telegraph system at the outbreak of the First World War. Who were the modern versions of those British censors, though?

The absence of a handle on the door to Room 641a on the sixth floor made Mark Klein suspicious. 'It looked kind of odd,' he thought. Klein was a technician at a San Francisco facility run by the telecoms company AT&T. In January 2003 he saw a new room being built by workmen. Afterwards, it seemed they had installed special equipment inside. When an air-conditioner was leaking downstairs from the secret room, regular staff had to wait until special security-cleared technicians could go in. A curious Klein followed the physical path of the cables going into the room and realised they were connecting up to the seventh floor where AT&T handled internet traffic.

The building was a key junction, or choke point, where the communications of AT&T customers were switched and connected to other networks. It was like the building in Hong Kong where the telegraphs came in and the censors situated themselves to read what passed through in 1914. But this time it was a place where fibre-optic circuits connected up. Klein realised that the light carried on the fibre-optic cable was being split – effectively copied – and sent into the secret room. Measuring about twenty-four feet by forty-eight, the room contained several racks of equipment, including a large box. Klein learnt that this was not the only such room in America. Other boxes were being installed in other cities, more than a dozen in all. AT&T responded at the time by saying that it was not permitted by the government to respond to Klein's allegations but that it vigorously protected its customer's privacy and only shared information as specifically authorised by the law.[14] Some NSA staff, unhappy at developments, have said they wondered why choke points at domestic exchange facilities were chosen as opposed to the landing places for foreign cables. The reason, they thought, was a desire to capture all the traffic – including domestic – rather than just focus on foreign collection.[15] These boxes are the heart – or perhaps they are better described as the brains – of modern computer espionage.

When bags of international letters were taken to a building on the Strand in the First World War, 300 examiners would then beaver away in a large, well-lit room. They would examine the address and then if necessary open and read a letter with reference to the current set of instructions – a watchlist of names and addresses – looking for a match. There were strict rules about the process the examiners were supposed to follow. The idea was to make this as 'mechanical' as possible to speed up the flow, while ensuring nothing was missed – the kind of repetitive action based on instructions that Alan

Turing had in mind when he first conceived of machines acting with instructions on data.

Today, big black boxes sit at internet and communications exchange points in America and around the world. These modern, specialised computers do the job of those First World War censors in an instant. They ingest information at a rate of 10 gbps, much faster than the poor people in the sorting office could ever manage. But they can also conduct Deep Packet Inspection to read the contents of the mail. A machine rather than a person can be instructed either to read the 'metadata', the address on the envelope of whom it is going to or who it is from. Or the machine can try to open it up and – if they can overcome the tricky task of reassembling distributed data packets – read what is inside. They can match the address against what are known as 'selectors' – the equivalent of the blacklists the human censors operated. If there is a match, then the message will be flagged up for the next stage, perhaps a person or perhaps another machine. A selector might be an email address or a phone number (in certain cases, it could be a keyword in the content or the signature of a malicious piece of code engaged in a cyber attack). In the First World War, certain items were immediately flagged – for instance, if they were in code or were suspected of using secret inks. The modern equivalent may be a message using encryption, taken by authorities to suggest it is hiding something. There is one other difference: in the First World War, cleared letters were placed in a tin at the end of the room to be taken away by Girl Guides. They are no longer needed.

In the days of the telegraph, Britain had savoured the global reach of its own companies, but also cannily encouraged other countries and companies to have new routes land on British territory scattered over the oceans. Remote, often barely inhabited rocky outposts like Ascension Island in the South Atlantic became crucial stopping-off points for traffic, providing Britain with a huge strategic advantage in intercepting communications. America's global intercept system relies less on traditional geography and more on the reach of its companies.

The installation of US taps in places like San Francisco was just one part of a broader programme of bulk access to communications, the origins of which pre-date 9/11 and which focused on foreign communications. The heart of modern global signals collection system lies in accessing the global telecommunications backbone through which data flows. Known as 'upstream' collection, it has three components. One is access through telecoms companies.[16] Companies (acting under legal compulsion) provide large-scale access to fibre-optic cables, switches or routers around the world allowing their traffic to be searched for selectors. One programme, codenamed Stormbrew, for instance, involved seven access sites at international choke points for global traffic which passes through the US, as well as two submarine cable landing sites in the US, one on the East and one on the West Coast.[17]

The Fairview programme dates back to 1985 and is described as being 'aggressively involved in shaping traffic to run signals of interest past our monitors' – implying that communications traffic is deliberately redirected so that it can be intercepted. Blarney goes back to 1978 and provides access to cables, switches and routers around the world (focusing originally on diplomatic traffic). The codename is perhaps a little joke by Irish-Americans since it is the successor to the Shamrock programme that collected cable traffic from the 1970s onwards. Why do companies co-operate? A mixture of reasons, ranging from patriotism to the hard-nosed knowledge that government is often a major customer (especially in defence). Two telecoms companies had contacted the NSA in the days after the attack and asked 'What can we do to help?' One company said it had noticed odd patterns in domestic calling records around 11 September 2001. Some companies were less willing, while a few wanted letters from the Attorney General – just as had happened after the Second World War. A 2008 law would provide new legal authority that meant that if the role of companies was revealed or challenged they could respond by saying that they were simply doing what was required of them. Licences for cable operators may also include requirements that if foreign operators buy up companies, then special cells of security-cleared Americans will continue to run secret programmes.[18]

A second method of access is when other governments hand

over the traffic or allow access to cables on their turf.[19] The NSA
would often build a facility that does the initial processing and then
share the results with the host. That country gets the benefit of US
technological know-how and massive computer processing power,
while the NSA gets its copy of the data for analysis. This is normally
done with the condition that neither side will use the information
to target the other. The NSA is said to work with thirty countries
including Germany, Israel, Japan, Jordan, South Korea, the UAE and
Saudi Arabia, the US often providing direct funding (the highest recipient being Pakistan, followed by Jordan).[20] These are known as
'third-party' partners. Second-party partners are those members of
the Five Eyes club who already share under their own rules.

The final method of collection is unilateral and clandestine
cable-tapping. This might involve covertly getting inside a country's
infrastructure to siphon off data. The US has built submarines with
special pods to tap undersea cables (although getting the information out was much harder). Overall, 'upstream' collection provides
bulk access to global traffic through the communications infrastructure, maximising the amount of communications that the NSA
can access and run its selectors against to look for matches (the US
argues this is not the same as actually collecting the communications in bulk). The selectors are usually things like phone numbers
and emails (not keywords like 'bomb' which would be too broad).
Upstream collection focuses on gathering information on non-US
persons based abroad as part of the search for different categories of
foreign intelligence. How much traffic does NSA access? Overall, it
says it 'touches' 1.6 per cent of data on the internet and only selects
0.025 per cent of that data for review, which is to say that analysts
only look at 0.00004 per cent of total internet traffic. The meaning
of figures is often opaque, though, much internet data being worth-
less anyway (consisting of music and films), and there are issues
over definitions: intelligence agencies often use the term 'collect'
to mean a person reading data rather than a machine siphoning it
into a database. But another programme made use of home-field
advantage differently – using the consumer power of American tech
companies and exploiting the way in which data had become the
currency of the information age.

*

The casinos of Las Vegas and Atlantic City and the credit card companies rather than the tech companies of Silicon Valley were among the first to understand the value of data.[21] From the early 1990s they were at the leading edge of a business trend called 'database marketing' or, as it later became known, 'database mining'. This grew out of direct marketing and a desire to understand customers by collecting information about their preferences. It has evolved into the modern concept of data mining or 'big data', which aims to extract useful knowledge hidden within large amounts of information. Casinos wanted to understand what got people gambling – identifying those people for whom a free room or a drinks voucher might lead to greater spending. Tracking someone's behaviour by collecting data meant a casino pit-boss no longer needed to remember their name or face to give them the special treatment. Instead a computer knew who they were and what they liked. But casinos wanted to do something else. They also wanted to look for suspicious patterns. So companies tried to build software to find blacklisted gamblers who had been banned but were trying to use fake identities. They also wanted to look for people who might be hiding a relationship with a member of the casino staff which they might abuse to manipulate the system. These early systems proved less capable of plucking people out of thin air than of investigating those who had already come under suspicion – a key distinction that would also exist when the practice transferred over to counter-terrorism.[22]

American Express was another early adopter of data. It bought large computers from a company called Thinking Machines to crunch through purchasing patterns and understand customer habits. From 1993, the UK division of American Express led the way in placing targeted offers on people's billing envelopes which led to an increase in card spending.[23] It might use computers to find the people who had bought jewellery on a business trip abroad in the last month, for instance. That was suggestive of a certain propensity to spend which could be useful to know about. This quickly became highly targeted and the company would sometimes send offers to as few as twenty individuals out of its millions of customers. At first it took days for computers to process all the data, but soon they speeded up and more and more companies like large retail stores followed.

Companies were learning about moving advertising and market-ing away from what was called 'non-productive reach' and instead directing it more towards a specific person's interests (dealing with the famous adman's saying, 'Half my advertising is wasted, I just don't know which half'). The big supermarkets realised the power and began to set up 'loyalty card schemes'. This would allow habits, preferences and the minutiae of purchasing patterns to be analysed in bulk. Promotions and discounts were the reward for handing over information which could help a smart company – with the aid of computers – to understand its customer base better. 'Consumers appear to be responding to the precision marketing,' *Businessweek* noted in 1994, 'but of course this private intelligence-gathering gives some people the creeps.' A credit card statement or a shop-ping bill could, it emerged, shine a light into almost every part of people's lives. 'This was a gold mine,' one executive said. The pro-cess of having computers churn through all the information was known as 'drilling down'. It was a world reliant on computing power and hungry for data – which might be bought and sold by markets and brokers – and which people willingly but perhaps un-thinkingly provided when they signed up for cards and filled in their details in return for convenience and discounts. Companies were now doing what the spies had done with large-scale traffic analy-sis of opponents' communications in the Cold War – building up a large enough data set to extract patterns and meaning and look for irregularities. In the Cold War, only the spy agencies had the com-puting power, knowledge and motivation to engage in this kind of large-scale computer-based data collection, and did so in secret. But from the late 1990s, as computing power moved out into the busi-ness world, it was becoming possible for others to follow suit and adapt one of the key intelligence collection techniques to the wider world. The spies would then wake up to what was happening in the private sector and the value of the data held not in their own secret basements but in the outside world. In Smiley's day, the information he needed to piece together clues was mainly in the hands of the government or on paper. Now it was in the hands of companies and on computers. At first this brought opportunity for spies. Only later would it bring tension.

As the new millennium arrived and the internet roared into life, tech companies would learn the value of data better than anyone because they were collecting more than anyone else, and one company came to epitomise this. Back in 1936, Turing had talked of algorithms providing a set of step-by-step instructions for how a machine should treat data. In the late 1990s Google powered into life on the back of its own unique algorithm to offer people a tool to search the ever-increasing mountain of information on the World Wide Web. The key was the ability to scan the entire web and then organise information and retrieve it based on an understanding of what people were looking for – ranking a page's relevance not just by what was written on it but its relational value based on what other pages were linked to it. This offered a means of establishing the relative significance of different types of data. The company's Stanford alumni founders, Sergey Brin and Larry Page, worked relentlessly to improve the system by studying user behaviour. Could you learn from people's searches whether they had found what they had been looking for and so improve the system for next time? Other companies like Amazon were also understanding that you could build a business around using customer data to make recommendations.

Google soon realised that the data it collected provided a unique insight into people based on their web searches. There were riches in this. In their desire to establish themselves, Google and other tech companies offered their services for free (perhaps intending to charge eventually). And consumers liked that and soon came to expect it. But someone had to pay for it all. And that was advertisers. And what the data offered was a chance to take the database marketing techniques to a new level in understanding consumers to target advertisements towards their interests. This was first done with the web searches, and the company realised it was gold dust. People's searches opened a window into their lives – their interests, aspirations and fears. The next step came when Google moved into developing its own email communications system. The company explored using its computers to scan people's communications to target advertising better. When a Californian politician proposed a law to opt out of such targeted advertising, she was approached by Page and Brin. She later recounted the encounter in a PBS TV documentary:

> All of a sudden, Sergey started talking to me. He said, 'Senator, how would you feel if a robot went into your home and read your diary and read your financial records, read your love letters, read everything, but before leaving the house, it imploded?' And he said, 'That's not violating privacy.' I immediately said, 'Of course it is. Yes, it is.' And he said, 'No, it isn't. Nothing's kept. Nobody knows about it.' I said, 'That robot has read everything. Does that robot know if I'm sad or if I'm feeling fear, or what's happening?' And he looked at me and he said, 'Oh, no. That robot knows a lot more than that.'[24]

This was the new world of computers automatically analysing the vast data flows of individuals – it was machines talking to machines. But what if the robot did not explode and some of the data could be retained? The more data you have and hold on to, the more you can do with it. This was becoming the business model of the new internet era. As time went on, other technological trends aided the process. The cost of data storage fell, making it economic to keep more information; the relative cost of computing power fell as well, making it economic to process it; and the places in which data was being collected expanded as our use of the internet and computers moved beyond just email and web searches into a much more integral part of our daily lives. A new world of 'apps' would emerge which would collect vast amounts of user data, both for their own functions but also for advertisers. As early as 2004 Google executives were reported as nervous about what would happen if the NSA realised how much they had data they could have access to.[25]

The fact that companies were collecting rich data on people was not a truth likely to be lost on the state. For a start there was the principle of it: if a company could automatically scan communications to sell advertising, why should the government not do it to protect national security? they argued. But there was also a practical point. If the companies had all this data, why not piggyback on their collection and try to get access to it? After all, the targets of intelligence agencies used the internet like everybody else. And so corporations would become the (not always willing) agents of state intelligence collection.

The first pillar of America's home-field advantage lay in the way it had built much of the infrastructure and controlled the digital world's piping and cabling – this facilitated upstream collection. The second lay in the fact that it was also American companies that people used to send communications and data through those pipes. 'The majority of known terrorist email addresses that NSA has tracked are hosted on US-based providers or foreign-managed providers hosted on servers in the United States,' a draft NSA report said.[26]

The volume of individual requests (in the form of legal demands) for data on foreign suspects served on companies like Microsoft, Google and Yahoo began to grow from 2001. Information from the accounts of suspects proved important in counter-terrorist inquiries in the UK, such as those of a known Al Qaeda operative, Dhiren Barot, picked up in 2004 as well as the men arrested for planning a fertiliser bomb the same year. But as the burden on companies increased so did their discomfort. This led to demands for a more streamlined process and new legal authorities, which came in the 2008 Foreign Intelligence Surveillance Act (FISA). PRISM was the result. This compelled companies to help the government acquire 'foreign intelligence information' on non-US targets believed to be abroad.[27] Where Upstream demanded access to the telecoms companies carrying global traffic, PRISM used the same provisions to demand help from the companies who provided the services that people used – companies like Microsoft, Google and Yahoo. PRISM was not a secret 'backdoor' way for the NSA to break into tech companies (although there was that as well). Companies insist they only allow 'front door' access – in other words legally enforceable demands for their users data. PRISM was in effect more of a discreet side entrance that the companies had to provide so that the NSA could enter, but without the public seeing them come in through the front (and it was the FBI who did the entering on behalf of NSA). The system was designed to facilitate the transfer of material that the US government was asking for on specific foreign suspects and to do so by increasing the level of automation in that transfer – in a sense by outsourcing some of the effort to where the data lay. This involved providing the companies with selectors – for instance an email address – which they had to run in house across their data, offering up the results. The companies insist they had to approve these requests individually.

The spies were tapping into the global ascendancy of American technology firms and the companies' hunger for people's data. This secret was kept incredibly tight even within the companies, with only a tiny handful of people aware (and they did not know it was called PRISM). PRISM did not just provide access to communications as they moved but also data as it was stored ('data at rest'). The trend in recent years has been to give your data to the big companies like Microsoft and Google to store and process on their servers (known slightly misleadingly as 'the cloud'). That meant this treasure trove was also becoming accessible. More than nine out of ten internet communications that the NSA acquired each year were obtained through PRISM.[28]

Caspar Bowden was not one of those in on the secret when he worked at Microsoft. He had fought on the British front of the 1990s crypto wars to push back against state control before becoming Chief Privacy Adviser for Microsoft. It was a role that involved advising the different National Technology Officers for the company who operated in individual countries. He had been one of the few to suspect what the 2008 legalisation might mean and became worried about how it might secretly be implemented when it came to a broad category of data belonging to foreigners that touched the US in some way. During an internal strategy conference in Europe in 2011, he says he voiced his concerns. 'If you sell Microsoft cloud computing to your own governments then this law [FISA 2008] means that the NSA can conduct unlimited mass surveillance on that data,' he told the assembled group. A senior executive 'turned green' and the room fell silent, he later recalled. During the coffee break, he was warned about speaking out. Two months later Bowden left the company. He would go on to campaign vocally on the issue.[29]

PRISM's access to the tech companies' data could provide

everything from surveillance of live email and chat through to files and photos and even real-time notification when someone logged in or sent a message. PRISM became the leading source of material for the NSA, accounting for nearly one in seven intelligence reports and, along with Upstream collection, contributing to more than a quarter of all reports on terrorism. As of September 2012, 45,000 selectors were going through PRISM, up 32 per cent on the previous year. The acquisition of data begins with the targeting of a known individual rather than trawls using key words to find suspects (forms of key-word searching are more possible in Upstream but not PRISM). But the scale of what was now possible was transformative. Together, Upstream and PRISM allowed the US government to target a much wider range of foreigners than was possible in the past and to do it with much greater flexibility. Those individuals did not need to be involved in terrorism but just have some knowledge of a suspected terrorist or fall into other broad categories (certified by a court) such as someone reasonably believed to be likely to communicate desig-nated types of foreign intelligence. The number of targets steadily increased from the passing of the 2008 law. By 2013, the programmes targeted 89,138 people.[30] This was home-field advantage.

The information collected by these systems flows into databases accessible not just by the NSA but also by the CIA and FBI (who can also make requests to 'task' certain selectors for collection). All of this data can be kept and then searched or 'queried' by ana-lysts. 'The NSA's intelligence analysts conduct at times complex queries across large data sets,' an oversight report notes.[31] Data in itself is not enough without the tools to analyse it and these were evolving. Hunting terrorists was less like breaking Enigma or per-forming traffic analysis against Soviet codes and more like the kind of spy-hunting of the Cold War, seeking out fragments and piec-ing them together. This process of investigating connections and looking for links has been transformed by computers. The prob-lem spies faced in the digital age was not a lack of leads but a tidal wave of data. What was crucial was finding tools that would allow you to find meaning within it all. In a Cold War spy hunt, like a police investigation, everything was done with pen and paper and charts up on a wall – the type seen in crime procedural dramas.

In a murder inquiry, these would allow you to visualise the array of suspects and contacts and how they might be connected to each other and the victim, often through billing records and other associations. Complex engineering projects like the construction of nuclear missiles back in the 1950s pioneered the use of what were known as PERT (Programme Evaluation Review Technique) charts to visualise complex networks and events over time. These were then adopted by law enforcement agencies in the 1960s to try to bring together all the data about an event (one of the earliest uses was by LAPD in the investigation into the assassination of Robert Kennedy). The charts could be up to thirty feet long, and a new piece of information would require rubbing things out and moving bits of paper around.

An officer from the Metropolitan Police brought some of these ideas back to the UK in the 1980s and thinking began about whether computers could be used to visualise this data. In 1990 a British company called i2 was founded in Cambridge. It used graphic drawing tools to display a spider's web of contacts and associations on a computer. This network analysis might reveal new leads to follow or unravel a complex financial web. Almost immediately after the company showed off its product at conferences and exhibitions, the intelligence community in the US and the UK approached it for their own version. Their eagerness came because their own tools were good at processing 100 million records but not at extracting meaning in a user-friendly way. Both MI5 and the FBI had been wasting vast amounts of money on databases and computer systems that had never really worked in previous years and would continue not to do so. The i2 system did not just reveal static connections, but looked at how they evolved over time: it could answer the who, what and when questions an investigator might have, although perhaps not the why. Crucially, in the mid 1990s, digitisation meant that phone companies were for the first time able to keep detailed call records for itemised billing. This provided a gold mine for police that would transform their work. i2 would become the equivalent of Microsoft Word for the intelligence and law enforcement agencies who could plug into their databases. The world of computer analytics was now meeting counter-terrorism.

By 2000 the system had been able to look for patterns, but processing still limited its capacity to do complex work against the
largest data sets. But by about 2005 the kinds of systems i2 had built
were able to test out more advanced algorithms and patterns – to
do what is now called big data. For instance, a computer could spot
that if A calls B then B calls C and C calls D and this happens every
Tuesday afternoon, there is a pattern. This kind of pattern analysis
in turn can be predictive. If the data says A has called B every Tuesday afternoon, then they are likely to do the same again. And if they
do not, that in itself might be revealing. In 2011, IBM bought i2 for a
reported half a billion dollars.

The idea of correlating fragments of data and looking for patterns
was explored immediately after 9/11 in a controversial project run
by John Poindexter, a former US National Security Adviser. It had an
ominous name: Total Information Awareness (TIA). This was different from the initial post-9/11 NSA system, which was based around
getting hold of big troughs of domestic data like phone records and
then querying them with specific numbers ('link analysis' – uncov-
ering associations and connections). What Poindexter had talked of
was data mining – using algorithms to wade through combined sets
of data to find patterns and anomalies. This offered the possibility
that you could make predictions about behaviour, rather than just
follow a known trail, to find unknowns.[32] Poindexter had talked of
a 'Manhattan Project for Countering Terrorism' based on the idea
that terrorists emitted a characteristic data signature, like the acous-
tic trail a Soviet submarine might leave as it tries to move silently
through the oceans. TIA involved bringing together records of all
kinds of domestic data – like credit card billing and travel patterns
– which would be correlated to find a 'signature', for instance, what
kind of travel patterns might typically constitute reconnaissance by
a terrorist cell. These could then be applied back to the vast mass of
data to find others like them. The processing power did not quite
exist when it was conceived, though, and – crucially – it was not
clear that the signatures were going to be possible to find. The project was quickly junked after a public outcry over privacy when it
was revealed in 2003.[33] Poindexter resigned. TIA died but the idea
behind it had not.

One former NSA official recalled going to hear General Keith Alexander talk when he was running the army's 'Information Dominance Center' in the early 2000s. 'He had all these diagrams showing how this guy was connected to that guy and to that guy,' the former official later told a journalist. 'Some of my colleagues and I were skeptical. Later, we had a chance to review the information. It turns out that all [that] those guys were connected to were pizza shops.'[34] But slowly the spies learnt the lessons. After Alexander became head of the NSA he would push harder.

It is a lot easier to build a network chart of someone's contacts if you can start with their Facebook profile. This is all a reflection of the new world in which so much valuable intelligence is 'open source' – in other words it is not secret or classified. Its power comes when it is cross-referenced with other data – just as it was in the days of Cold War spy-hunting. The difference is there is much more data and it is computerised and therefore easy to collate (although one challenge was that it was in the hands of the private sector). Analysing all this open source information is almost a new intelligence function in itself but has fallen to the secret intelligence agencies. Is accessing this material spying or not? A rich picture could be built up on someone without ever having to do anything that used to be considered spying. But if you take open source and add in secret information, then you have even more. In 2010 the NSA received permission to build analysis of American people's contacts through combining its metadata with other publicly and commercially available sources (like Facebook or bank details), allowing domestic contact-chaining as long as there was a foreign intelligence purpose. Previously this had been barred as too intrusive.[35]

At first computers had struggled to cope with the sheer volume of data on offer. The link charts that came out the other side were meaningless spaghetti rather than anything more instructive. The NSA again learnt from the tech companies in dealing with the mass of information. The ability to search across multiple databases and combine the results in a meaningful way was the next crucial step. From 2007, after in-house attempts proved less successful, the NSA began using a distributed system (Accumulo) that could analyse trillions of data points from different sources. It could then look for connections and establish the significance of these. This was based around Hadoop, which grew out of Google's work and was used by Facebook and other companies – Visa had used it to reduce the time needed to process 73 billion transactions from one month to just thirteen minutes. And so the idea of finding meaning in the vast sea of data began to turn into something more than an elusive dream. Spying was meeting the new world known as big data.

CHAPTER SEVENTEEN

BRITAIN

The first reports on the morning of 7 July 2005 were that a power surge was responsible for the people emerging coughing and blackened from smoke at underground stations in central London. It took only a few minutes to understand this was something far worse. The roof of a bus was ripped off by an explosion in Tavistock Square, the last of four bombs wreaking carnage on the capital's rush hour.

At MI5, Eliza Manningham-Buller gathered her staff in the atrium of their headquarters and told them their worst fears had been realised. They should brace themselves for questions as to whether they had missed anything, but they also needed to get on with finding out who else was out there. Over at Cheltenham, Iain Lobban returned from London to GCHQ. 'I walked back into this building,' he later said, 'and everyone was looking at me to see was I relaxed or confident, resolute, anxious, nervous, biting my nails, frowning, walking with my head down a hundred miles an hour.'[1]

For the first few days, no one had any idea who had carried this out. The speculation was that a team had come into the country to carry out the attack. And most likely they had left their explosives and escaped. But as the police combed through the fragments found at the bomb sites, they came across pieces of personal identification and other clues that suggested a different possibility. It emerged that the perpetrators were four Britons willing and able to be suicide bombers in their own country. No one had predicted or prepared for this, including the spies. For all the trauma their American allies had experienced on 9/11, it was only on 7/7 that the reality of what they were facing hit home for Britain's spies. Then two weeks later it nearly happened again.

The failed attack on 21 July was, intelligence officials of the time say, almost more terrifying than the successful ones two weeks earlier. Suddenly they were faced with the possibility of wave after wave coming at them. There was also the problem that the 21 July bombers had escaped after their bombs fizzled and failed to detonate. The fear was they would strike again. A manhunt ensued. In its initial stages it went terribly wrong when an innocent Brazilian was shot dead on the tube by police who believed he was one of the bombers. But after that, technology played a key role. Crises often lead to new capabilities being deployed. The police and MI5 needed to find the suspects but they had abandoned their existing mobile phones and practised good operational security in their communications. The authorities believed, though, that they might be in touch with each other or other contacts. So how could they find them? The answer lay in contact-chaining phone records – the technique the US had begun using after 9/11 with its bulk collection programme. This required access to records of UK phone calls on a large scale (the 1984 Telecoms Act likely provided the legal authority). Officials will not comment, but this is believed to be the first time such bulk analysis was used. Investigators began hunting for possible contacts. This led directly to working out what phones the men were now using within twenty-four hours. Two of the men were tracked to a flat in west London where they were arrested on 29 July. Another was arrested in Italy.

The attacks had hit home in a different way from the US. In America on 9/11, the hijackers were foreign (mainly from Saudi Arabia), some of them were known to US intelligence and they had travelled into the country to carry out the attack. That led to the conclusion that what was needed was to join the dots between domestic and foreign intelligence. In Britain on 7/7 the men were, to use the unpleasant jargon, 'home-grown' (although they had travelled abroad for training). They were not quite what were known as 'clean skins' (since two of them had been on the periphery of another counter-terrorist investigation the previous year), but no one had spotted how dangerous they were. So while the first question was the same as in the US – how many more plots and attackers? – the answer to the second question – 'how do we find them?'– was different. This was a problem of finding people within your own community rather than people who had come into your country from outside.

The twin attacks had led to a question: how did you find Britons willing to carry out such acts? To answer the question a new, powerful and secret capability employing data, telecoms and computing would be built at GCHQ in the coming years. It was classified to the highest level, with only named individuals knowing its full extent and authorised by the Prime Minister. One of the revelations that surprised British investigators who looked into the background of the four suicide bombers was how much of their lives had been lived online. In the confines of the 'Doughnut' at Cheltenham, GCHQ staff realised that theirs was an organisation that understood the bits and bytes of technology and the internet, but not necessarily how the world outside, and especially a younger generation, actually used it all – how they messaged each other, how they visited websites and used social media, how they planned and organised their lives. The irony is that for a technological organisation, GCHQ was pretty bad at understanding the way the internet and technology were integrating into people's lives and transforming them. That was because no one was allowed internet access at work or even to bring in a mobile phone (one veteran describes jumping the first time his mobile rang after he left government service as he had never heard it make that sound before). To find their new targets, the experts believed they needed to understand the way they lived their lives and find patterns and connections. Two of the bombers might have been to Pakistan, where they were trained and tasked. But how had they got there? It looked as if the initial radicalisation and recruitment that drew them over there had taken place on the internet. So to find people like them you needed to look online. What was wanted was a system that allowed you to investigate the richness of the trail people left in their digital lives – even perhaps to build a pattern of terrorist behaviour and then ask a computer who else matched it.

Could you take a huge pile of data and ask it: who was communicating between Britain and Pakistan? That sift would take you down from the whole population to still potentially tens of thousands or hundreds of thousands of people. But then ask: 'who of these people has also visited certain extremist websites?' – now you were down to perhaps a few hundred. And who of those had social connections to known extremists? Perhaps now you were down to a handful. This could all be done by looking not at the content of their communications but types of metadata. Having narrowed the list down to a few, the names of any people in the UK on that list could then be handed over to MI5 for investigation and correlation with their existing files. GCHQ remains foreign-facing while MI5 leads on investigations at home and would go to a government minister and ask for a warrant to intercept the actual content of those people's communications to establish if indeed they were involved in terrorism.

Stovepipes were broken down between the three British intelligence agencies – GCHQ, MI5 and MI6 – so that small joint teams would work together on terrorist targets and share leads. In the past, GCHQ was highly protective of its intelligence, fearing its use by the others could compromise the source – MI5 staff used to call the GCHQ officer dealing with their requests Dr No because of his predictable response to any request. But that changed, although GCHQ's role was often hidden. 'Quite often other people get credit for our work and we smile wryly and get on with it,' Iain Lobban said in a 2010 interview. 'Whether it be a very clever sniffer dog. Or a sister agency. Or an overseas intelligence agency. You just smile and think that's the price for staying in business sometimes.' Intelligence officials are adamant that what they built worked, and point to the fact that in the nearly ten years from 7/7 to the start of 2015 only one person was killed by terrorism in the UK (Lee Rigby in Woolwich in 2013).

The point of computers – as in the past – was not to replace people, but to be able to reduce the workload so that it becomes manageable for a finite number of people to focus their time on the priorities. 'Because I have a certain number of analysts . . . the more I can throw away the better,' Lobban argued in 2010. Computing power made this possible to a new degree.[2] GCHQ remained foreign-focused but, like the NSA, was drawn into the connections between Britain and overseas – after 7/7 specifically people who had trained or been in touch with Al Qaeda in Pakistan. A decade later it would be the same for Syria.

During the war Bletchley Park used data and decryption to find German submarines cruising the Atlantic looking for their prey. Now it was using data and decryption to hunt for terrorists hiding in the online space that everybody inhabited. And to perform this kind of data analysis, you need data. And not just some of the data. The point was that if you want to do it effectively, you need as much data as possible. The analogy of finding a needle in a haystack is sometimes used to explain the idea of finding a terrorist hiding in the population at large. But in some ways this does not explain how data sifts work. If you were really looking for a needle in a haystack then you would want as small a haystack as possible to increase your chances of spotting the needle. But in data mining to discover unknown targets, you want to have as much hay as possible – firstly to be sure that your target is in there and not in another data set, and secondly to make it easier to do the kind of pattern analysis that helps you understand what normal looks like and what makes a terrorist look different to everyone else, even when they are trying their best to blend in. The problem comes when, to totally over-extend the analogy, a needle is doing its best to look like hay and so you need to know how needles behave as opposed to hay, so that you can get better at spotting them over time.

The building of this capability relied on a number of key developments. One was the legal authorisation to undertake this kind of work and get hold of the data. Spies in the UK and the US have been careful to get legal cover for their work, but the histories and frameworks of the two allies are different. In 1763, an angry King George III issued a general warrant. Its aim was to find the authors, printers and publishers and anyone else involved in publishing a damning critique of the King's speech opening Parliament. Searches led to forty-nine people being arrested including John Wilkes, the radical author of the critique.[3] The case proved controversial because, rather than issue a warrant to search for a specific person, a blanket warrant had been used to find persons unknown. In modern parlance, this was the difference between targeted surveillance involving intrusion against a known suspect and a broader trawl of everyone. The controversy was most keenly felt in the British colony of America. Similar general warrants were used to trawl houses and premises for goods on which tax had not been paid to the Crown. After the revolution, this led directly to the Fourth Amendment to the Constitution that prohibited the new state from unreasonable searches and seizures without 'probable cause'.

Constitution-less Britain never saw quite such a restriction on general warrants and they continued to be employed, sometimes with mixed results. In the mid 1840s a secret parliamentary commit-tee found that six or seven individual warrants were in force to open people's mail, but on average two general warrants a year were being issued.[4] General warrants formed the basis for the First World War intercept system in the hunt for German spies. These covered categories of communication, sometimes as broad as anything 'sus-pected to contain matter of a dangerous tendency'. At times this could mean any letter addressed to France, Flanders or Holland, or even an envelope suspected to contain such a letter. General war-rants tended to be used to comb through communications to try to find people who might be German spies. Individual warrants were used against those already suspected. In the era of pervasive com-puting and data mining, these distinctions could also be applied to finding terrorist suspects. General warrants would allow bulk access to data travelling along the global cable system and a clause of the 1984 Telecoms Act would allow a minister to demand a network provider carry out any act deemed necessary for national security – and keep it secret. Forms of bulk access have actually taken place in the past – for instance, the blanket collection of calls between the UK and Ireland in the search for intelligence on the IRA during the Troubles in Northern Ireland (when briefed about it at the time, one American told a British colleague the NSA would never have been able to do that). But this had never been done on so large a scale and never combined with the kind of computer analytics that became possible after 7/7.

The other development was technology. It was only in the year or so before 7/7 that systems first became available to carry out the kind of data mining and analysis that was wanted. The initial work to master – rather than be defeated by – the internet had started in the late 1990s under a programme called SINEWS. It involved mapping the path of data around the world and using advanced algorithms to work out how then to piece together individual packets so you could reconstruct a message. Those involved say that this involved a 'Bletchley level' of ingenuity over many years. Around the time of 7/7, GCHQ then undertook a huge modernisation process (known as SIGMOD). This took up a 'significant proportion' of the entire UK intelligence budget. The results were staggering. Over two years from 2005 it delivered a twenty-fold increase in GCHQ's ability to access, process and store particular types of communications.[5]

This involved returning to the First World War business of cable-tapping. But this time it meant plugging into those fibre-optic cables that landed in places like Cornwall carrying international data. A splitter could now be attached to a fibre-optic cable to divert the light into a box and a probe or bearer could then take it onwards, as found in the American junction points. Unlike satellite communications, where you could just put up your own dish to pull down the signals, with cables you needed physical access. As in the past, this was done with the co-operation of telecoms companies with obligations under law (although willingness to work with the system may still vary from company to company).[6] This provided a fire-hose of data to work on. It particularly focused on Middle East, North African and European data – because this is where the data transiting Britain through cables was heading.

The 'take' of the intelligence systems was staggering. One (leaked) presentation from August 2012 said that over the previous five years, GCHQ access to 'light' (meaning the fibre-optic traffic) increased 7,000 per cent and the amount analysed and processed by 3,000 per cent.[7] One document talked of receiving upwards of 50 billion events (the many different outward details of a phone call or email being sent) per day and noted this was growing.[8] The legacy of Britain's imperial telegraph system meant a significant part of the world's cable traffic still flowed underneath those Cornish beaches. How much? A lot passes by the censors, but certainly not everything. Even against critical targets, the coverage of their communications is normally only partial. Tempora – GCHQ's filtering system – does not have access to all traffic. At one point it had access to around two hundred internet links but only the capacity to analyse forty-six at a time. And the system is only able to process a portion of that. But it appears that the UK may still have access to a far greater amount of traffic from cables than the US, thanks to geography (although the US has the advantage when it comes to data held by its companies).

The size of global data flows still poses a problem as well as an opportunity for spy agencies. It is the equivalent of placing a fire-hose to your mouth and trying to take a drink. The volume is enormous. If you are after something specific (as opposed to broad data mining), how do you find what you want in that torrent? The trick is to filter. Some of this can be done automatically (getting rid of all the video and music being streamed and downloaded). That still leaves a lot. But then you set up filters to reduce it down further and further to the point where it is manageable. Spies say their aim is to throw away as much as possible and claim this is no different from the days of Bletchley when people out in stations around the world would listen in to huge amounts of radio – most of it innocent – to find the German signal which could then be sent back to Bletchley for decryption. The difference is that now this can be done with smart programming through computers. 'We access the internet at scale so as to dissect it with surgical precision,' Iain Lobban would argue. 'Practically, it is now impossible to operate successfully in any other way. You can't pick and choose the components of a global interception system that you like (catching terrorists and paedophiles), and those you don't (incidental collection of data at scale): it's one integrated system.'[9]

Tempora slows down the fire-hose of global data flows by buffering some of it. The content is reported to be kept for three days and the metadata for thirty days, which gives analysts time to search through it.[10] This is done by using 'selectors' – say an email address or a phone number – which can be searched for within the buffer of data so that the relevant messages are then kept. Tempora is said to provide ten times more information than the next-largest database, according to leaked documents. A portion of traffic can also be searched through with bespoke queries based on three or four terms – 'Show me anyone in Russia communicating with Iran using encryption,' or 'Find anyone using a particular type of encryption and who is also searching for information about bombs or visiting extremist websites'. According to reports, GCHQ was running about

40,000 selectors and the NSA 31,000 across one data set.[11] The black box might be in Cornwall, but these can be added remotely by an intelligence analyst sitting at his or her desk hundreds or even thousands of miles away. A search needs to be proportionate: pull up too many results and it cannot be justified. As well as legal limits, there is also the simple restraint of manpower to work through the results. In a world of bulk intercept and computing power, the need for human eyeballs and ears has become one of the key constraints and restraints on spying. Millions of communications pass by GCHQ sensors and through its systems each day but only a few thousand are examined by analysts who are presented with an index – like the results of a web search – from which they pick the items that appear relevant. These can then be read or stored.

Modern communications intelligence systems do not always rely on reading the content of someone's phone call or email, but often on analysing the metadata – defined sometimes as the information generated as people use technology (formally called communications data in the UK). In communications this would be the fact of a message going from one person to another, and when, but not what was written or said. When it comes to bulk access as opposed to targeted intelligence, collecting metadata can be more useful than content. This is partly because it is much easier to store than content due to its size, but primarily because of its relatively defined or structured nature. It is what is known as 'machine-readable' – in other words, machines can process it with their algorithms. If someone presents you with a whole stream of content data, it will have all kinds of things mixed up together (video and music on different formats, for instance). Some of it might be encrypted. Also the language, even in written content, may include slang, nuance and code-words which computers struggle to understand. But metadata is much cleaner and simpler and therefore much easier for a computer to sort through and analyse. Increasingly, even though it is not as revealing as content, metadata itself is rich enough to answer many queries. Listening to phone calls requires real people and takes up a lot of their time. Analysing metadata to look for connections takes a computer a moment.

The controls covering metadata are also lighter. A general warrant allows bulk collection of traffic moving internationally which can then be sifted, a process often known as 'trawling' – the equivalent of a fisherman sweeping up everything in his vast net. Some domestic traffic may well be swept up since so much of the data goes abroad (for instance, to a Google server for a web search). This then becomes available for metadata analysis. However, the next stage of actually reading the content of someone in the UK's communications still requires an individual warrant or similar authorisation, personally signed by a minister. The lighter controls on metadata reflected the era in which noting down whom a letter was being sent to was seen as far less intrusive than opening it and reading the content. But modern metadata can be much more revealing about a person, since people communicate much more and leave more of a digital trail in their lives which can be reconstructed using metadata alone.

Everyone recognised this was a hugely powerful capability. Senior officials decided to limit how they would use this powerful new tool, resolving initially only to use it for terrorist-related searches (although the same global collection system is also used to search for traces of cyber espionage). Analysts who accessed the database were given instructions on how to use it. Any misuse might be initially punished with a serious warning and a fine; a second misuse would lead to an analyst being marched out of the door. Any search leaves an audit trail that can then be examined as part of the oversight process (including by an independent commissioner, normally a judge).

Targeted surveillance – following someone you know – was now joined in the computer age by 'target discovery' – finding a target you do not know but believe is out there. Terrorists do not always fit a profile, and seek ways deliberately to ensure they do not, so there is a technological cat and mouse game to try to find those in hiding amid all the noise. The desire for target discovery has driven this system – the fear of the unknown in the era of international terrorism and the desire to find terrorists before they do something. And those targets have become harder to find, spies claim. For instance, a large group of plotters planning something ambitious like 9/11 offer many opportunities when they talk to someone already being investigated or do something suspicious. But so-called lone wolves radicalising themselves online leave few traces – except perhaps in their online activity. The politics surrounding terrorism means the pressure on the state and intelligence agencies to prevent attacks is enormous (and the consequences of failure high), but the complicated issue is how far to go in trying to reduce risk. That is a profoundly political question, complicated by the fact that the capabilities involved have often been kept secret for fear of compromising their value by exposing them.

Some officials say that the new capabilities are valuable as part of a tool kit rather than as a silver bullet to replace other traditional techniques like running agents and human or individual technical surveillance. They say that they have been most useful for establishing connections and unravelling a network once you had already identified someone rather than finding complete unknowns. Digital intelligence, they say, can be used to narrow a large group down to those you really should be focusing on so that in turn you can target finite resources on them and not others. The problem on 7/7 was, after all, not that the men were truly 'clean skins' but that two of them had not been sufficiently highly prioritised for further investigation after a previous plot and their changes in behaviour over time had not been understood. Prioritisation is vital since MI5 has only a limited number of surveillance and investigative teams to deploy. 'We are not the Stasi, we can't cover everyone,' one British intelligence official said in the wake of 7/7 to explain what had happened.[12]

The rich trail of data and communications – and the power of computers – was changing spying again. As well a new way of finding targets, it offered new ways of following those who had already been identified. In the past, targeted surveillance was labour-intensive: following someone home or listening to their phone calls was possible but took a large team. Now, with a click of a mouse an analyst can carry out a form of electronic surveillance in an instant, capturing the data surrounding someone's life. If they are British, this still requires a ministerial warrant, but if they are foreign just the tick of a box on a computer screen to justify the need from a pull-down menu of options relating to the Human Rights Act. The vast data trail of modern life, the 'digital exhaust' that we all leave as we move around online, can then be accessed. The state is also able to acquire what are called 'bulk data sets' to use in its analysis. This could be a list of everyone who has a fire-arms or pilot's licence, for instance, but could potentially be much broader as well since some of the data sets are said to cover millions of people. These can be acquired through overt means (a demand to a company or public body) or covertly (effectively stealing them, especially if they are located abroad) and they could be acquired and retained with relatively little authorisation and, until revealed in 2015, no real oversight.

A person's digital exhaust is almost unique, meaning they can be identified unless they are very skilled at hiding. Mobile phones – which are more tied to a person than a PC – have aided that pro-cess enormously. These are usually tied to an individual in a way a computer, which can be shared in an office or internet café, is not. In the past, putting a tracking device or a bug in a car or a house was a risky task. Now, people carry just such a multi-purpose device around themselves that can provide many more details than just their location or what they are saying. Using tools, analysts can build up a digital fingerprint of a target based on how they move online and what devices and identities they use. In turn this can be used to track them in real time, much as a surveillance team would follow them around in the real world – just with a lot less people and a lot more computing. In the past, it would take dozens of people to follow one suspect around all the time. Now, electronic coverage can provide something close at a fraction of the cost. Again, com-puting has lowered the practical cost of surveillance. This is just one of the ways it has transformed spying.

In the late 1990s, an old-hand cryptologist made the thirty-five-mile drive from the NSA to go to work at Langley for the CIA. He found a different culture. He noted that NSA staff liked to think their work was 'cleaner' compared to the 'moral ambiguity' of the CIA's efforts, which involved manipulating or blackmailing people to betray their own countries and break their own laws. 'The intercept floor resembles a laboratory or high-tech "clean room" with lots of gizmos and people in clean military uniforms listening intently to radios. It is far from a back alley in a dirty foreign capital that reeks of discarded vegetable peelings,' he wrote. It meant sitting in a station bristling with antennae and never meeting your target face to face, just perhaps listening to them. 'On occasions when I have questioned NSA audiences, most refuse to believe they are engaged in spying.' But that officer noted that the days of 'splendid isolation' for the two sides were passing. They were going to have to work together. 'Moral ambiguity, meet Mr Clean, meet your saviour,' he told the CIA.[13]

Computers and data have changed the dynamic between GCHQ and MI6 and between communications and human intelligence as a whole. The techniques of target discovery and digital intelligence can now be used to support what might be thought of as 'old-fashioned' human intelligence. There was a point, former British spies say, when MI6 looked lost and a little vulnerable in the new digital world and feared it might be washed up. In the old days, if a document had been sought then they would have been the ones tasked with getting hold of it – recruiting a spy in the form of a diplomat or a clerk, perhaps in the foreign ministry of the target country. But from the 1990s onwards, GCHQ was often able to deliver the document far more easily. And perhaps even offer previous drafts of the document as well while they were at it. What makes cyber espionage so powerful was the ability to extract information on a totally different scale to old-fashioned espionage and with a lower level of risk. Spies like Kim Philby and Oleg Gordievsky had to risk their lives to carry out secret documents and hand them over to their controllers at clandestine meetings. Now someone can sit in London or Moscow and download thousands of pages of material in a few moments with little risk (although the most sensitive material that Philby or Gordievsky had access to is unlikely to be online).

The saving grace for the old human spies was that access to the hardest targets still often required a human agent. Sometimes it was because you needed a human agent to enable the act of cyber espionage to steal a document. Perhaps a person was needed to get into the foreign ministry to plug in a USB stick and gain access to the classified network that was air-gapped from the internet. In these cases MI6 might become the enabler of GCHQ cyber espionage operations by providing the human access point into a network.

Sometimes people were needed, though, because only a human being and not a computer held the answers you were after. This can apply even to technical tasks like finding out who was behind a cyber attack on your country that had been routed around the internet to cover its tracks. A spy inside the PLA might still tell you who was behind it. A human spy may also provide the first lead, which is then acted on with the communications capabilities – providing a suspicious name or number or perhaps a profile of someone that can be fed through the databases and systems.

Computers and data increasingly helped MI6 find its targets. GCHQ in the past was often the enabler for human intelligence operations by, for example, providing intercepts of the phone calls of someone that MI6 might be trying to recruit so they could learn about them and perhaps their foibles. Now computers might provide a richer trail of data to identify those worth approaching, so you can work out how to persuade or pressure them to spy. This can even be automated to some extent. MI6 might want to get into the Iranian nuclear programme and so may use computer databases to ask who has access to the programme through contractor companies, what their links are to the outside world (including any to the UK), whom they know on social networks and how they might be contacted. Having run that through global data mining, a small group of people might be found who fit the profile. Alerts could be then set so that if any of those potential targets are travelling out of the country or checking into a hotel room, then the human spy service can be informed immediately and can be waiting at the hotel in order to approach them. In this way, computers have allowed human espionage to become more targeted. In March 2015, the Director of the CIA announced a major shake-up of the agency that involved integrating cyber and digital intelligence much more closely with traditional human intelligence operations. Our online and offline worlds are merging to the point where distinctions will become increasingly meaningless, including in spying.

Computers were changing intelligence by making it harder to hide. Osama bin Laden managed it for years, but only by never connecting to the internet and using human couriers physically carrying USB sticks. Computers and digital surveillance can help spies find their targets, but of course the converse is also true. It makes it easier for other countries to find your spies.

In October 2011, German police commandos burst into a family home in Marburg. They found a Russian deep-cover spy (known as an 'illegal') communicating with Moscow. The woman promptly fell off her chair in shock. What was so telling was that her computer was receiving encoded data over a shortwave radio frequency rather than through the internet.[14] In other words, the Russians knew that connecting up to the internet might give them away. The Russians were also reported to be returning to typewriters in their embassies. Old-fashioned techniques like secret ink are making a return, and British spies talk of having to 'go medieval' to hide. This was a sign of how the techniques of data mining and bulk intercept can be used against spies and not just by them. Spy services are all too aware that data can be used for counter-intelligence. Italian investigators found CIA officers involved in a rendition operation in 2003 by combing the trails of things like phone records and hotel bookings they had left in Milan in order to identify who had been involved. Dubai police surprised many spy agencies by using similar techniques and CCTV to find a Mossad hit team which killed a Hamas operative in a hotel room. Open-source, publicly available data was also used to track the movements of CIA planes involved in rendition operations. Other countries will be able to spot a British MI6 officer working undercover much more easily now by looking at the plausibility of the digital trail he or she leaves behind. A faked birth certificate or passport might have been enough in the old days to build a back-story (a 'legend', in the parlance of John le Carré and the Cold War), and an officer could travel into a country on one passport and then assume three different identities in one day to meet three different agents. But now he or she might have to get through biometrics and database checks to enter a country, and their fake identity will need to have a presence online with its own history. Without that, judicious use of a search engine can be enough to show that someone is not who they say they are or that, suspiciously, they only appeared online recently. Spies used to be paranoid about their personal details emerging in public. But now, the very fact that you have no personal data trail may well mark you out as a spy. When MI6 ran a test to see how long an officer's cover could stand up against Google and a suspicions foreign state, the answer was about a minute. Using more advanced data mining, spy-hunters will be able to find 'unknowns' by looking for behaviour among the masses that exhibits the typical signature characteristics of an undercover operative. Real spies have always sought not to stand out like James Bond but to blend in and be the 'grey man'. That will require a different skill set in the future. Spying has been enabled but also challenged by computers. A digital data trail works both ways. Tradecraft online is as important as offline. Spy services in the future will rise and fall, succeed and fail, based on their ability to master these techniques to find their opponents and hide themselves in the digital world.

So what of the special relationship founded at Bletchley after the journey over the Atlantic? Britain was the senior partner during that war, but the scale of American resources – expressed in money, people and computing power – was confirmed from the start of the Cold War and persists beyond it. The US intelligence budget in 2012 was $54 billion; that of the UK was about $3 billion. 'They are obviously much bigger than us,' Iain Lobban said of the NSA in a 2010 interview, 'in terms of money, in terms of numbers of people. They rate our geography, our people, our expertise, our analytic contribution. And we are not an uncritical partner as they are not an uncritical partner. If we think they have got something wrong we will say so.' He argued that the relationship was based on sharing where it was possible to do so, but with limits based on international law. Are they equal partners? 'I think we hold our end up,' he replied. In the high-end maths of cryptanalysis that Turing once pioneered the relationship is still close, but always with an edge of insecurity in Britain, knowing that it has more to lose by the relationship weakening.

The geography of Empire and its legacy used to provide Britain with something to take to the party: Hong Kong, for example, had been of strategic importance for the interception of communications until it was handed over in 1997, and the overall importance of other foreign satellite and radio intercept stations diminished. In the early 1990s much of the intelligence that came out of GCHQ actually came from the US. But then the fibre-optic cables began to be laid, and here Britain continued to have unique access thanks to the global web that still followed many of the paths of the old imperial system. This is one reason why a British base at Bude in Cornwall was modernised, in part with millions of pounds of American funding. Thanks both to the history and geography of the cable system, as well as its different regulatory environment, Britain had access to more global traffic than the US. 'We are in the golden age,' one person from GCHQ noted.[15]

The Five Eyes alliance had deepened, dividing up the world for coverage. In the Cold War this might have been in order to look for the signals of Soviet submarines; today it is for data.[16] UKUSA made it clear there would be a presumption to share, but some categories of information could be withheld (this included intelligence Britain was collecting on Ireland and Northern Ireland during its struggle with the IRA). Officials maintain that members of the Five Eyes club do not spy on each other as a matter of course (although there are provisions to do so if deemed necessary for a country's own security) and are not allowed to evade their own country's restrictions by asking another to do something it is not supposed to do. Leaked draft documents do indicate, though, that the countries retain the right to break the agreement in circumstances where they feel it necessary. (A leaked diplomatic cable suggests that UK intentions and activities at the United Nations with regard to International Atomic Energy Agency [IAEA] policy towards Iran were on a human intelligence collective directive.)[17]

There are still cultural differences. A former Western intelligence official described what he calls a 'typical' Five Eyes meeting. 'The Americans look a bit bored. The British try and sound attentive. The Aussies get drunk. The Canadians are normally reeling from the latest scandal and the New Zealanders are desperately scribbling notes.' However, that person remembered that everything stopped when General Keith Alexander from the NSA talked. Everyone paid attention then.

When he became head of the NSA in 2005, General Alexander, a hard-charging army man, was already deeply versed in technology and what it could do but he would become a hugely powerful advocate for bulk collection and analysis. A brutal insurgency was growing in Iraq as he took the helm. American casualty figures were spiralling as soldiers were being killed or having their limbs blown off by improvised explosive devices that seemed to be everywhere. The traditional role of the NSA in supporting combat forces came under pressure, with commanders saying intelligence was too slow in reaching the people who needed it. Alexander responded with a new programme which sucked up all the communications metadata flowing through Iraq. This might allow you to find an individual target, but it also allowed you to do a lot more. 'Rather than look for a single needle in the haystack, his approach was, "Let's collect the whole haystack,"' a former senior US intelligence official told the *Washington Post*. 'Collect it all, tag it, store it . . . And whatever it is you want, you go searching for it.'[18] This allowed you to perform pattern analysis across the data.

The NSA developed techniques for analysing all cell phones that fitted the particular profile or signature of a phone typically used to detonate improvised explosive devices. The location of these phones could then be plugged back into the bulk analysis system to show where the bombs might be. The analysts next realised that if they found a cluster of phones all showing the same characteristics, then they may have found a bomb factory. This type of intelligence was pushed down much faster to military teams on the ground so they could avoid bombs and take out the factories, reducing the death toll. 'We successfully reduced that disconnect from sixteen hours to around one minute,' Alexander later said.[19] NSA computer hackers were also employed to get inside insurgent computer networks to disrupt their plans and deceive them, even to try to work out where videos of attacks might have been uploaded from using new forensic tools to try to overcome the attribution and anonymity problems associated with the internet.[20] The NSA's support extended to the most controversial tactic of counter-terrorism: drone attacks. 'We track 'em, you wack 'em' was the motto adopted by one NSA unit to describe its role. GCHQ also began to adapt to supporting the military, particularly in Helmand, Afghanistan, deploying its staff to work close to the front lines where they could supply tactical communications intelligence as fast as possible. For the NSA, the drive was to apply the same techniques more broadly – getting it all, so you could see what you could do – in terms of bulk collection and analysis. Alexander in particular was said to be a man who liked to push to do more.

The Cold War traffic analysis was designed to prevent the surprise of a nuclear attack. But counter-terrorist analysis is much harder. You are looking not for a signal sent from a known commander to missile silos telling them to launch. You are looking for perhaps an email from an Al Qaeda commander in Yemen telling someone in east London in guarded language to move ahead. Or perhaps a man in New York downloading bomb-making instructions and posting his intent on a closed web forum. Terrorists are more adaptive than the old Soviet military as they seek to hide. What this means is that the kind of assurance that Cold War signals intelligence provided – and for which a huge enterprise was built – is not easily replicable in looking for the modern terrorist equivalent of a 'launch signal'. And even just trying to replicate that level of assurance requires gathering many more signals, which are now blended with those of ordinary people in the vast tide of the internet, including those of your own population. That has implications which societies desperate to stop terrorism are only just thinking about.

Avoiding another 9/11 or another 7/7 was an imperative for spies and their political leaders. This meant that powerful capabilities were built up, and in secret. These were built with the agreement and sometimes at the instigation of political leaders and stemmed from a desire for a zero-risk approach to terrorism. Was it too much? Over the years, one or two on the inside dissented about what was being done by the spies (a GCHQ analyst called Katherine Gunn leaked details of a spying surge on the UN as part of the drive to war in Iraq in 2003). There were not many out of the thousands who inhabited the secret world. But one individual would grow to loathe the system, especially the one built by Alexander in America, and would try to bring it down. For GCHQ the legacy of Bletchley was utter secrecy: even so much as a whisper risked compromising capability by letting the other side know what you could and could not do. That secrecy was about to be blown apart.

CHAPTER EIGHTEEN

EXPOSURE

At the end of May 2013, a young computer programmer called Ladar Levison found a business card on his door in Dallas. On it was a number and a message asking him to call. That led to a visit from the FBI. Levison was one of the successors of the cypherpunks like Phil Zimmermann, who had created the PGP encryption system in the 1990s. Levison had built his own secure email service called Lavabit, which used encryption to keep messages away from prying eyes. Now the FBI were very interested in one of his users. Levison knew that all kinds of people with varying motives were attracted to a service that guaranteed anonymity, but it was a principle he believed in. 'The definition of bad guys is certainly a relative term that changes over time,' Levison argues in response to the idea that those with malicious intent gravitated towards his system. 'You know back during the American revolution the American revolutionaries were referred to as terrorists by the British.'[1] That was the language of a man unafraid of taking on the state.

Levison is barred from confirming whose email the authorities were interested in. One of his users had created an account with the address Cincinnatus@lavabit.com – the first part of the address referred to a statesman who had chosen to relinquish authoritarian powers once he had used them to defend Rome. The FBI seemed to know something was up. But how much did they know? The power of encryption was still strong enough to require them to approach Levison for the keys to unlock the messages they were interested in.

That email account had mysteriously summoned a group of journalists to Hong Kong, where they arrived as May turned to June. They found a man much younger than they had expected toying with a Rubik's Cube in a hotel lobby, an old-fashioned recognition signal that was more out of a Cold War spy novel than the modern techno age. As they spoke to him, the journalists learnt that the man passionately espoused a worldview that epitomised the way in which computers and the internet had become an indispensable part of the lives of some members of a new generation. 'The Internet allowed me to experience freedom and explore my full capacity as a human being,' he told them. 'For many kids, the Internet is a means of self-actualization. It allows them to explore who they are and who they want to be, but that works only if we're able to be private and anonymous, to make mistakes without them following us. I worry that mine was the last generation to enjoy that freedom,' he told his visitors.[2] That belief had driven him to betray the place where he worked and challenge the power that the state had built. 'For me, it all comes down to state power against the people's ability to meaningfully oppose that power.'

On 5 June, General Keith Alexander was (perhaps ironically given what was coming) in Germany meeting with allies, when he received news that a story was about to break. Alexander later said that the NSA had no idea where it had come from. Within a few hours the story had appeared in the *Guardian*. It published a court order to the company Verizon that demanded that it hand over details of every phone call in America – the programme created after 9/11. The revelation that the NSA was collecting domestic information which included that of everyone in the country was so stunning and so out of the blue that at first it was met with an eerie silence in the wider world, like that in the moments after a bomb blast, as everyone struggled to understand what had just happened.

In Cheltenham, Iain Lobban, Director of GCHQ, was a worried man. 'When I heard the news,' he later said, 'I lay awake saying to myself: "I hope this isn't a Brit".' He even went round colleagues in the 'Doughnut' to ask if there was anyone in their teams who had taken an unusually long holiday and who might be the culprit. He thought an employee gone rogue would have been spotted in the much smaller, more tightly knit British community, but he knew that if he was wrong and that was the case, the results would be disastrous, not least for himself. 'That would have been the end of me,' he later told a journalist.[3]

In London a few hours after the story first broke, Britain's Intelligence and Security Committee was holding a previously scheduled press conference to publish a report on 'Foreign Involvement in the Critical National Infrastructure'. The title was a little misleading because it was really only about the involvement of one company – the Chinese telecoms giant Huawei – in Britain's communications. It examined how that had come to pass without what the committee thought was sufficient oversight and looked at whether its presence created risks of Chinese espionage against the UK.

The next day the PRISM programme was exposed to the world – initially misinterpreted as some kind of secret hack into tech companies' systems. The companies responded by saying they had never heard of such a programme (technically true, since the codename was classified), and stressed in carefully worded statements that they simply abided by the law. By now, the counter-intelligence people at the NSA knew who was behind the stories. But what they did not realise was that the rest of the world was also only a few days from finding out as well.

As the shockwaves of the PRISM story began to ripple out in government, in public and the technology sector, President Obama was making last-minute preparations for one of the most important summits of his presidency to date. At the 200-acre Sunnylands Ranch in California, he was due to meet the new Premier of China. It was the moment a co-ordinated US campaign to corner Beijing over cyber espionage, encompassing the revelations by Mandiant about PLA Unit 61398 hacking American companies, was supposed to reach its crescendo as America's President sat down with China's leader and told him it was time to stop. 'We were spring-loaded,' one former US intelligence chief said of that moment. But hours before the summit opened, a Top Secret document was published which outlined America's doctrine for its own offensive cyber operations. Along with PRISM, it made it all too easy for the Chinese to ask who was *really* the most aggressive player in cyberspace. The plan to pressure Beijing fizzled out.

Within hours of the summit wrapping up, the mystery of who was behind the revelations was over. On Sunday, 9 June Edward Snowden went public to identify himself in an interview recorded in a hotel in the semi-autonomous Chinese territory of Hong Kong.[4] As if trapped in a Jason Bourne film, he warned darkly that the CIA might subject him to rendition or they might pay the Triad gangs of Hong Kong to come after him. Why had he done it? 'I do not want to live in a world where everything I do and say is recorded,' he explained.

The next day, Ladar Levison received a court order demanding he hand over details of the email account the FBI had been interested in. A six-week legal battle ensued. Levison resisted because he said that handing over the encryption keys would compromise the accounts of all his users. When he was forced to comply, he did it in his own way. He handed over the keys, not on a CD or USB stick or via email but in an envelope written out in tiny four-point type on paper. Each of the five key chains contained 2,576 characters – making up eleven almost totally illegible pages. 'I was giving them exactly what they asked for. They just didn't know what to do with it.' The FBI agent dutifully wrote out a receipt for one sealed white envelope. Levison knew that the FBI would have to input the thousands of characters by hand. One mistake would render the whole enterprise meaningless. And that gave him time to shut down his entire system and wipe all the data. He had named each of his servers after his ex-girlfriends and so pulled the plug on one relationship after another.[5] Levison does not see the NSA as the enemy: he is happy for them to protect his country against foreign threats and spy on foreign governments. But he knew which side he was on in the new round of the crypto wars. 'In the same way I don't think police should be rolling through our streets in tanks, I don't believe a military organisation with the capabilities of the National Security Agency should be spying on its own citizens,' he argues.

Seven months earlier, Edward Snowden had used his Lavabit email account to organise a 'crypto party' in an art space at the back of a Hawaiian furniture store. These parties are grassroots gatherings in which volunteers teach interested people how to use advanced encryption like PGP and TOR to protect their communications. Attendees at that December 2012 event recall that the quiet young man introduced himself as Ed, but seemed reluctant to tell people where he worked.[6] That was because his workplace

was an NSA base called Kunia – originally a massive bunker underneath a pineapple field built after Pearl Harbor, later a warehouse
for torpedoes before it finally became an NSA facility. Snowden
had taken a job in Hawaii (with a pay cut) to get access to the final
trove of secret documents that he wanted.[7] In the Cold War, spy
agencies always dreamt of recruiting a code clerk because all the
traffic of an embassy passed through their hands as they encoded
it to be sent. The modern equivalent in the computer world is a
systems administrator involved in maintaining computer networks.
As far back as 1991, the NSA had recognised that these individuals'
unique access to classified material made them a top target for foreign intelligence agencies since they could 'so easily, so quickly,
so undetectably, steal vast quantities of information'. However,
for all the effort it made in getting inside other people's systems,
the NSA seems to have been slow to think about securing its own
from within.[8]

In his final job choices Snowden may have sought maximum
access to damaging information, but attempts to push his story into
that familiar Cold War notion of espionage do not quite fit. He may
have received sanctuary and help from other states after he fled, but
no evidence has been produced to prove that he was acting out of
anything other than his own deeply held ideological beliefs when he
made his decision to reveal the NSA and GCHQ's secrets. He was, as
one NSA person puts it, 'self-radicalised'. Snowden was of a generation for whom computers were not just machines in the office you
worked in but things through which you lived your life and inter-
acted with people, sharing intimate details with friends and building
relationships. Snowden also valued anonymity; web postings would
emerge in which he used that anonymity to express himself and his
political views in a trenchant manner, which suggested he felt he
could take on a different identity online. He also explained to the
journalists that he had learnt from video games the importance of
individual protagonists fighting against powerful forces. Snowden
was from a different generation from the Haydens and the Alex-
anders, for whom computers were a simply a tool for their work.
Computers were his world.

Snowden was himself an exemplar of how computers had changed spying. Spy agencies needed people with his skills for the new world of espionage – not martini-swilling, Aston-Martin-driving James Bonds who could navigate a casino, but Diet Coke-drinking hackers who knew their way round a network. Older managers struggled to relate to these new recruits who, in turn, were often contemptuous of the lack of technical knowledge of their seniors. And it should be little surprise that one or two of these hackers may have infused the libertarian ideology of the cypherpunks, who believed information needed to be free and – at the further end of the spectrum – believed government could be the enemy of liberty.

Barack Obama fought to keep his BlackBerry when he assumed the presidency rather than have the security officials take it away. But, to Snowden's disappointment, the new President seemed to heed his securocrats' advice when it came to keeping the intelligence programmes he was inheriting from George W. Bush. After taking office Obama continued with the bulk data programme – especially after Umar Farouk Abdulmutallab nearly blew up a plane with a bomb in his underwear on Christmas Day 2009. That highlighted the political dangers of being seen once again to have failed to connect the dots. The persistence of these programmes led to further disillusionment from Snowden, who says a final breaking point came when he saw Director of National Intelligence James Clapper deny to Congress that the NSA was collecting data on Americans (Clapper would later apologise for this after the details of the phone metadata programme were revealed).

Snowden, like Bradley (now Chelsea) Manning, who downloaded a mass of State Department and Pentagon files and gave them to Julian Assange's Wikileaks website a few years earlier, highlighted two truths. First, that the US secret state had grown so large that it had lost the ability to keep its own secrets. Around 1.5 million people had Top Security clearances and many worked as contractors, part of the military-cyber complex. Snowden himself worked for a private company but said he was tasked by the NSA. The emphasis on sharing post-9/11 to connect the dots had provided too many people too much access to too many secrets. The other truth was – as every spy agency had learnt – that it had become a lot easier to gather and exfiltrate a ton of secret material. Manning transferred diplomatic

cables onto a CD while pretending he was listening to music by
Lady Gaga. Snowden was a touch more sophisticated, using accu-
mulated passwords and special privileges to 'scrape' all the files he
could get hold of. No one is quite sure how many he took. Perhaps
over a million, some claimed; Britain thought at least 58,000 of its
GCHQ files that had been accessible to the NSA. For an organisation
whose raison d'être is secrets, the NSA had failed when it came to
protecting its own.

Snowden argued that the data collection and programmes instituted after 9/11 constituted a threat to privacy. 'Every time you pick
up the phone, dial a number, write an email, make a purchase, travel
on the bus carrying a cell phone, swipe a card somewhere, you leave
a trace and the government has decided that it's a good idea to collect it all, everything, even if you've never been suspected of any
crime,' he said. He also argued that the danger was that a capability
so powerful – and which had also been secret – could in the future
be turned on the population without them knowing. 'The NSA is
surely not the Stasi,' Snowden told *Time* magazine, in reference to
the East German security service, 'but we should always remember
that the danger to societies from security services is not that they
will spontaneously decide to embrace mustache-twirling and jackboots to bear us bodily into dark places, but that the slowly shifting
foundation of policy will make it such that mustaches and jackboots
are discovered to prove an operational advantage toward a necessary
purpose.'

A powerful capability had grown up, and it had grown up in
secret. But was it actually being misused to spy on ordinary people?
Snowden and his supporters argued it was. Supporters of the NSA
argued it was not. The agency said that over the previous ten years
there had only been twelve cases of intentional misuse, such as
people looking for information on current and former partners,
for instance when an NSA employee owned up to looking at her
husband's phone calls because she suspected he was being unfaith-
ful. GCHQ in 2014 also sacked one employee for gross misconduct
after unauthorised searches but said this was the first such case.[9]
But Snowden and his supporters argued that the act of gathering up
everyone's data to be processed by machines was itself an intrusion

into privacy. 'The abuse doesn't occur when people look at the data; it occurs when people gather the data in the first place.'[10]

This goes to a fundamental question of what constitutes privacy in the digital age. Does the bulk collection of data and its analysis by computers constitute spying and surveillance? Back in the pre-computer First World War, the task of censoring the post had been undertaken by hand. Each examiner read on average 110 private letters each day. Even an official report acknowledged the distastefulness of this. 'No one employed in the work could fail to experience an involuntary and deep-seated disgust when he first broke the seal of an unknown person's correspondence,' it noted. The knowledge of surveillance also created a kind of self-censorship: 'men and women have been oppressed when writing by the fear of the unknown eyes that would scan their correspondence; they have feared to put their correspondences on paper, have felt a not unnatural dislike to utter their true desires and their genuine feelings,' the internal official history of postal censorship recorded. It called the practice a 'tolerable evil'.[11] But is a computer collecting and reading your email the same kind of intrusion? The spies maintain it is not but realise they may have made a mistake by not making their case before it was exposed. 'Government should, long ago, have explained that meeting legitimate demands for digital intelligence for law enforcement, as well as for their main national security mission, means bulk access to the Internet,' argues former GCHQ Director Sir David Omand. 'Our communications do all already pass through many computers in the course of their delivery, just as all our financial transactions pass through the banks' audit systems. Computers are not conscious and I do not think we should worry that these intimate details of ours are subject to the security algorithms.'[12]

Privacy activists questioned many of the specific details of what was revealed – the adequacy of legal authorisations, the rules for sharing intelligence, the rigour of oversight process, the weaker controls over metadata and for foreign intelligence compared to domestic. But they also maintain that the act of gathering up communications and then having computers filtering them en masse is in itself a violation of privacy. They argue the new systems represent a fundamental shift in the relationship between the individual and the state which will lead people to alter behaviour and self-censor – to become more conformist and be afraid of experimenting for fear it will be recorded and be accessible in the future. They believe a shift towards identifiability online will aid surveillance that crushes dissenting voices and lead to control or self-censorship to reinforce existing power structures.[13] The privacy activists' argument necessarily challenges not just the state but also technology corporations who carry out similar automated collecting and scanning and also the whole future of big data analysis. One of the main differences is that you can opt out of Google, but not GCHQ. The question therefore becomes whether GCHQ's use of data is necessary, proportionate and effective in supporting national security (rather than selling advertising) to justify its employment.

All spying is invasive of someone's privacy – its point is finding out something that someone else wants to keep secret. In that sense it always raises ethical issues. Some oppose it entirely, but most believe it comes down to the details of who is being spied on and why and with what controls over the spies' behaviour. Judging how much spying you want relates to how far you balance the need for security and the cost of privacy. That may change over time as different security threats emerge. But it is also ultimately a personal judgement. It is one that in democratic societies – unlike authoritarian ones – can be subject to debate. Snowden's supporters argued that he did a public service by kick-starting that debate. His critics contend that the way he did so damaged security.

The sudden revelation of huge capabilities (and on the terms of the agencies' critics) left GCHQ and the NSA reeling. Senior officials acknowledge they could have been more transparent in the past, and that perhaps even programmes like the US gathering of domestic call metadata might have been able to withstand the light of day. But no one had tried. GCHQ argued that the issue of what could be done with the capability was confused with what it was actually being used for. However, the organisation had barely moved beyond the reflexive position of total secrecy it had inherited from Bletchley Park. The vacuum of information was filled by critics who claimed it was spying on ordinary people. A panel of judges found that the collection system did not breach human rights but that there had been a lack of failure over the specific issue of a lack of transparency in how it worked and what safeguards operated. GCHQ maintained it was simply doing what it had been tasked to do by the government under the law and that it had contributed to saving many lives in stopping terrorist attacks (although a 2015 oversight report redacted all the details of how bulk collection may have actually made a difference). Staff were left bruised. 'We don't employ the type of people who would be prepared to intrude into the private lives of ordinary people,' Sir Iain Lobban told an oversight committee in a 2013 hearing of British spy chiefs which was televised for the first time. 'If they were asked to snoop, I wouldn't have the workforce, they'd leave the building.' He echoed that claim in his departing speech. 'We don't suddenly lose our souls the moment we swipe into the Doughnut,' he said. What was clear was that the old days in which signals intelligence was the most secret part of the secret world was passing. This was beginning anyway in a world in which computers and surveillance and large-scale data processing had moved well beyond the purview of spies alone and overlapped with the private sector and cyber security. But now the spies were going to have to come to terms with living and working in a very different environment, one that David Cameron christened 'living in Snowdonia'.

In the US there was a more heated reaction, partly because of the stronger libertarian streak that encompassed both the left and right of the political spectrum. There was also more debate over the utility of the programmes. Supporters said they were vital. 'I can't think of any terrorist investigation where the NSA was not a pre-eminent or central player,' Michael Leiter, who ran the US National Counterterrorism Center, said.[14] The harder question is the extent to which the specific new bulk analysis powers acquired after 9/11 were more valuable than the traditional route of intercepting the specific communications of a target. Talk of more than fifty 'plots' having been stopped using call data was challenged and downgraded.[15] Those examining the details have said they have found evidence that PRISM had been considerably more useful than domestic call metadata collection, pointing, for instance, to a case in which PRISM targeted the email address of an Al Qaeda courier in Pakistan. This led to emails being intercepted from an individual in the US urgently seeking advice on mixing explosives. The individual, Najibullah Zazi, was then put under intensive FBI surveillance and eventually convicted. The issue of the proportionality and effectiveness of bulk access and data mining is hard to judge from the outside since so much is secret. Supporters say it was highly classified for a reason: if the person you are trying to find knows you can do this, and if they know the parameters you use to try and find patterns and signatures of behaviour, then they will change the way they operate in order to minimise their chances of being spotted. Privacy activists contend that this argument has allowed the state to expand its intrusive powers without sufficient public understanding or accountability and that what was created was disproportionate to the threat. Judging the impact of secret intelligence is always challenging, leaving critics asking for proof and supporters maintaining that secrets need to stay secret to be effective. That problem also related to the debate over the damage caused by Snowden's disclosures.

Intelligence officials in the US and UK maintain that Snowden did real damage to intelligence-gathering capabilities. They say he heightened awareness of communications security issues among their targets, leading them to change their behaviour. They say they saw advice going round those they were monitoring telling them to change communications providers (most likely away from American companies) or switch encryption systems (although some observers reckon these may now be easier to crack).[16] They claim coverage of their targets was lost. Those who worked inside the agencies say they watched sources go dark. Privacy activists remain sceptical.

Officials also said they had to work on the assumption that other states had got access to Snowden's documents. They focus on the fact that Snowden had found refuge first in China and then in Russia, the two countries most eager to understand Western technical capabilities (neither of which, ironically, are proponents of the kind of internet freedom that Snowden himself espouses). 'Not even the KGB in its heyday of Philby, Burgess and Maclean in the 1950s could have dreamt of acquiring 58,000 highly classified intelligence documents,' said Sir David Omand.[17] Snowden and his supporters reply that there is no evidence that these states have got access to the material and say steps were taken to prevent that by ensuring Snowden himself was no longer able to provide the keys. 'He's a smart fellow,' Chris Inglis, Deputy Director of the NSA until early 2014, has said of Snowden. 'He knows something about security and encryption. But what we have determined over seventy years of cryptologic history is that single minds never prevail against a diverse set of minds. The idea that a single person could secure information against the dedicated efforts of intelligence services that are quite capable is asking a lot. There's some amount of hubris in that.'[18]

The Snowden revelations seemed shocking because, while most people know that spies spy, few had given much thought as to how they might do that in the digital world and thought about whom they might spy on. And it was not just on terrorists. The counter-terrorist mission had become the overt justification for modern intelligence-gathering, but the traditional mission of securing national advantage remained. This might mean collecting intelligence on other states or issues like the proliferation of weapons of mass destruction. It might mean using intelligence to answer questions such as who gave the orders when Syria fired chemical weapons in 2013 or when a missile shot down a Malaysian airliner over Ukraine in 2014. But securing national advantage can also mean spying on those closer to you. This is what Britain did in the First World War when it intercepted American diplomatic traffic to obtain the Zimmermann Telegram from Germany's foreign minister. Countries know this happens, but the first rule of espionage is 'don't get caught'. And thanks to Snowden, the US and UK got caught. International bodies like the UN, IAEA and EU had been targeted both by old-fashioned bugging of offices as well as getting inside computer networks.[19] At one point in 2011, the NSA stumbled upon the Chinese who were also spying within the United Nations. Spying on allies has been normal practice within Europe for many years, including ahead of major European summits. (Breaking French codes was a priority when Britain was trying to join Europe decades ago, and more recently MI6 provided intelligence to policymakers ahead of every major treaty negotiation during the 1990s, as the French did against Britain.)[20]

NSA officials maintain they do not conduct commercial espionage in the way the Chinese do – stealing information from foreign companies to give to domestic companies. But the NSA does conduct economic espionage. This may involve spying on companies for national security reasons, for instance to see if they are breaking sanctions or selling items for use in a country like Iran's nuclear programme. It might involve spying on Gazprom because the company is closely aligned with Russia's national security policy, which can use energy as a tool to squeeze other states. But economic espionage may also involve spying on companies and countries for economic gain on a national level. American officials argue that spying on Japan's trade position for the benefit of the US government in its negotiations is not the same as stealing secrets from a Japanese company and giving them direct to a US company. But not everyone may accept the distinctions.

Countries responded differently to the revelations, based on their own history and culture, including the US and UK. 'Brits tend to trust their government and mistrust their corporations. Americans tend to mistrust their government and trust their corporations,' Sir Tim Berners-Lee said of the differing reactions.[21] Distinctions were also evident in Europe. France has always been more laid-back about espionage and state power. Perhaps the idea of America stealing French secrets made for less fuss since the French elite knew they do pretty much the same thing to the Americans. France is also reported to carry out the kind of bulk collection that the US undertakes, sucking up international metadata flowing between France and abroad, and is reported to have similar arrangements with some of its telecoms and technology companies to provide bulk access through Mediterranean cables.[22] What is different is the sheer size of the US capability and its ability to exploit control over the infrastructure and the dominance of their companies at every level. 'They have wiretapped the entire world,' one former French spy argues enviously. 'When the Chinese try to steal information from our servers, we can spot and block them because they use hacker methods. However, the United States does not need to hack us because it controls the internet. It is at the heart of the network.'[23]

In Germany the legacy of the Gestapo and the Stasi makes espionage a toxic and emotive subject, in stark contrast to the British reference point of Bletchley Park. Mass surveillance – based on human informers – is something many in East Germany can remember in their own lifetimes. 'You know, for us, this would have been a dream come true,' a former member of the Stasi said of the revelations, pointing out that in his day only forty telephones could be tapped at one time because of a lack of equipment.[24] An interest in economic and financial stability earned Germany a ranking of three out of five on the NSA scale of interest (with China, Russia, Iran, Pakistan and Afghanistan rating a one).[25] Chancellor Angela Merkel had been targeted by the NSA perhaps as far back as 2002, one of around thirty-five world leaders.[26] But Germany is both a target and a partner: as well as being spied on, it also seems to provide bulk access to the NSA.[27] Both German and American officials say co-operation on counter-terrorism surveillance has helped stop terrorist plots in Germany. 'If the NSA were to provide us with no more information, it would be a disaster for us,' a German intelligence officer told Die Welt as the Snowden affair began.[28] This dual aspect of being a target and a partner reflects the tension in international operations like finding terrorists who cross borders, dealing with the proliferation of weapons of mass destruction or supporting a joint military engagement in Afghanistan – and the more traditional espionage missions involved in securing national advantage by spying on the diplomatic or trade talks of the same countries you are co-operating with in other fields. France was also both a target and a partner, with reports that it had passed NSA information from undersea cables which carried traffic from Asia and Afghanistan in exchange for intelligence about parts of the world where France has little visibility.[29] Israel likewise partners America against Iran's nuclear programme, but still spies on American foreign policy.[30] The Germans pushed to have a deal similar to the one Britain enjoys with the US which would limit spying. Negotiations began, but one report says the Germans 'blanched' when they were told that, in return for joining the club, they would have to take on responsibility for intelligence collection and cyber operations in certain parts of the world – a sign of just how the Five Eyes really works.[31]

*

It was telling that Snowden had spent his spare time in Hawaii organising a crypto party. His revelations reignited the crypto wars
that had started in the 1970s with a blazing intensity. In the 1990s,
we left the intelligence agencies abandoning hopes of stopping the
spread of public key encryption. And so, as the twenty-first century began, many people thought that code-making had reached
ascendancy over code-breaking and those who had sought to get
inside secure communications had simply given up in the face of the
power of encryption. But, as they always try to do, spies found new
ways to reach their targets. Cryptanalysis – using supercomputers
and people to break codes – had remained a core part of the mission.
It is costly in terms of time and resources, however, both for the few
human brains capable of working out possible attacks and for the
vast supercomputers. That cost – and the challenge of public key
and the like – has meant that spy agencies have had to look for ways
around encryption rather than face it head-on in order to gain access
to communications.

There are a number of possibilities. Domestically, a court order
to a communications provider might work to get the keys to a message. In Britain, after the crypto wars were lost in the 1990s, the
government included a provision in the 2000 Regulation of Investigatory Powers Act that allowed law enforcement to demand keys
from people. Lawful interception was easy when people simply used
UK or US phone companies, but once they started using international internet firms to communicate then it became much harder.
PRISM was particularly valued since it provided a means of getting
hold of foreign encrypted communications from a US-based provider without having to intercept and decrypt the messages. But
other countries like China increasingly also began to demand keys
and access to American technology products for their law enforcement, a move met with horror (including by the US government
who had demanded the same thing themselves). If getting the keys
through the front door was not possible then sometimes you might
hope to pick up a message through the intelligence machine (which
provides only partial rather than total access). But what if a message
you intercepted was encrypted?

One of the most controversial revelations from Snowden's cache was the claim that the agencies had undermined encryption. Precisely because modern encryption is commercial, it offers spies opportunities. They can get inside the process of code-making in a way they could not have done in the past – the equivalent of having influence on the design of Enigma machines. That might be very useful, not just in understanding the workings of a machine but also in messing with its settings to reduce the randomness. There were reports of a $250-million campaign to 'influence' as well as 'overtly leverage' commercial product designs through sensitive relationships with 'industry partners'. This included, it was reported, work to 'insert vulnerabilities into commercial encryption systems' which make them exploitable – in other words, trapdoors, following on from covert work in the Cold War. In an echo of Diffie and Hellman's battle in the 1970s, there were claims that the NSA had influenced an international Random Number Generator standard. The standard was eventually withdrawn. If a random number is not quite as random as you thought, then knowing the weakness might allow you to crack it.

The fundamental tension of the crypto wars remains, but it is now writ large by the spread of communications and computing and the desire to protect the growing pile of personal data. In the days of Bletchley Park, Alan Turing's and his colleagues' battle against encryption was waged against custom-built systems like Enigma used almost exclusively by enemy armed forces. By the seventies, encryption was starting to spread to large companies and institutions. Next, in the nineties, to determined individuals (including criminals). But now modern encryption is commercially available for everyone to use. So deliberately weakening (or, more subtly, failing to point out a weakness) has a much broader impact.

This creates far more complicated quandaries in the relative balance between offence and defence. The public may want spies to be able to read the messages of those who wish to do them harm, but what if those same systems are used to protect innocent people's personal information from criminals and other states? The tension applies not just to encryption but to all forms of computer vulnerabilities. What the spies call the 'equities' problem – balancing offence and defence – becomes acute. Should inadvertent weaknesses in computers and communications be patched up to protect or left open to spy? 'It is false to imagine or say that NSA broadly has the capability to decrypt most of the encryption used by any citizens in the world but in particularly US citizens to access their web browsers or financial systems or things of that sort,' says Chris Inglis.[32] He argues that the NSA finds ways of keeping US citizens' communications safe while exploiting those of adversaries. 'We do in fact find ways to do precisely that,' he says, while avoiding specific detail. In practice this may parallel the system developed from the first battles in the crypto wars, in which focus is placed on exported systems (although the distinction is harder now) and in which trapdoors are well hidden. Critics claim that this position is a dangerously arrogant one. Who is to say that not just China or Russia with their crypt-analytical expertise but soon perhaps the most advanced hackers and criminals will be able to find and exploit the weaknesses? 'We are not dumb enough to think we are the smartest kid on the block,' is all one American official will say, but critics charge that too little emphasis has been placed on defence.

The revelations about encryption were a '9/11 moment' for those involved in computer security outside of government, says Ross Anderson of Cambridge University. This was not because they were surprised by what the spies were trying to do but by the sheer scale of it (although naming programmes to defeat encryption after battles in your country's respective civil wars perhaps did not help – Bullrun for the NSA and Edgehill for GCHQ). Systems like TOR which provided anonymity online were also targeted to try to understand who might be using them. 'They've pushed it even further than we thought they would,' Anderson argues. 'The surprising thing to us was that there appear to have been occasional pockets of competence within NSA and GCHQ – many of us had for many years thought that the real secret was that, like other public-sector IT projects, it didn't work and there was really nobody there. But to find they had built this machine and got it working was an eye-opener.'[33] This is true of much of what Edward Snowden revealed. No one who thought about it should have been surprised that code-breaking communications intelligence agencies tried to break codes or collect communications. The surprise was not the attempt but the success and scale of what they achieved. 'The details were surprising, the sheer magnitude of the programmes was surprising, the amount of money, the effort, how broad they were,' says Bruce Schneier, who examined some of the Snowden cache. 'Because we never really thought about the details that much. So while not a lot was surprising, in the end a lot of it was surprising.'[34]

Until they appeared all over the internet and on the front pages of the newspapers, these capabilities to target encryption were among the most closely guarded secrets. 'Do not ask about or speculate on sources or methods underpinning Bullrun,' one document instructs the reader. 'These capabilities are among the SIGINT community's most fragile, and the inadvertent disclosure of the simple "fact of" could alert the adversary and result in immediate loss of the capability,' a GCHQ document said, echoing the Bletchley fear of opponents abandoning a system they had thought secure.[35]

'Always look for plaintext' was one of NSA veteran Robert Morris's golden rules, meaning that, rather than break the code of a message, find a place along its path where it was not encrypted. The Tempest attacks of the Cold War – picking up electromagnetic signals – were one way of getting hold of text as it was typed into a machine and before it was encoded. Today, there are modern methods – these include what are called side-channel attacks – for instance, analysing the amount of power consumed by a computer making its code to work out the setting. And if you were really desperate, there was always the oldest technique of all: human spies. Agents could be run within the telecoms industry or a company to compromise a system or an individual account, another example of how human and electronic intelligence can be interwoven. Why expend all the energy of a supercomputer to try to break in if you can bribe your way in? This might mean a black-bag job – breaking into a foreign embassy or mission in Washington, New York or London – or it might mean physically intercepting an encryption machine being sent to a foreign capital and inserting a weakness or backdoor into it that can be exploited. Computer servers and routers being sent abroad were also intercepted en route and redirected to a secret location where implants could be installed.

Computer espionage – hacking – offers the other way around

even the most advanced encryption by targeting the endpoint – in other words, someone's computer where the clear text of a message is first written before being encrypted or read after it is decrypted and where data is stored. Just as China might place an implant in a Tibetan activist's computer to act as a spy, so the US and UK could do the same. In the NSA, the emergence of fibre-optic cables and encryption in the nineties is believed to have led directly to increased funding to support a new emphasis on hacking into computers to gather data which might otherwise be harder to access.[36] This activity was undertaken by the Office of Tailored Access Operations (TAO), formed in 1997.[37] 'In the early days TAO used to be just a bunch of hackers!' one member of the department wrote in 2012. 'We did things in a more ad hoc manner . . . one guy did it all. Now we're more systematic.' TAO now consists of separate teams for each target, one looking at China and North Korea together, another Iran, another Russia, as well as cyber counter-intelligence and counter-terrorism. In the teams are developers to create software and hardware tools alongside analysts who plan operations. A planner would take an operation to a mission director who would assess the risks, and then finally, if approved, hand it to the elite team of hackers in the Remote Operations Center (ROC). In 2004–05, the hacking team expanded rapidly into a 40,000-square-foot office housing 215 personnel able to undertake at least a hundred operations a day. 'What if your job was to exploit a target's computer, collect voice cuts from an adversary's phone system, use a terrorist's web-based email account to infect them with a Trojan horse, and assist the military in locating a high-value terrorist target for capture – all in a day's work? Then you would be working in the Remote Operations Center!' an internal 2006 note explains.[38] ROC's role at the time was described as to collect data, geo-locate individuals, provide real-time support in rendition of 'high-profile terrorists' and manage a global covert infrastructure. Their motto was: 'Your data is our data, your equipment is our equipment – any time, any place, by any legal means.'

These US teams had been getting into Chinese systems from the late 1990s and into those of Al Qaeda in Iraq from 2004, according to reports.[39] They sometimes work in real time, waiting for a target to go online, which might be in the middle of the night, and then moving fast. 'We try to grab all of it for analysis later. We don't take time during the operation to sort out the good from the bad,' a hacker noted.[40] The NSA is a place, insiders claim, in which a military organisation values its geeks. There are plenty of people with pink and blue hair, one NSA official explains, using his slightly dismissive shorthand for counter-culture hacker colleagues; one likes to go round the offices on a skateboard, another on a unicycle, juggling while he rides. When they are testing the government's own systems, they are told to spend $1,000 on software and are then put in a room 'with Twinkies and Red bull'; 100 per cent of the time they get into their targets, it is claimed, even without using NSA techniques. And 98 per cent of the time the other side does not know it. That is because it is not the tools but the knowledge that counts. The NSA needed hackers and its staff went around their conventions donning black T-shirts and handing out business cards to recruit them.

TAO has developed a vast array of spy equipment – the modern, technical equivalent of the kind of things that Q gave James Bond in the films. Catalogues which have been leaked show that this ranges from rigged computer monitor cables which allow TAO to see what is displayed to modified USB sticks which can send data via radio links. Designers have thought about every which way of getting into modern technology.[41] 'What I took away from reading the Snowden documents was that if the NSA wants into your computer, it's in. Period,' says cryptographer Bruce Schneier.[42]

Initially computers had been targeted in the same way as traditional hackers – for instance through emails. But increasingly the NSA and GCHQ used more sophisticated techniques by getting in through the infrastructure, for instance through a compromised router which directs traffic around the internet, and fooling the computer to send it to a fake site owned by the NSA or GCHQ at which malware can be installed in order to take control of the computer. In 2009, researchers who had attended an international scientific conference were sent a CD after they arrived home. The CD contained pictures and materials but also malware. This appears to have been inserted onto the disk when it was intercepted en route to its destination. The team at Kaspersky Lab would find that the group behind the spying operation (whom they called 'Equation') were operating as far back as 1996 and had targeted forty-two countries including Iran, Russia, Pakistan, Afghanistan, India, Syria and Mali topping the list. Although not directly named, the NSA was widely suspected as the culprit, not least because this was judged to be the most sophisticated attack seen, getting deep inside computer systems, including the hardware. Kaspersky's discovery also indicated how even the act of spy-hunting has now been outsourced to private firms.

The scale as well as sophistication of cyber espionage had grown. According to leaked documents, by the end of 2013 the NSA's aspiration was to have 85,000 implants around the world.[43] A system called Turbine promised to allow this procedure to be scaled up to handle 'millions' of implants through automated control. Computer espionage itself could now increasingly be taken out of human hands.[44] Again, computing has changed the scale of what was possible.

The elite hackers of GCHQ and the NSA moved beyond simply gathering information. Covert action teams – like Britain's hackers at GCHQ's Joint Threat Research Intelligence Group (JTRIG) – have the capacity to deny, disrupt, degrade or deceive a target online.[45] This might involve taking an opponent's computer offline, or it could involve manipulating information using all the old-fashioned tricks – spreading misinformation or leaking damaging information, stings, infiltrations, ruses. This is what spies have always done. But now they do it online. JTRIG offered a catalogue of 'effects', saying it could develop others on request. There was apparently also considerable research into psychology to understand how best to influence people online.[46] Whom were these tools used against? One of the few examples to come to light was 'Operation Cupcake' in 2011, when an Al Qaeda publication offering instructions on bomb-making to would-be jihadists had its content replaced with garbled code which proved to be recipes for cupcakes.[47] Documents suggest some of the techniques – such as discrediting individuals – were for use against people who might be involved in radicalising others towards violent jihad. Those targets were thought to be particularly vulnerable when their private and public behaviours were not consistent – for instance, by viewing pornography. When exposed, that might undermine their authority and so their ability to spread their message.[48] The 'honey trap' is a staple of espionage. The online version appears to involve getting an individual to go somewhere on the internet or perhaps in real life in pursuit of someone, perhaps a pretty face or the suggestion of one.[49] If spies did this offline in the past, are we comfortable with them doing it online now? Some may recoil from spies ever using these techniques on anybody, online or offline. The question for many others may well come down to who the target is and whether a technique is considered acceptable and proportionate for the kind of danger they represent. That is the central question for all espionage, online or offline.

One of the controversial aspects was that some of these techniques were reported as being considered not just against terrorists and the like but also against groups like Anonymous – a loose collective of hacktivists who disrupt or steal data from corporations and governments they dislike. The members might occasionally be engaged in criminal activity, but they often turn out to be teenagers. The use of high-end techniques against a wider range of targets is one of the issues facing intelligence agencies: for instance, they may be best placed to go after paedophiles who hide on what is known as the 'dark web' using anonymising tools like TOR and encryption. That may be popular with the public and politicians, but it also takes spies and their advanced capabilities out of the realms of national security.

An angry Mark Zuckerberg of Facebook told President Obama that his administration 'blew it' when it defended PRISM and other programmes by saying they were only used to spy on foreigners. At a White House meeting, Zuckerberg and other tech bosses told the President that they stood to lose billions of dollars. The President had publicly said that the NSA did not spy on Americans. But this did not help much when the majority of your customers were foreigners who had just been told they were fair game.[50] The appearance of their logos on the leaked NSA slides about PRISM was a catastrophe for American technology companies, exposing the tension between their global aspirations and American roots. Being an American company was no longer an advantage when your country's spy agency was perceived as exploiting its home-field advantage to spy on the world. The revelations were deeply uncomfortable, partly because they showed the companies' working to the state (under compulsion) through programmes like PRISM, but also because of the questions asked of their wider business model of collecting customer information for their own uses like advertising – two problems which interrelated. Like the state's spying, that process had not been transparent. The public had traded their own privacy for the convenience offered by tech companies, but were now learning how companies were using this data and also how the state was taking advantage of corporate acquisitiveness to get hold of the information itself. If the government asked people to carry around a tracking device, they would never do it. But they will happily carry around a phone that can act as a tracking device thanks to the data it emits and which spies can then make use of. How far most people worry about this is debatable, but a push-back against the companies began by privacy activists and some customers. The alliance between companies and cypherpunks that had existed uneasily since the 1990s was broken. 'Surveillance is the business model of the internet,' cryptographer Bruce Schneier says of the tech company system. 'Your private thoughts and conversations are the product they sell to their customers.' Schneier compared consumers to tenant farmers in medieval Europe who are working on land owned by the big companies to harvest data to make them rich.[51]

There was more to come. The anger of tech bosses at the exposure of PRISM was serious. But relations nosedived even further when it was revealed that, as well as demanding information from a side door, the NSA was also hacking into the companies' internal data links to steal data out of the back. 'We were attacked by the Chinese in 2010. We were attacked by the NSA in 2013,' Eric Schmidt, Chairman of Google, said.[52] A GCHQ operation codenamed Muscular, reported as starting in 2009, collected internal traffic of Yahoo and Google from a point in British territory. Those revelations caused surprise, not just among the tech companies. 'Why in the world would we burn a relationship with Google by breaking into a data center?' one former US intelligence officer said to journalists.[53]

Companies felt they had been helping at the front door while being burgled at the back.

'We've created a Huawei problem for these companies,' said one US official. In other words, American companies would now be met around the world with the kind of suspicion that the Chinese telecoms company encountered in the US. And if spies had inserted vulnerabilities in encryption, then people wondered if they might have done the same in other fields of computer technology – whether in software programmes or the hardware of chips inside a computer or the routers through which internet traffic passes, something China and Russia had long feared. After all the talk of Chinese backdoors, it was suddenly imported American technical equipment that was being looked at with suspicion. Cisco began to notice an impact on its international sales, including in China. It came out strongly denying that it had worked with the US state to place backdoors or vulnerabilities in its systems that could be exploited. More countries started demanding access to the source code for devices and software used in critical industries, in the way Britain had from Huawei. And what of Huawei? There were some in the company who struggled to hide their satisfaction at the fact that the US had been caught out doing exactly what their company had been accused of, although this also increased overall awareness of the possibilities of espionage. In addition it emerged that the NSA had itself hacked Huawei extensively from at least 2007, obtaining details about its routers and switches. The NSA's TAO was reported as gaining access to Huawei boss Mr Ren's communications.[54] One of the aims of the spying was to try and establish what links might exist between the company and China's PLA. But another was to be able to carry out surveillance on other countries which were using Huawei kit and to understand how that equipment worked (understanding how domestic kit worked so it could be spied on was much easier).

The revelations provided the perfect justification for China to crack down on US tech companies operating in China, leading to an even greater drive to indigenise software and hardware. Companies like Cisco, IBM and Microsoft came under renewed pressure as Beijing tried to reduce dependency on their products and pushed new regulations and investigations. Chinese media called on the state to 'punish' American firms for their alleged role in facilitating espionage.

Trust in the neutrality of technology began to evaporate as suspicion spread. This was all part of a trend towards what supporters of a free and open system feared was a 'balkanised internet' – companies and countries wanting to wall themselves off and control access. Countries talked about building their own cables and infrastructure – Germany's Deutsche Telekom was one of those involved in trying to domesticise as much traffic as possible to create a 'German internet' with packets travelling within the country. Brazil talked of building its own fibre-optic cables, which would not go through America or involve American companies. The point of the global internet was that anything could connect to anything else by any route. But that is not set in stone. The internet was beginning to segment into walled gardens run either by companies or by states in which borders were patrolled and those inside needed to identify themselves. An age was passing as a single World Wide Web began to fade.

Vladimir Putin described the internet as a 'CIA project' in April 2014, signalling his desire to break up US dominance.[55] The global debate on internet governance was shifting. The Russian-Chinese position of more state control was strengthened as other countries came to see the American vision of a multi-stakeholder web as a means for America to maintain dominance, partly through its stakeholders in the form of corporations. Many countries thought their own walled-off, protected internet was preferable to one in which American spies and their corporate allies had free rein. The reaction of many other governments to the PRISM revelations was to say 'we would like a piece of that'. Russia has always had less access to global traffic flows than the US and UK because cables do not pass through it – this is one reason it has focused more on cyber espionage to gather intelligence. However, it does have strong control over the domestic environment. The FSB's SORM (System of Operative-Investigative Measures) has its own direct access to phone and internet providers in every region, allowing information to be collected and stored.[56] Putin criticised a major Russian search engine for basing its servers overseas and Russia's parliament passed a law in 2014 requiring foreign social media websites to keep their servers in Russia and save user information for six months. This would allow the Russians to demand access to systems in the way PRISM had done for the US.

Localising data storage was something states pushed for since it gave their law enforcement and spies more chance of getting at the information. India's $4 billion Central Monitoring System provided direct access to phone and online activity, reportedly including automatic voice-print recognition against the traffic. One blog said the Indian system made PRISM look like the 'paragon of restraint', especially since there was reported to be relatively little oversight over the way it was used. German intelligence told its parliament it needed more money for its real-time and automated monitoring of the internet so that it could 'catch up' with the NSA and GCHQ, and that if it did not get it it would 'fall behind' others like Spain and Italy. Countries approached technology companies and telecoms providers and said that if they did not co-operate, then laws would be passed forcing them to collect and retain data and provide it if required. Companies take different positions about how far they co-operate with this based on their corporate position, relative power and view of a country's behaviour but they found more and more countries demanding more and more data. Rather than hasten the demise of the system he despised, Edward Snowden's disclosures may have inadvertently led to the acceleration of its spread in other countries seeking to emulate the US and UK.[57]

British and American spies talked about a 'second golden age' for signals intelligence in the post 9/11 era. But was it a short-lived one? Edward Snowden may have exposed their power, but deeper trends were also complicating their work. The challenges are manifold. Home-field advantage is not what it was, as the centre of gravity for the internet shifted. Asia was rising. By 2014, the amount of international internet traffic touching the US and UK had nearly halved.[58] 'We won't see this traffic crossing the UK. Oh dear,' one person wrote in a paper before asking whether GCHQ should do the same

as the US, which involved 'buying up real estate in these places' –
apparently Asia.[59]

As well as cable traffic, concerns over PRISM meant countries
were less willing to store their data in the US or with US companies.
China had already built up its own consumer rivals to Google and
Twitter. Playing a leading role in developing the internet allowed
the US to export its values, import other countries' information
through spying and make a lot of money for American corporations
along the way, but that era may now be passing. Companies like
Huawei were on the rise. Then there was the speed of technological innovation. During the Cold War, the communications devices
the West was trying to intercept changed very infrequently. Keeping
up with slow Soviet development and procurement cycles was easy.
Keeping up with the pace of commercially driven innovation was
far more challenging.

Issues of accountability and transparency are also challenging espionage. It may be possible to get companies to be more open about
how they use data in order to ensure informed consent from the
public, but that is trickier with spying since revealing its capability
can undermine its effectiveness. But this creates a complex quandary
for spying in the world of computer-based espionage. 'I used to say
when I was Director of NSA that to be effective NSA really needs to
be only two things – powerful and secret,' Michael Hayden observes.
'And we exist inside a political culture that frankly only distrusts two
things – power and secrecy.' It was precisely the desire for power
and secrecy that worried critics of the spy agency. There had been
too much of both, they felt. After he left the NSA, Hayden ran the
CIA. He asked his advisory board to look at a number of questions,
including one broad issue: 'Can America continue to conduct espionage in a society that every day demands more transparency and
more public accountability from every aspect of national life?' The
study remains classified but Hayden did give a short summary of its
conclusion. 'The answer is we don't know.'[60] How much risk do we
want and what level of security do we consider proportionate to the
threats we face? That trade-off is a deeply political, even personal
judgement. The harder question is how to even make that decision
when so much of the information is secret, because making it public

would risk compromising its effectiveness. Squaring this circle in democratic societies is a major challenge.

The conundrum of how to watch the watchers has always been a problem for the secret world. Optimists believe transparency and data will mean that the screens Orwell depicted in *1984*, his dystopian novel about authoritarian power, could be used not just to surveil us but could also become two-way. Government secrecy is under pressure in a similar manner to individual privacy (as Snowden himself exemplified). Despite being subject to hostile cyber attacks, the tiny Baltic state of Estonia shows some signs of how this might work. It has pioneered moving government services online. Using a national identity card and a digital signature, citizens can access some 4,000 services from voting through banking to filing taxes, as well as accessing medical records and ordering prescriptions. The benefits are clear for the public, who can access all their data at will, and also for the state. But it can only query the databases that it needs and not cross-reference them. The key safeguard is an audit trail. Every query about someone's data is recorded on a log file that the individual can see, with criminal liability for anyone who accesses data they should not. 'I feel more like I am the Big Brother who is watching what the state is doing,' an Estonian claims. Estonia points to a form of mutual surveillance in the world of pervasive data in which citizens are able to use technology and surveillance to watch their state as it watches them. Just as police now regularly film protesters at demonstrations to track them, so protesters now also film the police to catch signs of abuse. The 'if you have nothing to hide, you have nothing to fear' argument is being turned back on the state. A 'panopticon' in which everyone watches everyone else is one possible future, although some believe that the mere possibility of being watched in itself leads to changes in behaviour and will undermine the freedom of both state and individual. Audit trails may work for most types of data, but are harder for national security and intelligence information agencies for whom the problem of not tipping off a target remains. Spies say secret services need to keep some secrets or else they are useless. But which ones? And who gets to decide? Transparency and accountability in these fields requires different and technically savvy mechanisms of oversight. But people's attitude will often revolve around a simple question: how much do you trust your state? Your answer may be very different depending on who you are and in which state you live.

In his farewell speech at GCHQ, Iain Lobban launched an attack on the ideology that suffused the thinking of cypherpunks, Snowden as well as parts of Silicon Valley. For its most ardent supporters, the internet was a symbol of freedom and liberation from the confines of traditional authority. They had argued that the state – and especially spies – were the serpent in the Garden of Eden. Lobban demurred. 'We all know that the beautiful dream of the internet as a totally ungoverned space was just that – a beautiful dream,' he argued. 'Like all utopian visions, it was flawed because it failed to account for the persistence of the worst aspects of human nature. Alongside the amazing benefits . . . there are the plotters, the proliferators and the paedophiles. From what we know of ungoverned spaces in the real world, do we really believe that the world would be a better place if the internet becomes an ungoverned space where anybody can act freely with impunity?' For law enforcement agencies and spies, the internet was no paradise but a place, like any other, where the bad congregated and communicated. It was an ungoverned space which they felt could not be left alone, which meant piercing the veil of anonymity that the internet and encryption offered. The advocates of privacy and internet freedom counter that the ecosystem of the internet is far more fragile than is often understood and that the spies have done more than police the Garden of Eden and have too often been the snake.

A dozen uniformed police in riot gear smashed their way into a terraced house on a quiet street in Shepherd's Bush, west London, in late 2005. After a scuffle with a young man, they found his laptop was still switched on at his desk. He was logged in under the name IRH007 and working on a website called Youbombit. It took a few weeks for detectives to realise they had caught one of the world's most wanted cyber jihadists, a man whose real identity had been unknown but who went by the name Irhabi – or terrorist – 007, a strange tribute to James Bond that reflected the mixed-up pop-culture references of a new generation of extremists.[61] YouTube had not been founded when Younes Tsouli was first talent-spotted by Al Qaeda in Iraq to spread its videos. He still needed to convert formats and then hack websites in order to upload and host films showing their violence. But a decade later, the next generation fighting in Iraq and Syria would take the use of social media and the internet to a new level. That would heighten the tension between spies and companies.

On his first day in the job in November 2014, the new GCHQ Director, Robert Hannigan, launched a blistering salvo. The group calling itself Islamic State (ISIS or ISIL) was, he wrote, 'the first terrorist group whose members have grown up on the internet' and whose members were adept at using technology to communicate and organise, as well as exploiting social media to spread fear and radicalise and recruit. The rise of ISIS and its aggressive use of social media threw many of the tensions over the internet and espionage into sharp relief and with it the question of whose priorities and values – the state or the tech companies – would predominate. Hannigan accused American technology companies of being 'in denial' at the way they had 'become the command-and-control networks of choice for terrorists and criminals'.[62] He called for greater support and co-operation from those companies, a move that was greeted with some incredulity from parts of the tech sector who argued they were doing and would continue to do what they had to do by law. But, in the post-Snowden world, those companies were becoming more resistant to government demands to hand over customer data, positioning themselves as champions of the user and privacy and increasing their use of encryption. This led to bitter rows.

In 2011, GCHQ had flagged up an individual showing interest in extremist material online. The lead was assigned to an MI5 team whose job was to investigate further, using digital intelligence to try to work out who they were, what they were up to and if they were a threat to national security. The power of jihadist media material made those kinds of leads increasingly important. The online Al Qaeda magazine *Inspire*, for instance, which called for lone-wolf attacks and provided instructions, was, MI5 claimed, read by those involved in at least seven out of the ten attacks planned within the UK since it first appeared in 2010. The person who had been spotted in 2011 – Michael Adebowale – was assessed as not being a major threat. Eventually a warrant for more intrusive surveillance went up to a minister in May 2013. But within hours of the warrant being sent, Adebowale and another man who had also crossed MI5's radar mowed down a soldier, Lee Rigby, with a car in Woolwich and brutally killed him with a knife.

A British intelligence oversight report into the killing in November 2014 lashed out at American tech companies, arguing the only lead which might have stopped the killing of Rigby was a message in which Adebowale had talked graphically with a known extremist about wanting to kill a soldier. The committee said this was not passed on by the company, in this case Facebook. But whose job was it to spy on the content of people's private messages sent on Facebook? The company's or the state's? The companies were nervous of a system in which they would have to scan all their customers' data and then proactively report suspicious content to governments. This would effectively mean the act of spying (and not just the collection of data) had been outsourced to the private sector. That was not just technically challenging but also awkward when it came to winning over their customers' trust. And which states would they report this to and how would they define extremist content? The UK government pushed for companies to retain data so it was available for the state, and also for companies to agree to implement warrants for people's data and communications even if those companies were based abroad or kept the data abroad. The US pushed for the same. Companies resisted co-operating on anything more than a voluntary basis, fearing Russia and China would be next to make the demand. In the first half of 2014, Google received almost 15,000 government requests for user data from around the world, complying in 65 per cent of cases. It increased its use of encryption as well to ensure governments had to approach at the front door to get hold of user-data.[63]

In terms of challenges for the spies, the oldest remained the most fundamental: secret codes. An encrypted future may be arriving as companies and people seek to protect their communications and data from hackers and prying eyes. This is becoming the central battleground, and again raises tensions between the corporate world and governments. Robert Hannigan of GCHQ and other spy chiefs have begun sounding the alarm on the spread of encryption. Companies are deploying encryption both on consumer products and across corporate networks, positioning themselves as protecting their customers' data against a range of threats. Apple in 2014 made a point of saying it did not hold the keys to decrypt its new iPhone, leading to a row with the FBI. Ubiquitous encryption can protect information from criminals, hackers, spies and also states, including your own. 'There is no encryption system that keeps the Chinese government out, but that lets local law enforcement have access to that data,' Christopher Soghoian of the American Civil Liberties Union argues. 'With encryption, you either keep everyone out, or you keep no one out.'[64] Is that a good thing or a bad thing? The inherent tension of the crypto wars remains unresolved. Can the defensive value of guaranteed protection and privacy be offset against the cost to law enforcement and spies of not being able to reach their targets (whoever they might be)? Encryption might force spies to do more 'traditional espionage' in terms of bugging, burgling and endpoint computer attacks, rather than relying on passively intercepting data using their global system. But is that a good thing? The familiar fear of the spies that they are about to go dark and that encryption is going to super-empower individuals is matched by the fear of privacy activists that the state is taking us to an Orwellian future of pervasive surveillance.

'Let us speak no more of faith in man, but bind him down from mischief by the chains of cryptography,' Edward Snowden proclaimed. For him, encryption had the power to bind the state and to empower the individual, altering the fundamental balance of power between the two. The state and its spies, unsurprisingly, did not want to be bound.

TO INFINITY AND BEYOND

If you journey beneath GCHQ's 'Doughnut', down a few flights of stairs (assuming you can get through the access control system), past a mini-underground road that delivers heavy equipment, you find yourself in a cavernous computer hall, stretching to 10,000 square metres. Although large, it is not as big as the secret facilities run by Google and other tech companies. Ear protectors are required for visitors by health and safety rules to hush what sounds like a constant electronic waterfall. No food or coffee cups are allowed. A visit is carefully managed. The exact names of some of the computers – a mix of general-purpose and special-purpose – are secret, as is the amount of their processing power. 'I'm not at liberty to tell you that,' a manager replies when asked about storage capacity, before adding as an aside, 'it is not full yet'.[1]

Bletchley's great innovation had been the melding of human ingenuity with the kind of processing power that only a computer could offer. That remains the model today. Above the computer hall, inside the main 'Doughnut', the mathematical heirs to Turing sit in front of screens with ones and zeros rather than the pads of paper with letters that made up the Enigma code. But it is the same task of understanding a system and its weaknesses. 'My job is a combination of maths and computer programming and just being crafty at problem solving,' a female twenty-something mathematician explained in 2010.[2] 'You know why it is important. And that is what spurs you on,' she said. 'There are all sorts of clever techniques that you have to do. And it is not really about solving hard sums. It is more about trying to come up with clever ways to solve the problem. And every time you do it it is different. And that is why it is interesting.' Breaking codes remains about understanding how the 'keystream' is generated – the random element added to a 'plaintext' message to hide it. This is not, she says, a solitary enterprise, since it requires working with technologists who understand the systems as well as analysts who offer insights into how people use it.

Cracking Enigma – and the Tunny machine even more so – required finding chinks in the armour that a machine could then work on. This remains the case. 'Usually the only time you can get anywhere is because people make mistakes,' says the code-breaker. 'So you have to be particularly crafty and you have to kind of get inside the mind of somebody perhaps setting something up and where they might have slipped up. Maybe a particularly clever computer could do it. But we haven't invented it yet.' The modern cryptanalytic special-purpose machines that hum away in the basement of GCHQ and in the specially built computer centre at Fort Meade do what Colossus did – just much, much, much faster. Cray Computers are building computers capable of 'sustained multi-petaflops' of calculations, according to the company's 2014 claims about its commercial machines (and spy agencies' special-purpose models are likely to be even faster).[3] A petaflop means a machine can undertake a thousand trillion calculations a second. Colossus could read 5,000 characters per second, a slightly different measure but, even a decade ago, it could perform its purpose-built statistical attack as fast as a Pentium II laptop.[4] China, Russia, Europe, Japan and the US are all in a race to build an exaflop computer that can undertake a quintillion (1,000,000,000,000,000,000) operations per second.[5]

What does it feel like to break a code? 'It feels amazing, really. You feel like you've won,' the GCHQ code-breaker explains. 'You are doing it for a reason. Because someone has said this is something they want to get into. Then you ring them up and you say: "I've got it. I've got it." And they are absolutely astounded because to them it's magic.' This approach to code-breaking – mixing human inventiveness, mathematics and supercomputers – is one of the threads that connects Bletchley Park to the present.

Governments rely on their own specialist crypto systems to command their military (as with Enigma); to communicate at the highest levels of leadership (as with Tunny); to control nuclear weapons launches and to keep secret what their spies are doing.

Cryptography – knowing you have a secure system and being able to break into someone else's – has since the Second World War been a fundamental, but little commented on, aspect of national sovereignty. It is now more important than ever, with the modern crypto war playing out on the internet as well as in the real world between states. If you want to be what used to be called a Great Power, then code-making and code-breaking are indispensable. 'In the future, superpowers will be made or broken on the strength of their cryptanalytic programs,' a 2007 NSA document argued. 'It is the price of admission for the US to maintain unrestricted access to and use of cyberspace.'[6]

The hot, open desert of Utah is a world away from the manicured lawns of Bletchley. But out in Bluffdale is another sign of where the history of spies and computing has taken us. Here it is all about data rather than cryptanalysis and code-breaking. An army of cranes toiled away for years constructing a gigantic new NSA data centre with its own water and power systems and four vast halls. Everyone is building vast centres to store data. But some have claimed that the aim behind Utah was to build a warehouse big enough to store *all* the data. 'The capacity of NSA's planned infrastructure is consistent, as a mathematical matter, with seizing both the routing information and the contents of all electronic communications,' one (disgruntled) former NSA staffer has claimed. The possibility of doing that is disputed by others, including the NSA, who say it is not possible and nor is it the purpose of Utah.[7] But whatever the reality of Utah's role, the spies' ambition to keep pace with the explosion of data is clear. Their fear of going dark drives them onwards.

Tommy Flowers' Colossus – the first computer – was the size of a large room, all hot valves and whirring tape. The valves of the telephone exchanges were a hint that computers and communications were going to merge, a process that is approaching completion in the twenty-first century. We carry computers around in our pockets that almost incidentally act as phones. These computers have become increasingly central to our lives. We use them not just to communicate but also to pay for things and tell us where we are. That makes them mobile data generators, pinging out information every second. In the past, spies would have dreamt about their targets carrying around a tracking device through which they could not just be located and identified but also their social connections and behaviour analysed. Now we provide this information for free to companies in return for the services they offer.

Soon we will perhaps wear computers under our skin as the virtual and physical worlds merge. A billion users connected to the internet through the 1990s, another 2 billion joined them in the following decade, largely with mobile devices. But in the coming 'internet of things' at least another 40 billion devices will be connected up over the internet, from our fridges to our cars, talking direct to each other with minimal human intervention. Sensors collecting data will be ubiquitous. We will live in 'smart homes' that aim to know what we want before we even tell them but in which some of our most intimate details will be picked up. A monitor attached to your watch checking your condition could give you advance warning of a heart attack and send for an ambulance. But that monitor would also know where you were all the time and have a pretty good idea of what you were up to. This rich digital trail and the connectivity created by it can be used, like any technology, for good or for ill. It can be used to make us healthier, keep us safe, sell us products or to spy on us.

The internet of things is, like the whole internet, built on weak security foundations. The lessons of the past about the vulnerabilities of computers apply starkly. The fundamental designed-in weaknesses of computers and the internet are, in many cases, the same as those the Anderson Report pointed out in 1972. Attackers have an advantage over defenders because of the openness and complexity of systems, making it easier to find a way in than to close every possible vulnerability. Companies continue to be torn apart by hackers seeking money or secrets. The internet of things provides an internet of things to be hacked to disrupt our lives, whether by criminals or hostile states. Even as Western states' cyber espionage programmes are exposed, other states are doing all they can to catch up with that capability and there is little sign of the most aggressive cyber espionage players outside the West – China and Russia – adjusting their posture. But the cyber espionage skills that were once the preserve of the select few are now available to the many.

The tools of hacking have been industrialised, commoditised and even commercialised, leaving our technologically dependent world insecure, vulnerable to both espionage and sabotage from an ever-greater range of actors.

The type of information-gathering that used to be considered part of spying is now something undertaken not just by the state. It is rarely called spying, not just because that would offend but also because it relies on collating scraps which individually are not secret. 'Open source' intelligence is increasingly valuable and the spies have often lagged behind in exploiting it because of their focus on the 'secret' – the private sector remains in the lead. The race is on among advertisers to have the best data and become an 'intelligence broker', one person in the industry explains. Consumers used to be placed in large groups to think about what they might buy, but they can now be segmented down to almost the individual level. 'This is an information war,' one of those on this battlefield told *The Economist* in 2014.[8] That person's company had a billion or so profiles of potential customers around the world, each with an average of fifty data points. The data that we produce is increasingly stored, thanks to both the decreasing costs of doing this and the increasing value of what can be done with it. Data, unless people want a right to be forgotten, may be kept indefinitely, leaving a permanent record of our actions (posting on social media should be thought of as like getting a tattoo, one person reckons). Companies often say they do not need to know people's names when their computers collect and analyse this information, but identifying someone has become much easier – especially if you aggregate different data sets.

It's not just companies acquisitively eyeing our digital exhaust who are engaged in accessing this information. It is all of us. Google says its mission is to organise the world's information and make it universally accessible and useful to everyone. Not just a public library, it can also serve as something like the Registry that MI5 used to run which was full of files about people and organisations. Social media search tools now allow users to pull up a list of anyone who lives in a certain town, of a certain age and with a particular interest, carrying out the kind of analytics spies were only dreaming of a decade earlier. You can delve into someone's digital past and

trail them from your laptop. A worried parent can purchase a programme that tracks a family member's phone as it moves around.
The use of surveillance software by abusive spouses to monitor
their partners has reached 'epidemic proportions,' a charity warned
in 2015, as companies marketed their products on the ability to track
partners through their phones and covertly switch on a microphone
to listen in.[9] In a sense, computers have allowed us all to become
spies. Intelligence, in the form of information and data, has been
commoditised and commercialised and the advanced tools that
spies only once deployed have now been democratised. Computers
are now fusing all the different sources of information together –
challenging our very notions of what constitutes secrecy, privacy
and the act of spying itself.

If you take this ever-growing pile of information and add in the
decreasing cost of storage and the increasing ability to process it,
you have big data with its promise – perhaps real – of predicting
behaviour by looking for patterns and correlations. This takes the
Cold War traffic analysis of data to study the normal to the next
level. The ability to use data in innovative ways (often not for the
purpose for which it was collected or for which consent was given)
and combine it with other data sets may be hugely powerful but, as
one study argues, 'it renders ineffective the core technical and legal
mechanisms through which we currently try to protect privacy'.[10]
Aggregating health data may predict early signs of illness and save
lives. But could your insurance company have access to it and use
it to adjust premiums? And could the state use it? If a credit card
company can tell if someone is likely to get divorced before they
take the decision based on their spending patterns and adjust their
credit rating accordingly, can a spy agency tell if someone is likely to
become a terrorist? And if so, would it be useful for early intervention or are we moving towards the world of pre-crime featured in
the film *Minority Report*, in which suspects are arrested before they
have actually done anything? In the future we may not just ask how
effective an algorithm is at answering a question but how ethical that
algorithm is. Will it be transparent enough for a person to challenge
why they were refused a mortgage or placed on a terrorist watchlist?
People have talked of the risks of a 'dictatorship of data' in which

computers make all the decisions. Real dictatorships may also be able to cement their power by exploiting data.

Alan Turing used to ponder what free will meant in a world of machines. Sceptics are unsure whether computers and big data can really be predictive about the complexity of human life – especially when it comes to something as complex as radicalising towards terrorism – but advocates of 'machine learning' and 'artificial intelligence' claim we are approaching a new age. In a 1950 paper, Turing proposed changing the question 'Can machines think?' to 'Can machines do what we (as thinking entities) can do?' 'Machine learning' holds out the promise (and the fear for dystopians) that computers will be able to learn automatically from data and their own mistakes to improve their work and become truly intelligent and independent. Computers began as people. Then they became ways of helping people, doing things they were not capable of such as calculating, then they began to help us communicate. Now they communicate with each other. The next challenge for spies and society may come over autonomy. Computers were built to help people spy. Then they became the targets of spying. Soon they may be able to spy all by themselves. If computers do make spying easier, one of the limiting factors is the role of people in actually reading emails and the like. But if the process becomes increasingly automated even that constraint may be removed.

In the First World War, the people called secret censors sat at the telegraph stations and sorting offices. A century later, black-box computers act as secret sentries on data and communications, filtering and analysing at the borders where the internet comes into their countries or companies. Deep Packet Inspection provides a powerful, semi-automated monitoring capability and, allied with data retention and mining, is transforming spying. Looking for the abnormal among the normal can be used by a state to find signs of malicious foreign cyber espionage against its companies and citizens or the first signs of a cyber attack so that it can be stopped in its tracks. But cyber security tools are also dual-use. Searching through data for signatures can also be used to hunt for terrorists at home and abroad and to search for the signatures of political dissent or other behaviour that a state deems right or wrong. The issue is not the technology – it is neutral – the issue is the state and how it exercises power. In authoritarian countries, the public has no choice. In democratic societies, the public *may* have some say.

There is a parallel between data collection in cyber security and counter-terrorism. In cyber security, the vast complexity of code means that it is impossible to spot every vulnerability and close it off from attacking some point of your system. That means the trend is towards monitoring to make sure you spot anything anomalous. 'With total surveillance, and total surveillance alone, it is possible to treat the absence of evidence as the evidence of absence,' argues Dan Geer, a veteran thinker on the subject. 'Only when you know everything that did happen with your data can you say what did not happen with your data.' In counter-terrorism a parallel problem exists: spy agencies, tasked with the mission of 'never again', are looking for someone hiding in the population at large who could attack at any moment and strike a 'soft target'. And so, if you want zero risk from terrorism (which the public often demands and politicians struggle to resist), it may be tempting to monitor people more closely or at the very least to collect more and more data to reduce the chances of missing something.[11] Many of those who inhabit the spy agencies of democratic countries appreciate the dangers of un-hindered technology, but a real debate about trading off risks has barely begun. And the demand for secret intelligence will likely intensify as the post-Cold War world proves eminently unpredictable – with old-style crises emerging suddenly in places like Ukraine and new threats such as cyber attacks on companies like Sony hitting the news, as well as continuing concerns over terrorism, whether hostages in Syria or gunmen in Europe.

The fundamental questions of the crypto wars – privacy versus security, anonymity versus identifiability and the place of encryption – remain unanswered. Some people ask if we should be more scared of our governments or those that they are there to protect us from. The answer to that may well depend on where you live and what your politics are. Security and privacy are sometimes portrayed as two competing poles. In some cases that is correct, but the relationship is often more complex, especially in the global tangle of cyberspace. You might want your data to be secure in order for it to be private from prying eyes. A company may encrypt it for you but also scan it themselves to sell you things. A state may demand the data for its definition of security. But you might rely on the same state to protect your data from cyber criminals and foreign cyber spies. And to do that, the state may want to scan information going in and out of the country to spot them. It might then use the same system to search for signatures of terrorists who threaten your security or it might also use the same system to look for you. The choices may not be simple in a global interconnected world. But they are important.

Encryption will be the central battleground. From being a subject discussed in hushed tones in Bletchley's huts, it is now moving to the centre of political debate. As has been the case since Bletchley, spies fear going dark. Today, they see themselves battling against an encrypted future in which they will be drowning in unreadable data. Their critics fear that as our lives move online they will become too powerful and there will be no dark, only bright light. The sources may be fragile but history suggests the phenomenon of communications intelligence is likely to be robust, thanks both to the ingenuity of spies at overcoming the obstacles placed in their way and to the continued, exponential growth in the amount of communications and data that we produce.

There was a time a few years ago when American and British spies talked of 'mirroring the internet' – in other words, creating a perfect copy of the digital world which could be preserved, interrogated and analysed at will, a kind of massive, endless filing cabinet to rifle through. One former spy said this idea had been abandoned. But privacy activists fear Utah is part of a project to do something close to that ambition. How do you find a signal in a growing sea of noise? How do you connect dots you don't know you have? How do you know what information you will need in the future? What if the value of one dot only becomes clear much later when you can connect it to another you collect? These questions have led some in the spy agencies to an ambitious conclusion in their quest to live up to the zero-risk demands of the public and politicians. 'Since you can't connect dots you don't have, it drives us into a mode (where) we fundamentally try to collect everything and hang on to it for ever,'

Ira Hunt, then Chief Technology Officer of the CIA, said in 2013. In Turing's 1936 essay 'On Computable Numbers', he imagined an infinite tape containing data that would be fed into his 'universal machine'. In a modern echo, Hunt made his own dramatic statement: 'I think we are at high noon in the information age . . . It is really very nearly within our grasp to be able to compute on all human generated information.'[12]

A phone turns on and connects to the internet, a tower run by a Chinese company carries the data into a network monitored by a black box as it enters an undersea cable and travels across the world to a private company who process it, analyse it, store it and combine it with a billion other pieces of data to sell to an advertiser. Meanwhile a spy service demands or just steals the data, perhaps deploying their supercomputers if it is locked with encryption. The secret history of computers and spies began with Tommy Flowers in his workshop, paper tape flying around him as he built the first computer, Colossus, for Bletchley to break codes. The world has followed his lead and gone digital, computers becoming all-pervasive and all-powerful in a way the East End bricklayer's son could not have imagined. And the vast halls of Utah and Cheltenham may be a symbol of the ambition of spies not to be defeated by the computer age but to master it. Computers have changed our world – as they have changed spying – and will continue to do so in ways we cannot predict. 'There is a technical solution to every political problem,' one former inhabitant of the secret world argues.[13] That kind of technological utopianism is evident in America, not just in the thinking of Snowden but also in that of the engineers of the NSA and the tech bosses of Silicon Valley – a belief that technology, whether encryption or big data, is the answer. But often technology simply offers up new problems to grapple with. A fundamental question is not what *can* you do with technology, but what *should* you do with technology. It is a question that extends beyond espionage and is at heart a political question too important to be left to a select few. This question, as Sir Tim Berners-Lee said of the World Wide Web, is for everyone.

ACKNOWLEDGEMENTS

This book has grown out of reporting on issues related to intelligence and cyber security over a number of years. It based in part on interviews with many individuals who have worked in the field and I am grateful for their assistance. Endnotes indicate where people have spoken on the record but there are many others who have provided advice and thoughts who would likely not thank me for naming them here. But they too have my gratitude.

I am particularly indebted to those who read drafts of parts of the book and provided comments. Other people and institutions I would like thank for help in my research include: the National Museum of Computing, the National Cryptologic Museum and its librarian René Stein, Bletchley Park, the Imperial War Museum and its sound library, the National Archives at Kew, the Charles Babbage Institute at the University of Minnesota, Satu Haase-Webb for research assistance in the US National Archives, Charlotte Dando at the Porthcurno Telegraph Museum, David Hay at the BT Archive, Erich Schmidt-Eenboom in Germany and my colleagues at the BBC, especially Mark Savage. My agent Georgina Capel has provided support and encouragement, while this book would not be what it is without the guidance and patience of my editor Bea Hemming. My greatest debt is owed to my family.

Prologue

1 This account is drawn from Bourdeaux's report in POST 56/55 held at the Post Office archive. All the cables connecting Germany to the outside world would be cut – save two which went to Sweden and Norway. See also POST 30/4304 in the BT archive.

2 Most of the information on cable censorship comes from the official report, National Archives DEFE 1/130. There were one or two gaps in the coverage, for instance in Spain and Portugal. Porthcurno is mentioned in *The Invisible Weapon: Telecommunications and International Politics 1851–1945* by Daniel Headrick (Oxford University Press, New York and London, 1991), p. 146; the 'Fixity London, Fixed' reference comes from National Archives CAB 17/92.

3 Great War Postal Censorship Report, National Archives POST 56/57 (held in Post Office archive)

4 Occasionally even purely domestic mail was blanket-intercepted on a local basis. In October 1914, fears that material for enemy submarines was being stored on the western coast of Scotland led to censorship of all letters and telegrams in the Hebrides and Isles of Skye and Mull. But a trial of broad-blanket domestic censorship in 1915 found the huge volume was too much. The 'nothing of any value' quote comes from p. 64 of the Postal Censorship report.

5 The account is drawn from Chapter 4 of *Wireless: From Marconi's Black-box to the Audion* by Sungook Hong (MIT Press, Cambridge, MA, 2001) and 'Dot-dash-diss: The gentleman hacker's 1903 lulz', Paul Marks, *New Scientist*, 27 December 2011. During the recent Boer War a trench had been bombarded by British artillery and the soldiers inside were asked how they felt about the awesome power of British shells raining down on them. They sent back the answer 'rats'. The term came to signify a warning against overwhelming pride.

6 *The Codebreakers* by David Kahn (Macmillan, New York, 1972), p. 266, http://marconiheritage.org/ww1-intel.html

7 *Room 40: British Naval Intelligence*, Patrick Beesley (Hamish Hamilton, London, 1982), p. 52

8   http://www.cryptomuseum.com/kits/enigma/manual/pdf/p52.pdf, *Cryptologia*, vol. 26, no. 1 (NSA declassified journal).

Introduction

1   Job advert is in National Archives, CSC 6/33, also see http://www.london-gazette.co.uk/issues/33549/supplements/7079/page.pdf/ Other references go back even further.
2   http://www.nsa.gov/about/_files/cryptologic_heritage/publications/misc/tiltman.pdf
3   The machine is called Frostburg and is now on display at the NSA's public museum.

Chapter One: Birth

1   B. Jack Copeland et al., *Colossus: The Secrets of Bletchley Park's Code-breaking Computers* (Oxford University Press, Oxford, 2006), pp. 281–9
2   Imperial War Museum Sound Archive Accession Number 18332
3   *Alan Turing: The Enigma of Intelligence* by Andrew Hodges (Counterpoint, London, 1983), p. 93
4   'On Computable Numbers' accessed at http://classes.soe.ucsc.edu/cmps210/Winter11/Papers/turing-1936.pdf
5   National Archives CAB 23/55 and *Churchill's Man of Mystery: Desmond Morton and the World of Intelligence* by Gill Bennett (Routledge, London, 2009), Chapter 5
6   'The Spanish Link in cracking the Enigma Code', Gordon Corera, BBC News, 23 March 2012. Available at http://www.bbc.co.uk/news/magazine-17486464 and José Ramón Soler Fuensanta, Francisco Javier López-Brea Espiau and Frode Weierud, 'Spanish Enigma: A History of the Enigma in Spain', *Cryptologia*, vol. 34, issue 4 (2010), pp. 301–28
7   Quoted in *The Evolution of British SIGINT: 1653–1939* (HMSO, Cheltenham, 1997), p. 58
8   There are lots of ways of calculating these figures. This estimate is taken from *The Cryptographic Mathematics of Enigma*, a NSA publication, published by the Center for Cryptologic History (2011).
9   See *Cryptologia*, vol. 27, no. 2. For Turing's description, see Jack Good, 'Early work on computers at Bletchley', *Cryptologia*, April 1979
10 Quoted in Copeland et al., *Colossus*, p. 30
11 Another important method of statistical attack that Turing developed called Banburismus, which looked for coincidences between different messages which, if they happened enough, might suggest they had been enciphered with the same setting.
12 Mary Stewart in an account recorded in a display at Bletchley Park
13 'Some Reminiscences' by Brigadier John H. Tiltman, NSA declassified document ID 3838686
14 Imperial War Museum Sound Archive Accession Number 18332

15 Report on Progress from M.H.A Newman to Commander Travis, 1 March
    1943
16 *Colossus: Bletchley Park's Greatest Secret* by Paul Gannon (Atlantic, London,
    2006), p. 246
17 National Archives DSIR 32/326
18 Copeland et al., *Colossus*, p. 367
19 Gannon, *Colossus*, p. 398
20 Flowers, Imperial War Museum Sound Archive
21 A CIA official quoted in 1971 in 'American Cryptology during the Cold War',
    Book 1 by Thomas R. Johnson (NSA declassified document), p. 1
22 Hodges, *Alan Turing: The Enigma of Intelligence*, p. 289
23 Quoted in 'Bletchley Park's Forgotten Heroes' by Helen Morgan, *Cybertalk*,
    Issue 3, September 2013
24 The loss of Flowers' and all original Colossus documentation is mentioned
    in National Archives HW 25/24. In DSIR 32/326, Flowers recounts his own
    mistake of having rejected the use of computers in the phone industry. Also
    see Flowers' comments in Copeland et al., *Colossus*, Chapter 6
25 Hodges, *Alan Turing: The Enigma of Intelligence*, p. 445
26 Copeland et al., *Colossus*, p. 172
27 Interview with John Cane, 2014. Bombes appear to have still been used in the
    US to attack Enigma machines after the war even though the only users were
    the East German fire and police service. They may have been monitored for
    signs that they had found the tunnel beneath Berlin built by the CIA and MI6
    to tap phone lines. 'The Last Days of Enigma' (NSA declassified
    document).
28 National Archives HW 25/24
29 Such as the Hagelin, used more in places like the Middle East than by the
    Soviets.
30 National Archives HW 25/24

Chapter Two: Marriage

1 'The Origin of US-British Intelligence Co-operation 1940–41' by Robert
   L. Benson, NSA http://www.nsa.gov/public_info/_files/cryptologic_
   spectrum/origin_us_british.pdf and SIS website
2 'The Secret Intelligence Service and the Origins of the Anglo-American
   intelligence relationship 1940–41' by Christopher Baxter, SIS website
3 Frank Rowlett, 1976 NSA interview, National Cryptologic Museum
4 Frank Rowlett's account from 'Frank Rowlett, A Personal Profile', NSA
   http://www.nsa.gov/public_info/_files/cryptologic_spectrum/frank_
   rowlett.pdf and Solomon Kullback NSA interview, NSA OH-17-82
5 'The Many Lives of Herbert O. Yardley' (NSA publication) available at http://
   www.nsa.gov/public_info/_files/cryptologic_spectrum/many_lives.pdf and
   'Yardley Revisited' by Nathan X. Woodeman, *Studies in Intelligence*, vol. 27,
   1983 (CIA declassified publication)

6　Solomon Kullback, NSA interview, NSA OH-17-82 and Abraham Sinkov, NSA OH-02-04-1979

7　Solomon Kullback, NSA interview, NSA OH-17-82 and Frank Rowlett, NSA interview

8　'The Many Lives of Herbert O. Yardley', p. 24

9　'The Origins of the US National Security Agency 1940–1952' (NSA declassified document 1990), p. 9

10　'My Purple Trip to England' by Prescott Currier, *Cryptologia*, vol. 20, no. 3 (1996)

11　Quoted in *The Secrets of Station X* by Michael Smith (Biteback, London, 2011), p. 148

12　'Brigadier John Tiltman: A Giant among Cryptanalysts' (NSA publication), Center for Cryptologic History, 2007

13　Baxter, 'The Secret Intelligence Service and the Origins of the Anglo-American intelligence relationship 1940–41'

14　Abraham Sinkov, NSA interview

15　US National Archives, RG 457, Entry A1 9032, HCC, Box 1413, NR 4565

16　Johnson, 'American Cryptology during the Cold War', Book 1, p. 7 and 'The Origins of the US National Security Agency 1940–1952', pp. 22 and 26

17　Alan Turing, Report from Washington DC, November 1942, accessed at www.turing.org.uk

18　National Archives HW 25/1 and 'The Achievements of the Signal Security Agency in World War II', 1946 (declassified by the NSA, 2010), p. 29

19　Smith, *The Secrets of Station X*, p. 241

20　*It Wasn't All Magic: The Early Struggle to Automate Cryptanalysis, 1930s–1960s* by Dr Colin Burke, 1994 (NSA declassified publication), p. 83

21　US National Archives, RG 457, NSA, Entry A1 9032, HCC, Box 800, NR 2305, RG 457, NSA, Entry A1 9032, HCC, Box 1296, NR 3873 and G 457, NSA, Entry A1 9032, HCC, Box 1328, NR 4009

22　Arthur Levenson, NSA Oral History interview, NSA OH-40-80

23　NSA Oral History interview with Howard Campaigne, NSA OH-14-83

24　At the end of the war, despite military budget cuts, the US Army retained about 1,500 staff and the navy about 700, including a significant technical capacity.

25　'The Achievements of the Signal Security Agency in World War II', p. 24

26　Burke, *It Wasn't All Magic*, p. 199

27　Ibid., p. 199 and footnote reference on p. 224, and Johnson, 'American Cryptology during the Cold War', Book 1, p. 16

28　Material drawn from *Before Bourbon: American and British COMIT efforts against Russia and the Soviet Union before 1945* by Michael Peterson (NSA declassified publication); Johnson, 'American Cryptology during the Cold War', Book 1, Chapter 4: The Soviet Problem, p. 160, and 'The Origins of the US National Security Agency 1940–1952', p. 25

29　National Archives HW 80/1-7

30 National Archives HW 80/9 June 1951

31 'Communication intelligence shall never under any circumstances or in any form be disseminated to any ministry, department, agency, organisation, office, or individual from which or from whom it might be reasonably expected to find its way, officially or extra-officially, into the possession of any person or group who could use it for commercial competition or commercial gain or advantage,' the documents read.

32 March 1946, Technical Conference for UKUSA implementation, National Archives HW 80/6

33 National Archives HW 80/7

34 National Archives HW 80/19

35 Burke, *It Wasn't All Magic*, p. 203

Chapter Three: Into the Cold

1   The details of Rowlett's visit are in Part 1 of the NSA's declassified history of Venona.

2   Burke, *It Wasn't All Magic*, p. 265

3   NSA Oral History interview with Oliver Kirby, declassified and available at http://www.nsa.gov/public_info/_files/oral_history_interviews/nsa_oh_20_92_kirby.pdf

4   'The Days of Bourbon' (NSA declassified document) http://www.nsa.gov/public_info/_files/crypto_almanac_50th/Days_of_Bourbon.pdf

5   Burke, *It Wasn't All Magic*, Chapter 2, and 'The Achievements of the Signal Security Agency in World War II'

6   'NSA before Supercomputers' (NSA declassified document)

7   *Alan Turing and his Contemporaries*, ed. Simon Lavington (British Computer Society, 2012), provides a detailed account of early British computing.

8   National Archives HW 76/1

9   National Archives HW 76/4

10 *Moving Targets: Elliott-Automation and the Dawn of the Computer Age in Britain, 1947–67* by Simon Lavington (Springer, London, 2011), pp. 79–89

11 'NSA before Supercomputers'. Also see Peter Wright, *Spycatcher* (Heinemann, London, 1987), p. 195

12 Wright, *Spycatcher*, p. 196

13 Ibid., p. 148

14 Johnson, 'American Cryptology during the Cold War', Book 3, p. 159

15 Burke, *It Wasn't All Magic*, p. 242

16 *Innovating for Failure: Government Policy and the Early British Computer Industry* by John Hendry (MIT Press, Cambridge, MA. and London, 1990), p. 30

17 NSA, History Today, 14 June 2013, DOCID: 4126050

18 'Early Days in NSA Computer' (author's name classified), *Cryptolog*, August 1977

19 The names of the projects and prototypes from this period are bewildering, which shows the degree of experimentation. There was a special-purpose

machine called DELLA in the early 1950s, general-purposes computers like the IBM 701 and 704 and BOGART and SOLO (the first attempt to build a computer entirely using transistors rather than valves). ROGUE was the first system to use remote terminals connected to a central processor.

20 *Electronic Brains: Stories from the Dawn of the Computer Age* by Mike Hally (Granta Books, London, 2006), and John Pinkerton, Oral History, Babbage Institute, OH149

21 Arthur Humphreys, Oral History, Babbage Institute

22 National Archives AB 16/2929, AB 16/38977 and AB 16/2046

23 US National Archives, G 457, NSA, Entry A1 9037, Project Harvest, Box 1, 158881 and Burke, *It Wasn't All Magic*, pp. 295 and 311

24 Burke, *It Wasn't All Magic*, p. 265

25 'Lightning' by Howard Campaigne (NSA declassified document), https://www.nsa.gov/public_info/_files/tech_journals/Lightning.pdf

26 NSA Oral History interview with Howard Campaigne, NSA OH-14-83

27 Burke, *It Wasn't All Magic*, pp. 205 and 286

28 Ibid., pp. 53, 313, 315 and 308

29 Quoted on p. 59 of *The Craft of Intelligence* by Allen Dulles (The Lyons Press, Guildford, CT, 2006) ; Security Compartmentation of Computer Data, 2 February 1964, Declassified and available on CIA website and Prospectus for Science and Technology, Albert D. Wheelon, September 1966, declassified CIA memo

30 Johnson, 'American Cryptology during the Cold War', Book 2, p. 293

31 Burke, *It Wasn't All Magic*, p. 277

32 Along with cryptanalysis and data processing, generating your own random number keys was the other main use for computers. Johnson, 'American Cryptology during the Cold War', Book 1, Chapter 4: The Soviet Problem, p. 219

33 Solomon Kullback, NSA Oral History

Chapter Four: Coming of Age

1 'The Cybernated Generation', *Time* magazine, 2 April 1965

2 US National Archives, RG 457, NSA, Entry A1 9032, HCC, Box 1296, NR 3873 and Currier, 'My Purple Trip to England'

3 1954 Report for Admiralty Signals and Radar Establishment in National Archives ADM 220/675. Britain began building its own computer-controlled fighter intercept system known as Fire Brigade using an Elliott computer at Borehamwood linked to the RAF in Norfolk – see Lavington, *Moving Targets*.

4 http://www-03.ibm.com/ibm/history/ibm100/us/en/icons/sage/

5 See for instance Roger Schell, http://conservancy.umn.edu/bitstream/11299/133439/1/oh405rrs.pdf. Also see Scott Sagan, *The Limits of Safety: Organizations, Accidents and Nuclear Weapons* (Princeton University Press, Princeton, NJ, 1993), Chapter 3

6 http://conservancy.umn.edu/bitstream/11299/133439/1/oh405rrs.pdf

7   Interview with Roger Schell, 2014

8   Some details of Anderson's background and the context are from Roger
    Schell's oral history, http://conservancy.umn.edu/bitstream/11299/133439/1/
    oh405rrs.pdf

9   A 1970 Defense Science Board Report identified this as the major problem in
    1970 but did not go into as much detail as the Anderson Report of 1972.

10 'Cybersecurity: A Pre-History' by Michael Warner, *Intelligence and National
    Security*, vol. 27, issue 5, pp. 781–99

11 Roger Schell's oral history, http://conservancy.umn.edu/bitstream/11299/
    133439/1/oh405rrs.pdf

12 'The Beginning of Packet Switching: Some Underlying Concepts' by Paul
    Baran, *IEEE Communications Magazine*, July 2002

13 *A Brief History of the Future* by John Naughton (Phoenix, London, 2000), p. 124

14 D. W. Davies interviewed by M. Campbell-Kelly at the National Physical
    Laboratory, Monday, 17 March 1986

15 Information from National Museum of Computer History display at
    Bletchley Park. Other accounts say Davies had not been allowed to go over to
    the US to get involved in their work ('So, who really did invent the Internet?'
    by Ian Peter, NetHistory).

16 http://nrg.cs.ucl.ac.uk/internet-history.html and Naughton, *A Brief History of
    the Future*

17 For a good description of this, see John Naughton's *From Gutenberg to
    Zuckerberg: What You Really Need to Know about the Internet* (Quercus, London,
    2012)

18 *Hackers: Heroes of the Computer Revolution* by Stephen Levy (O'Reilly,
    Sebastopol, CA, 2010) and http://www.wheels.org/spacewar/creative/
    SpacewarOrigin.html

19 Interview with Herb Lin, 2013

20 'Computer Security: the Achilles' heel of the electronic Air Force?' by Roger
    Schell, *Air University Review*, January–February 1979. Copy provided by Roger
    Schell.

21 http://en.wikipedia.org/wiki/The_414s

22 *Cached: Decoding the Internet in Global Popular Culture* by Stephanie Ricker
    Schulte (New York University Press, New York/London, 2013), pp. 21–6

23 Warner, 'Cybersecurity: A Pre-History'

24 Six months later a faulty computer chip (costing 64 cents) led to another
    false warning of Soviet attack. For detailed accounts, see Sagan, *The Limits
    of Safety*, Chapter 5, and *Command and Control: Nuclear Weapons, the Damascus
    Accident, and the Illusion of Safety* by Eric Schlosser (Allen Lane, London, 2013),
    pp. 365–8

Chapter Five: Spy-hunting

1   *Defence of the Realm: The Authorised History of MI5* by Christopher Andrew
    (Penguin, London, 2009), also http://www.stephen-stratford.co.uk/

janssen_and_roos.htm and http://journals.chapman.edu/ojs/index.php/
VocesNovae/article/view/634/869

2 Wright, *Spycatcher*, p. 153

3 Ibid., pp. 82–4

4 'Tempest: A Signal Problem' (NSA partially declassified document)

5 This was only in the early 1980s. Mark Urban, *UK Eyes Alpha: Inside Story of
British Intelligence* (Faber and Faber, London, 1996), p. 256.

6 Daniel Edwards interview, OH427, Babbage Institute

7 Johnson, 'American Cryptology during the Cold War', Book 2, p. 368

8 Report of the Computer Study Group, May 1973 (NSA declassified document)

9 One former senior NSA man wrote that conventional wisdom was that 'In
the early days, NSA and its predecessor organizations drove the computer
industry. In the 1960s we kept pace with it. We started losing ground in the
'70s, and in the '80s we struggled to keep up with the industry', quoted in
Burke, It *Wasn't All Magic*, p. xi

10 'Soviet Computing and Technology Transfer: An Overview' by
Seymour Goodman, *World Politics*, vol. 31, no 4, 1979, http://
www.eecs.berkeley.edu/~alanmi/publications/other/SEG_WP_
SovietComputing&TechTransfer.pdf

11 'A View of the Central Computer Complex in the late 1970s and early 1980s'
by Cecil J. Phillips, *Cryptolog*, March 1977

12 Burke, *It Wasn't All Magic*, p. 316, and interview with veteran engineer

13 The role of cryptanalysis in events like breaking Argentine codes during the
Falklands War in 1982 meant that the military and other officials were not
going to want to wave goodbye to the capability.

14 Full details of how the complex operated remain classified, but from what
can be gleaned a front-end complex received communications from the field
sites which did the intercepting (usually UNIVAC and Honeywell computers)
and then broke it down based on what target it was. Then came a processing
function – which batched the signals down and put them onto magnetic
tapes, carrying out traffic analysis; the content of the tapes was then every
few hours sent on a data link to the NSA. Then at Fort Meade, the RYE
complex processed it so that it could be sent out to analysts. Two complexes
were the home of early supercomputers like the CDC 6600 and CDC 7700
and then later the CRAY 1 purchased in 1976. A young systems engineer
suggested connecting together all six computer complexes with standard
connection protocols based on the technology developed for ARPANET but
with extra security (flaws in this system emerged though). The US military
also developed a secret device called Blacker to secure communications over
TCP/IP.
'The PLATFORM Network Evolution' (NSA declassified paper), available
at http://www.nsa.gov/public_info/_files/cryptologic_quarterly/The_
PLATFORM_Network_Evolution.pdf and Johnson, 'American Cryptology
during the Cold War', Book 3, p. 151

15 'The Yawn of the Computer Age – Or When your Terminal is Terminal' (author's name classified), *Cryptolog*, January 1975

16 'Some Reflections on the Reality of Computer Security' by Robert J. Hanyok, *Cryptolog*, June–July 1982. Also see 'Computer Security Policy for the Intelligence Community', *Cryptolog*, 2nd issue 1989 for reference to 1972 directive.

17 Interview with Victor Sheymov, 2014

18 Lord Curzon, quoted in Smith, *The Secrets of Station X*, p. 7

19 *Secrecy and the Media* by Nicholas Wilkinson (Routledge, London, 2009), Chapter 38

20 'Recollections from the Church Committee's Investigation of NSA' by L. Britt Snider, CIA website

21 *The Puzzle Palace: A Report on America's Most Secret Agency* by James Bamford (Sidgwick & Jackson, London, 1983), Chapter 6, and *Body of Secrets: Anatomy of the Ultra-Secret National Security Agency* by James Bamford (Doubleday, New York, 2001), p. 587

22 Frank Rowlett, NSA Oral History

23 'Unlucky Shamrock – The View from the Other Side' by James H. Hudec, *Studies in Intelligence*, no. 10 (2001), pp. 85–94. See also Bamford, *The Puzzle Palace*, Chapter 6. The companies had begun to pass traffic although nervous executives continued to seek assurances from the President and Secretary of Defense that in the first few years they would not be prosecuted. However legislation was never enacted. The assumption within the NSA was that presidential directives granted the authority to pursue targets as long as there was a foreign intelligence purpose, a view others would dispute.

24 Wright, *Spycatcher*, p. 360

25 Interview with Stella Rimington, see 'Book Tells of MI5's Secret Past' by Gordon Corera, BBC News, 2009 http://news.bbc.co.uk/nol/ukfs_news/hi/newsid_8280000/newsid_8289800/8289897.stm

26 *Memoirs of a Spymaster: The Man Who Waged a Secret War Against the West* by Markus Wolf (Pimlico, London, 1998), pp.112 and 331

Chapter Six: Crypto Wars

1 The quotes from the meeting in 1976 are drawn from a transcript, http://toad.com/des-standford-meeting.html, with additional information and recollections from interviews with Martin Hellman and Whit Diffie in 2014. *Star Wars* actually came out in 1977 once the battle had been fully joined with the NSA, so there may be a touch of hindsight in Hellman's reference. Further information from Stephen Levy's *Crypto*.

2 *Crypto: Secrecy and Privacy in the New Cold War* by Steven Levy (Penguin, London, 2002), pp. 37–49

3 Martin Hellman interview with Jeffrey R. Yost, http://conservancy.umn.edu/bitstream/handle/11299/107353/oh375mh.pdf?sequence=1&isAllowed=y

4 Interview with Whitfield Diffie, 2014

5   Johnson, 'American Cryptology during the Cold War', Book 3, p. 233

6   Ibid., p. 232

7   'Unclassified Summary: Involvement of NSA in the Development of the Data Encryption Standard', Staff Report of the Senate Select Committee on Intelligence, April 1978

8   Johnson, 'American Cryptology during the Cold War', Book 3, p. 239

9   Interview with Clifford Cocks by the author, 2010. Also see 'The History of Non-Secret Encryption' by J. H. Ellis, 1997

10 Levy, *Crypto*, p. 321

11 Johnson, 'American Cryptology during the Cold War', Book 3, p. 235

12 Ibid., p. 190

13 Interview with Admiral Bobby Ray Inman, 2014

14 Interview with Martin Hellman, 2014

15 Interviews with Martin Hellman and Admiral Bobby Ray Inman, 2014

16 Levy, *Crypto*, pp. 157–63

17 'Rigging the Game', *The Baltimore Sun*, 10 December 1995

18 Levy, *Crypto*, p. 196

19 Phil Zimmermann interview in 'Crypto Wars', BBC Radio 4, 2014

20 Levy, *Crypto*, p. 210

21 *This Machine Kills Secrets: How WikiLeakers, Cypherpunks and Hacktivists Aim to Free the World's Information* by Andy Greenberg (Virgin Books, London, 2012), pp. 119–21

22 *Cypherpunks: Freedom and the Future of the Internet* by Julian Assange (OR Books, London, 2012)

23 Stewart Baker interview in 'Crypto Wars', BBC Radio 4, 2014

24 'Don't Worry Be Happy', Stewart A. Baker, *Wired* magazine, June 1994

25 http://web.mit.edu/newsoffice/1997/morris-1126.html

26 Interview with James Lewis, 2013

Chapter Seven: Attack

1   One of the only connections was for the UK academic network JANET but that had an email-only gateway onto ARPANET. *A Pathology of Computer Viruses* by David Ferbrache (Springer-Verlag, London, 1992), p. 195

2   Quotes of conversation from *The Cuckoo's Egg: Tracking a Spy Through the Maze of Computer Espionage* by Cliff Stoll, (Pocket Books, New York, 2007), Epilogue; *New York Times* obituary of Robert H. Morris; Incident reports, http://foofus.com/amuse/public/Morris_Worm_Incident_Report_1.pdf; Cornell Investigation, http://www.cs.cornell.edu/courses/cs1110/2009sp/assignments/a1/p706-eisenberg.pdf; Technical report into the worm, http://www.academia.edu/2771263/The_Internet_Worm_Incident_Technical_Report_CSD-TR-933

3   Quoted in 'A youth's passion for computers gone sour', Michael Wines, *New York Times*, 11 November 1988

4   *Cyberpunk: Outlaws and Hackers on the Computer Frontier* by Katie Hafner and John Markoff (Touchstone, New York, 1991), p. 264
5   Dennis Richie published his recollections of the incident online at http://www.cs.bell-labs.com/who/dmr/crypt.html
6   Dennis Richie's account is at http://cm.bell-labs.com/who/dmr/crypt.html
7   http://blog.kaspersky.co.uk/morris-worm-turns-25/
8   Hafner and Markoff, *Cyberpunk*, pp. 262–3
9   'Robert Morris, a developer of UNIX, dies at 78', T. Rees Shapiro, *Washington Post*, 30 June 2011
10 Hafner and Markoff, *Cyberpunk*, p. 274
11 Quoted in 'A youth's passion for computers gone sour', Michael Wines, *New York Times*, 11 November 1988
12 Ferbrache, *A Pathology of Computer Viruses*, p. 3
13 http://www.nsa.gov/public_info/_files/cryptologic_quarterly/computervirus.pdf
14 Daniel Edwards interview, OH427, Babbage Institute
15 Ferbrache, *A Pathology of Computer Viruses*, Chapter 2
16 'A Lesson in Computer Security' (author's name classified), *Cryptolog*, 2nd issue, 1988; NSA documents list the attack as in April. In Hafner and Markoff, *Cyberpunk*, p. 79, Mitnick's attack is written up as the previous month but the method of attack matches precisely between the two accounts. Mitnick also refers to getting into the non-secure part of the NSA in *Ghost in the Wires: My Adventures as the World's Most Wanted Hacker* by Kevin Mitnick (Bay Back Books, New York, 2011)
17 See for instance Mitnick, *Ghost in the Wires*
18 Mitnick, *Ghost in the Wires*, p. 75
19 Ibid., p. 115

Chapter Eight: Enter the KGB

1   Hafner and Markoff, *Cyberpunk*, p. 221
2   Account primarily drawn from Stoll, *The Cuckoo's Egg*; also Hafner and Markoff, *Cyberpunk*
3   Interview with Jim Christy, 2014
4   The Farewell Dossier by Gus Weiss, https://www.cia.gov/library/center-for-the-study-of-intelligence/csi-publications/csi-studies/studies/96unclass/farewell.htm and *Farewell: The Greatest Spy Story of the Twentieth Century* by Sergei Kostin and Eric Raynaud (Amazon Crossing, Las Vegas, 2011), and *The Mitrokhin Archive: The KGB in Europe and the West* by Christopher Andrew and Vasili Mitrokhin (Penguin, London, 2000), p. 285
5   Andrey Zverev of Ruselectronics speaking on Ekho Moskvy Radio, 14 October 2013
6   The Soviet Bloc's Unified System of Computers, available at http://old.cistp.gatech.edu/programs/inter-diff-innov-info-tech/docs/The%20Soviet%20Bloc's%20Unified%20System%20of%20Computers.pdf

7   Andrew and Mitrokhin, *The Mitrokhin Archive*, pp. 285 and 245

8   Wolf, *Memoirs of a Spymaster*, p. 182

9   Hafner and Markoff, *Cyberpunk*, p. 173

10 Ibid., p. 193

Chapter Nine: Out of the Cold and into Cyberspace

1   http://info.cern.ch/hypertext/WWW/TheProject.html

2   There were reports in the US about a large-scale operation called Rahab by German intelligence but this does not appear to be correct according to German officials. There was more limited espionage in the 1980s according to Erich Schmidt-Eenboom and his book *Die schmutzigen Geschäfte der Wirtschaftsspionage* (Düsseldorf, Econ Verlag, 1994), pp. 90–93

3   *GCHQ: The Uncensored Story of Britain's Most Secret Intelligence Agency* by Richard Aldrich (London, Harper Press, 2010), p. 468

4   'Thoughts on a knowledge base to support information operations in the next millennium', *Cryptolog*, Spring 1997

5   'Cornerstones of Information Warfare' quoted in Warner, 'Cybersecurity: A Pre-History'

6   'Thinking out loud about cyberspace' by William B. Black, Director's Special Assistant for Information Warfare, *Cryptolog*, Spring 1997

7   Interview with Michael Hayden, 2013

8   'Information Warfare: The War of the Future' (author's name redacted), *Cryptolog*, Summer 1996

9   'Information Security: Computer Attacks at Department of Defense Pose Increasing Risks', GAO Report, 1996

10 Interviews with members of US Air Force investigative team. Also 'Information Security: Computer Attacks at Department of Defense Pose Increasing Risks', statement of Jack L. Brock, US Government Accounting Office, 1996, and 'Security in Cyberspace', US Senate permanent subcommittee on investigations appendix, http://www.fas.org/irp/congress/1996_hr/s960605b.html and interview with Jim Christy, 2014

11 Interview with Mark Morris, 2014

12 'The Schoolboy Spy' by Jonathan Ungoed-Thomas, *Sunday Times*, 4 April 1998

13 'UK hacker says he found anti-gravity engine file' by Matthew Williams, 2 July 1999, from Kuji Media website run by Mathew Bevan

14 Ungoed-Thomas, 'The Schoolboy Spy', *Sunday Times*, 4 April 1998

15 'Solar Sunrise' by Tim Mauer, in *A Fierce Domain: Conflict in Cyberspace 1986–2012*, ed. Jason Healey (Cyber Conflict Studies Association, Arlington, 2013)

16 Warner, 'Cybersecurity: A Pre-History'

17 John Hamre, PBS Frontline interview, http://www.pbs.org/wgbh/pages/frontline/shows/cyberwar/interviews/hamre.html

18 Interview with Richard Clarke, 2013

19 https://www.fas.org/sgp/library/pccip.pdf
20 PBS Frontline interview with John Hamre, 2003, http://www.pbs.org/wgbh/pages/frontline/shows/cyberwar/interviews/hamre.html
21 Warner, 'Cybersecurity: A Pre-History'
22 PBS Frontline interview with John Hamre, 2003, http://www.pbs.org/wgbh/pages/frontline/shows/cyberwar/interviews/hamre.html
23 'Private Information and Moonlight Maze' by Adam Elkus, in Healey (ed.), *A Fierce Domain*

Chapter Ten: Titan Rain

1 The story is recounted in a US defence publication, 'Foreign attempts to gain US technology from defense industry', *Targeting US Technologies 2012*, Defense Security Service. Further details at http://www.pbs.org/wgbh/theymadeamerica/whomade/slater_hi.html and http://www.bloombergview.com/articles/2013-02-01/piracy-and-fraud-propelled-the-u-s-industrial-revolution
2 *For All the Tea in China: Espionage, Empire and the Secret Formula for the World's Favourite Drink* by Sarah Rose (Arrow Books, London, 2010)
3 'Hitachi-FBI tapes are released', *New York Times*, 16 May 1983, and *Friendly Spies: How America's Allies Are Using Economic Espionage to Steal Our Secrets* by Peter Schweizer (Atlantic Monthly Press, New York, 1993), p. 50
4 Interview with Brian Shields, 2013, and 'Chinese hackers suspected in long-term Nortel Breach', *Wall Street Journal*, 14 February 2012
5 A participant's list from 2001 is at http://www.dhs.gov/sites/default/files/publications/Network%20Security%20Information%20Exchanges_2001.pdf
6 http://www.marbridgeconsulting.com/marbridgedaily/2008-08-19/article/18911/huawei_denies_low_bid_price_for_cdma_equipment
7 http://www.obj.ca/Technology/2014-03-18/article-3653809/Nortel%26rsquo%3BS-failure-was-years-in-the-making,-Telfer-study-finds/1
8 *Chinese Industrial Espionage: Technology, Acquisition and Military Modernisation* by William C. Hannas, James Mulvenon and Anna B. Puglisi (Routledge, London, 2013)
9 Quoted ibid., p. 189
10 Ibid., p. 13
11 '1998 report told of lab breaches and China threat', *New York Times*, 2 May 1999
12 'China's Economic Espionage' by James A. Lewis, *Foreign Affairs*, 13 November 2012
13 'The invasion of the Chinese Cyberspies', *Time* magazine, 29 August 2005, and *Surviving Cyberwar* by Richard Stiennon (Government Institutes, Lanham, 2010), Chapter 1
14 Plaintiff's Memorandum in Opposition to Summary Judgment, Shawn Carpenter v. Sandia Corporation
15 'Q&A: Reverse Hacker describes Ordeal', *Computerworld*, 26 February 2007

16 'The invasion of the Chinese Cyberspies', *Time* magazine, 29 August 2005

17 *China's Cyber Power and America's National Security* by Colonel Jayson M. Spade, United States Army and US-China Economic and Security Review Commission Report on the Capability of the People's Republic of China to Conduct Cyber Warfare and Computer Network Exploitation

18 http://www.esecurityplanet.com/trends/article.php/3513761/UK-Under-Cyber-Attack.htm

19 http://online.wsj.com/news/articles/SB124027491029837401

20 'Chinese steal jet secrets from BAE', *Sunday Times*, 11 March 2012

21 Documents published by *Der Spiegel*, 17 January 2015, http://www.spiegel.de/international/world/new-snowden-docs-indicate-scope-of-nsa-preparations-for-cyber-battle-a-1013409-2.html

22 Wikileaks, Diplomatic Security Daily, 3 November 2008, and Snowden documents released by *Der Spiegel*, January 2015

23 https://www.microsoft.com/en-us/news/features/2012/jan12/gatesmemo.aspx

24 The group was also known as the 'Comment Crew' by some who watched its work because it hid code in the comments on web pages, or as the Shanghai Group by others.

25 'Does China's New J-20 Stealth Fighter Have US Technology?' by James A. Lewis, Center for Strategic and International Studies Commentary, 26 January 2011

26 'Chinese Army Unit is seen as tied to hacking against US', *New York Times*, 18 February 2013

27 http://www.bloomberg.com/news/2011-02-24/exxon-shell-bp-said-to-have-been-hacked-through-chinese-internet-servers.html

28 Interview with Kevin Mandia, 2014

29 'China Hacker's angst opens a window onto cyber-espionage' by Barbara Demick, *Los Angeles Times*, 12 March 2013

30 'The Chinese Defense Economy Takes Off', ed. Tai Ming Cheung, iGCC, 2013, p. 20

31 Interview with David DeWalt, Chairman of Mandiant, 2013

32 Snowden documents released by *Der Spiegel*, January 2015

33 Interviews with Nigel Inkster and James Lewis, 2013

Chapter Eleven: In the Wires

1 National Archives CAB 17/92, official report into wartime censorship

2 Wikileaks cable, Huawei: The Human Resource Component of a Modern Chinese Company, 15 December 2006

3 Quoted in 'Just because Huawei crushes all comers in telecom doesn't mean China will conquer the world, does it?', *The Globe and Mail*, 25 November 2011

4 *Sinophobia: The Huawei Story* by Eric Anderson (CreateSpace, 2013)

5 'Technology Two-Timing' by Kelly Motz and Jordan Richie, *The Asian Wall Street Journal*, 19 March 2001

6   Court filing, Cisco v. Huawei, US District Court for the Eastern District of Texas, 22 January 2003

7   A few of the details of a neutral expert's report were selectively released by Cisco in 2012 in response to comments by Huawei that the American company thought misleading. These excerpts suggested that Huawei had access to Cisco code and it had been electronically copied. However, the full details remain closed. http://blogs.cisco.com/news/huawei-and-ciscos-source-code-correcting-the-record/ For Chinese narrative, see Wikileaks, Shanghai IPR Roundtable, 19 February 2010

8   'The AT&T Divestiture and National Security' (author's name redacted), *Cryptolog*, August 1984

9   http://www.zdnet.com/blog/projectfailures/fbi-counterfeit-cisco-routers-risk-it-subversion/740

10 Wikileaks, Huawei says 3Com plans still uncertain, 21 March 2008

11 Information drawn from interviews with a range of people involved on both the commercial and government side. Also Intelligence and Security Committee Report on Foreign Involvement in the Critical National Infrastructure, June 2013

12 Intelligence and Security Committee Report on Foreign Involvement in the Critical National Infrastructure, June 2013

13 'China could use BT network to cripple Britain', *Daily Mail*, 30 March 2009

14 'Spy chiefs fear Chinese cyber attack', *Sunday Times*, 29 March 2009, http://www.thesundaytimes.co.uk/sto/news/uk_news/article158319.ece

15 Huawei Cyber Security Evaluation Centre: Review by the National Security Adviser, December 2013

16 Intelligence and Security Committee Report on Foreign Involvement in the Critical National Infrastructure, June 2013, and private information

17 http://johnsuffolk.typepad.com/john-suffolk/2011/07/following-approval-from-the-uk-prime-minister-john-suffolk-to-join-huawei-as-their-global-head-of-cyber-security-reporting-to.html

18 Investigative Report on the US National Security Issues posed by Chinese Telecommunications Companies Huawei and ZTE, US House of Representatives Permanent Select Committee on Intelligence, 8 October 2012

19 'Huawei spied for China, claims ex-CIA head Michael Hayden', *Financial Times*, 18 July 2013

20 Anderson, *Sinophobia*

21 Interview with Mike Rogers, 2013

22 Intelligence and Security Committee Report on Foreign Involvement in the Critical National Infrastructure, June 2013

23 Huawei Cyber Security Evaluation Centre: Review by the National Security Adviser, December 2013

24 Huawei also claims that reports linking the Chairwoman of Huawei, Sun Yafang, to the Ministry of State Security are erroneous (Chinese media reports said she worked in the Communications Department).

25 Interview with Chen Lifang, 2013
26 Interview with Scott Charney, 2013
27 Quoted in 'New telecoms superpower Huawei looks West to build its dream', *Daily Telegraph*, 31 March 2013
28 http://blogs.cisco.com/news/comment-on-der-spiegel-articles-about-nsa-tao-organization/
29 http://www.bbc.co.uk/news/technology-27468794
30 Australian Financial Review, interview transcript: former head of NSA and commander of the US Cyber Command, General Keith Alexander, 8 May 2014

Chapter Twelve: Britain and the Cyber Spies

1 'How the internet came to the Tibetans', Tibet report, 14 July 2009, https://tibetreport.wordpress.com/2009/07/14/how-the-internet-came-to-the-tibetans/ and interview with Thubten Samdup and other activists
2 http://www.tibetangeeks.com/geeks/history/how_the_internet_came_to_dhasa_tibetans-dan_haig-20130927.html
3 'State hackers spying on us, say dissidents', *South China Morning Post*, 18 September 2002
4 'Google's claim is old news to China activists' by Barbara Demick, *Los Angeles Times*, 15 January 2010
5 The twenty-something was the author, who had to explain to the then Prime Minister how to use a mouse while producing a TV broadcast in Downing Street soon after the 1997 general election
6 'Titan Rain: How Chinese hackers targeted Whitehall', *Guardian*, 5 September 2007, http://www.theguardian.com/technology/2007/sep/04/news.internet
7 RSA posting: https://blogs.rsa.com/anatomy-of-an-attack/
8 Interview with Sir Iain Lobban for BBC Radio 4, 2013
9 'Industrial espionage: the American advantage', *Le Monde*, 31 October 2013. In 2010, the NSA had an operation called Ironavenger, in which it watched an ally conduct a spear-phishing attack on an adversary and monitored what was taken out and how their ally operated, gaining valuable insight. 'No morsel too minuscule for all-consuming NSA', *New York Times*, 2 November 2013. See also *Der Spiegel*, 'The digital arms race', 17 January 2015, http://www.spiegel.de/international/world/new-snowden-docs-indicate-scope-of-nsa-preparations-for-cyber-battle-a-1013409-2.html
10 Interview with MI5 head of cyber, 2013
11 'MI5 warns universities on cyber spying', *Financial Times*, 10 April 2013
12 'Significant Cyber Incidents since 2006', Center for Strategic and International Studies, http://csis.org/files/publication/140310_Significant_Cyber_Incidents_Since_2006.pdf
13 'Figures from Rio Tinto case highlight risks in China', *Financial Times*, 5 April 2010

14 'William Hague: UK is under cyber attack', BBC News, 4 February 2011, http://www.bbc.co.uk/news/uk-12371056

15 George Osborne speech at Google Zeitgeist 2011, https://www.gov.uk/government/speeches/speech-by-the-chancellor-of-the-exchequer-rt-hon-george-osborne-mp-at-google-zeitgeist-2011

16 'Hackers have breached Top Secret MOD systems, cyber security chief admits', *Guardian*, 3 May 2012, http://www.theguardian.com/technology/2012/may/03/hackers-breached-secret-mod-systems

17 'GCHQ: Inside the top secret world of Britain's biggest spy agency', *Guardian*, 1 August 2013

18 'Cabinet split over bullying by Beijing', *Sunday Times*, 17 February 2013

19 Wikileaks, Diplomatic Security Daily, 3 November 2008

20 'Digital spying burdens German-Chinese relations', *Der Spiegel*, 25 February 2013

21 'China blamed after ASIO blueprints stolen in major cyber attack on Canberra HQ', ABC, 28 May 2013

22 Interview with David DeWalt, 2013

23 'GCHQ: Inside the top secret world of Britain's biggest spy agency', *Guardian*, 1 August 2013, http://www.theguardian.com/world/interactive/2013/aug/01/gchq-spy-agency-nsa-edward-snowden

24 Interview with James Lewis, 2013

25 Interview with David Garfield of BAE Detica, 2013

26 http://www.telegraph.co.uk/earth/copenhagen-climate-change-confe/6746370/Climategate-was-Russian-secret-service-behind-email-hacking-plot.html

27 Interview with Bill Lynn, 2013

28 NSA declassified documents published by *Der Spiegel*, 17 January 2015, http://www.spiegel.de/international/world/new-snowden-docs-indicate-scope-of-nsa-preparations-for-cyber-battle-a-1013409-2.html

29 'NSA leaks make place for cyberdefense unlikely', *New York Times*, 12 August 2013, and 'US plan calls for more scanning of private Web traffic', email, Reuters, 21 March 2013

Chapter Thirteen: Dissent

1 Quoted in 'Cyber Spies', File on Four, broadcast on Radio 4, 20 September 2011

2 'Tracking GhostNet: Investigating a Cyber Espionage Network', Information Warfare Monitor, 29 March 2009, and interview with Thubten Samdup

3 Quoted in 'Cyber Spies', File on Four, broadcast on Radio 4, 20 September 2011

4 'Tracking GhostNet': Investigating a Cyber Espionage Network, Information Warfare Monitor, 29 March 2009

5 'The Snooping Dragon: social-malware surveillance of the Tibetan movement' by Shishir Nagaraja and Ross Anderson, University of Cambridge Computer Laboratory Technical Report, March 2009

6   Interview with Ron Deibert, 2013, and *Black Code: Inside the Battle for Cyberspace* by Ron Deibert (McLelland & Stewart, Toronto, 2013), pp. 148–51

7   Interview with Tenzin Seldon, 2013

8   'China hacking nearly destroyed our movie', *Hollywood Reporter*, 19 April 2013, and 'Tibet taboo leads to cyberattacks on film crew', *Washington Post*, 28 March 2013

9   Interview with Heather Adkins, 2013

10 'Occupying the Information High Ground: Chinese Capabilities for Computer Network Operations and Cyber Espionage', report prepared for the US-China Economic and Security Review Commission, 7 March 2012

11 'Enter the cyber-dragon' by Michael Joseph Gross, *Vanity Fair*, September 2011

12 'Google confronts China's "Three Warfares"', Timothy L. Thomas, *Parameters*, Summer 2010

13 Quoted in 'Enter the cyber-dragon' by Michael Joseph Gross, *Vanity Fair*, September 2011

14 'Google confronts China's "Three Warfares"', Timothy L. Thomas, *Parameters*, Summer 2010

15 http://googleblog.blogspot.co.uk/2010/01/new-approach-to-china.html

16 'Vast hacking by a China fearful of the web', *New York Times*, 4 December 2010

17 Wikileaks cable, Google China Paying Price for Resisting Censorship; Requests High-level Usg Help, 18 May 2009

18 Wikileaks cable, Google Update: PRC Role in Attacks and Response Strategy, 26 January 2010

19 Wikileaks cable, from US Embassy Beijing, 25 January 2010

20 Quoted in 'Enter the Cyber Dragon' by Tobias Feakin, Australian Strategic Policy Institute Report, June 2013

21 'China and the Internet', *Economist* Special Report, 6 April 2013

22 Urban, *UK Eyes Alpha*, p. 111

23 'You've got dissent', RAND Monograph, 2002, http://www.rand.org/pubs/monograph_reports/MR1543.html

24 Interview with Nigel Inkster, 2013

25 Deibert, *Black Code*, pp. 118 and 74

26 http://www.telegraph.co.uk/news/worldnews/1316520/Spy-plane-video-raises-tension-ahead-of-meeting.html

27 'Early Patriotic Hacking' by Jonathan Diamond, in Healey (ed.), *A Fierce Domain*. For Code Red Worm see http://www.scientificamerican.com/article/code-red-worm-assault-on/

28 Quoted in 'Operation Saffron Rose', Report by Fire Eye, 2014

29 In 1999, versions of Windows were found to have a piece of code called _NSAKEY which some believed was a backdoor for the NSA into the cryptographic systems in export versions, a possibility denied by the company.

30 Interview with Dr Huang Huikang, Beijing, 2013
31 'You've got dissent', RAND Monograph, 2002, http://www.rand.org/pubs/monograph_reports/MR1543.html
32 'Firms aided Libyan spies', *Wall Street Journal*, 30 August 2011, http://www.wsj.com/articles/SB10001424053111904199404576538721260166388

Chapter Fourteen: Sabotage
1 Interview by the author with Joachim Rønneberg, 2013, to mark the seventieth anniversary of Operation Gunnerside. http://www.bbc.co.uk/news/uk-22298739
2 'Stuxnet's Secret Twin' by Ralph Langer, *Foreign Policy*, 25 November 2013
3 'Stuxnet and the Future of Cyber War' by James P. Farwell and Rafal Rohozinski, *Survival: Global Politics and Strategy*, vol. 53, no. 1 (2011), pp. 23–40 and comments from Eugene Kaspersky regarding the Russian plant
4 Interview with Eugene Kaspersky, 2013
5 USA, Israel promoted retaliatory cyber attacks – Iran TV, BBC Monitoring translation of an Iranian TV documentary broadcast on 2 November 2012
6 'Obama order sped up wave of cyberattacks against Iran', David Sanger, *New York Times*, 1 June 2012
7 These details come from individuals with knowledge of the Natanz facility and the impact of Stuxnet. They also tally with other reports such as 'Basic Attack Strategy of Stuxnet 0.5' by the Institute for Science and International Security, February 2013. Also 'Stuxnet's Secret Twin' by Ralph Langer, *Foreign Policy*, 25 November 2013, and BBC *Horizon*, 'Defeating the Hackers', broadcast 19 August 2013. The most detailed account of Stuxnet so far is in *Countdown to Zero Day: Stuxnet and the Launch of the World's First Digital Weapon* by Kim Zetter (Crown, New York, 2014)
8 'A declaration of cyber-war' by Michael Joseph Gross, *Vanity Fair*, April 2011
9 'Obama order sped up wave of cyberattacks against Iran', David Sanger, *New York Times*, 1 June 2012
10 Interview with John Bumgarner, 2013
11 http://arstechnica.com/security/2012/06/flame-crypto-breakthrough/
12 'US, Israel developed Flame computer virus to slow Iranian nuclear efforts, officials say', *Washington Post*, 19 June 2012
13 It also says GCHQ at the same time was apparently pushing for a trilateral partnership with the US and Israel to work on Iran rather than for the UK and US to share with Israel separately. NSA briefing note for visit by Iain Lobban, reproduced by First Look, https://firstlook.org/theintercept/document/2014/04/30/lobban-nsa-visit-precis/
14 'Silent war' by Michael Joseph Gross, *Vanity Fair*, July 2013
15 For the most detailed account of 'Olympic Games', see *Confront and Conceal: Obama's Secret Wars and Surprising Use of American Power* by David Sanger (Crown, New York, 2012)
16 Interview with Michael Hayden, 2013

17 USA, Israel promoted retaliatory cyber attacks – Iran TV, BBC Monitoring translation of an Iranian TV documentary broadcast on 2 November 2012

18 'Iran's Cyber War: Hackers in Service of the Regime; IRGC Claims Iran Can Hack Enemy's Advanced Weapons Systems; Iranian Army Official: "The Cyber Arena Is Actually the Arena of the Hidden Imam"' by Y. Mansharof, MEMRI report, 25 August 2013

19 'Cyber threats to energy security, as experienced by Saudi Arabia' by John Roberts, The Barrel Blog, Platts, 27 November 2012

20 http://www.cybersecurity.ox.ac.uk/seminars/821.html – the speaker did not name Aramco or RasGas but it is clear those were the incidents he was discussing.

21 'Saudi official says cyber attack on Aramco planned, executed from "outside"', Arab News website, 10 December 2012. Also see Christopher Bronke and Enneken Tikk-Ringas, 'The cyber attack on Saudi Aramco', *Survival*, vol. 55, issue 2, pp. 81–96

22 *Cyberpower and National Security, Summary of a Roundtable Discussion of October 2012*, American Foreign Policy Interests

23 http://www.bloomberg.com/news/2012-10-25/code-in-aramco-cyber-attack-indicates-lone-perpetrator.html

24 Interview with Richard Clarke, 2013

25 Interview with President Ilves, Estonia, 2013 and other interviews with officials involved in responding to the campaign

26 'The Estonian Cyberattacks' by Andreas Schmidt, in Healey (ed.), *A Fierce Domain*

27 'Cyber Espionage Against Georgian Government', Ministry of Justice of Georgia, http://dea.gov.ge/uploads/CERT%20DOCS/Cyber%20Espionage. pdf

28 US Cyber Consequences Unit Report, http://www.registan.net/wp-content/uploads/2009/08/US-CCU-Georgia-Cyber-Campaign-Overview.pdf

29 'In bleak Russia, a young man's thoughts turn to hacking', *New York Times*, 29 June 2000

30 'The Russo-Georgian War 2008' by Andreas Hagen, in Healey (ed.), *A Fierce Domain*

31 'Chinese military mobilises cybermilitias', *Financial Times*, 12 October 2011, http://www.ft.com/cms/s/0/33dc83e4-c800-11e0-9501-00144feabdc0. html#axzz35O4jJovv and UK statement https://www.gov.uk/government/organisations/joint-forces-command/about/recruitment

Chapter Fifteen: The Lights Go Off

1 'The cyber attack threat to London's Olympic Ceremony' by Gordon Corera, 8 July 2013, http://www.bbc.co.uk/news/uk-23195283

2 *Cyber Security and Global Interdependence: What is Critical?* by Dave Clemente, Chatham House Report, February 2013

3 'Cyberattack leaves natural gas pipelines vulnerable to sabotage', *Christian Science Monitor*, 27 February 2013

4   One example of reconnaissance may be the targeting of Canadian firm
    Telvent, which provides systems for much of the energy industry.
5   Interview with Michael Hayden, 2013
6   Interview with Chris Inglis, 2013
7   Interview with Michael Hayden, 2013
8   'US spy agencies mounted 231 offensive cyber-operations in 2011, documents
    show', *Washington Post*, 31 August 2013
9   http://www.theguardian.com/world/2013/jun/07/obama-china-targets-
    cyber-overseas
10 *Washington Post*, 27 January 2013
11 http://www.sans.org/newsletters/newsbites/newsbites.php?vol=10&issue=5
    and http://www.whitehouse.gov/assets/documents/Cyberspace_Policy_
    Review_final.pdf
12 'Secret DTI inquiry into cyber terror', *Sunday Times*, 9 June 1996
13 Interview with Michael Hayden, 2013
14 'Sony Pictures CEO says he had "no playbook" for mega-hack on
    studio', Associated Press, 9 January 2015; 'NSA breached North
    Korean networks before Sony attack, officials say', *New York
    Times*, 18 January 2015; http://www.newsweek.com/north-korea-
    training-cyber-warriors-289414 and http://uk.businessinsider.com/
    north-korean-defector-jang-se-yul-trained-with-hackers-2014-12?r=US
15 http://www.wired.co.uk/news/archive/2014-08/14/monstermind

Chapter Sixteen: Rebirth – Cables
1   Most cables now come into Bude in Cornwall, but a few still land at
    Porthcurno including the cable 'Flag Europe-Asia' as of September 2014.
2   That estimate looks to be based on a 480gbps so has been roughly doubled
    based on new cable capacity, http://www.porthcurno.org.uk/userfiles/misc_
    pdfs/Cable%20Networks%20website.pdf
3   'The Business of Sigint' by Sir David Pepper, *Public Policy and Administration*,
    January 2010
4   'Information Technology: Into the Next Millennium' by William P. Crowell,
    Cryptolog, issue 1, 1995
5   'Production and Reporting in a Changed Environment', *Cryptolog*, issue 1, 1995
6   'Why we spy on our allies' by R. James Woolsey, *Wall Street Journal*, 17 March
    2000
7   'The Intelligence Gap' by Seymour Hersh, *The New Yorker*, 6 December 1999
    and Aldrich, *GCHQ*, p. 468
8   'Cryptologic Almanac, 50th Anniversary Series, The First Round: NSA's Effort
    against International Terrorism in the 1970s' (NSA declassified document)
9   *The Secret Sentry: The Untold History of the National Security Agency* by Matthew
    Aid (Bloomsbury, New York, 2010), p. 108, and https://www.eff.org/
    deeplinks/2007/10/qwest-ceo-nsa-punished-qwest-refusing-participate-illegal-
    surveillance-pre-9-11

10 General Keith Alexander, keynote at Black Hat conference, 31 July 2013, and the counter view in 'The Al Qaeda Switchboard' by Lawrence Wright, *The New Yorker*, 13 January 2014

11 NSA Inspector General report, 24 March 2009, working draft

12 Figures from draft NSA IG report, http://www.theguardian.com/world/interactive/2013/jun/27/nsa-inspector-general-report-document-data-collection

13 Interviews with Richard Clarke and Michael Hayden, 2013

14 http://www.att.com/gen/press-room?pid=4800&cdvn=news&news articleid=22372

15 'New details show broader NSA surveillance reach', *Wall Street Journal*, 20 August 2013

16 New detailed analysis of how NSA and its foreign partners intercept undersea fibre optic cable traffic, 18 January 2014, on Matthew Aid's website citing work by Peter Koop.

17 Glenn Greenwald, *No Place to Hide: Edward Snowden, the NSA and the US Surveillance State* (Hamish Hamilton, London, 2014), pp. 105 and 107

18 'Agreements with private companies protect US access to cables data for surveillance', *Washington Post*, 7 July 2013

19 Known as RAMPART-A and reported to provide access to over three terabits a second of data through seventy different cables. 'NSA "third party" partners tap the internet backbone in global surveillance programme', *Dagbladet*, 19 June 2014

20 Greenwald, *No Place to Hide*, p. 123. One issue Snowden raised was that even though a country like Denmark might say that its tap cannot be used to spy on Danes, and the Germans' not to spy on Germans, each other's nationals traffic will flow through the other's pipes without those restrictions applying. 'NSA "third party" partners tap the internet backbone in global surveillance programme', *Dagbladet*, 19 June 2014. See also Report on the Surveillance Program Operated Pursuant to Section 702 of the Foreign Intelligence Surveillance Act, Privacy and Civil Liberties Oversight Board, 2 July 2014 for Upstream details.

21 'Database Marketing', *Businessweek*, 4 September 1994

22 See *The Watchers: The Rise of America's Surveillance State* by Shane Harris (Penguin Press, New York, 2010), p. 223

23 'The Rise and Fall of Thinking Machines', *Inc*, 15 September 1995, http://www.inc.com/magazine/19950915/2622.html and see http://decisionstats.com/2010/10/05/interview-michael-j-a-berry-data-miners-inc/

24 http://www.pbs.org/wgbh/pages/frontline/government-elections-politics/united-states-of-secrets/transcript-61/

25 Harris, *The Watchers*, p. 268

26 Draft NSA IG Report http://www.theguardian.com/world/interactive/2013/jun/27/nsa-inspector-general-report-document-data-collection

27 Report on the Surveillance Program Operated Pursuant to Section 702 of the

Foreign Intelligence Surveillance Act, Privacy and Civil Liberties Oversight Board, 2 July 2014 and 'Secret to Prism program: Even bigger data seizure', Associated Press, 15 June 2013

28 Report on the Surveillance Program Operated Pursuant to Section 702 of the Foreign Intelligence Surveillance Act, Privacy and Civil Liberties Oversight Board, 2 July, 2014

29 http://www.networkworld.com/article/2866286/microsoft-subnet/former-microsoft-chief-privacy-officer-on-the-cloud-conspiracy.html

30 Report on the Surveillance Program Operated Pursuant to Section 702 of the Foreign Intelligence Surveillance Act and Greenwald, *No Place to Hide*, p. 111

31 Report on the Surveillance Program Operated Pursuant to Section 702 of the Foreign Intelligence Surveillance Act

32 Office of the Director of National Intelligence Data Mining Report, 15 February 2008. For instance in 2008, the US was looking at ways of finding patterns which would predict if there was going to be a 'change in the threat likelihood from a known individual' to provide advance warning.

33 Harris, *The Watchers*, provides a detailed account

34 http://foreignpolicy.com/2013/09/09/the-cowboy-of-the-nsa/

35 'NSA gathers data on social connections of US citizens', *New York Times*, 28 September 2013; *Big Data: A Revolution That Will Transform How We Live, Work and Think* by Viktor Mayer-Schönberger and Kenneth Cukier (John Murray, London, 2013), p. 46, and 'Under the covers of the NSA's big data effort', Gigaom, 7 June 2013, https://gigaom.com/2013/06/07/under-the-covers-of-the-nsas-big-data-effort/

Chapter Seventeen: Britain

1   2010 interview with Iain Lobban for BBC Radio 4 programme 'GCHQ: Cracking the Code', broadcast 30 March 2010

2   Ibid.

3   Skeleton Argument served by Privacy International and Bytes for All in front of Investigatory Powers Tribunal, July 2014

4   Report from the Secret Committee of the House of Lords relative to the Post Office, 7 August 1844

5   Intelligence and Security Committee Annual Report 2006/07

6   Most likely under Section 94 of the 1984 Telecoms Act, which allows the government to demand actions of telecoms companies and keep them confidential.

7   'GCHQ: Inside the top secret world of Britain's biggest spy agency', *Guardian*, 2 August 2013

8   Greenwald, *No Place to Hide*, p. 100

9   Iain Lobban farewell speech, October 2014

10 'GCHQ taps fibre-optic cables for secret access to world's communications', *Guardian*, 21 June 2013

11 'Mastering the Internet: how GCHQ set out to spy on the world wide web', *Guardian*, 21 June 2013

12 http://news.bbc.co.uk/1/hi/uk/6477777.stm

13 'A Cryptologist Encounters the Human Side of Intelligence' by Thomas R. Johnson, *Studies in Intelligence* (CIA declassified publication, undated) available at http://www.foia.cia.gov/sites/default/files/DOC_0001407027.pdf

14 'Russian couple's arrest could mar diplomatic ties', *Der Spiegel*, 25 October 2011

15 'Mastering the Internet: how GCHQ set out to spy on the world wide web', *Guardian*, 21 June 2013

16 *Canada and the Five Eyes Intelligence Community* by James Cox, December 2012, published by the Canadian Defence and Foreign Affairs Institute

17 See Wikileaks cable, Reporting and Collections Needs: The United Nations, 31 July 2009

18 http://www.washingtonpost.com/world/national-security/for-nsa-chief-terrorist-threat-drives-passion-to-collect-it-all/2013/07/14/3d26ef80-ea49-11e2-a301-ea5a8116d211_story.html

19 Interview transcript, *Australian Financial Review*, 8 May 2014

20 *@War: The Rise of the Military-Internet Complex* by Shane Harris (Headline, London, 2014), p. 19

Chapter Eighteen: Exposure

1 Interview with Ladar Levison, 2014

2 Quotes from Greenwald, *No Place to Hide* and documentary film *Citizenfour*

3 'This is not blitz Britain', *Daily Telegraph*, 11 October 2014, http://www.telegraph.co.uk/news/uknews/defence/11154322/GCHQ-This-is-not-Blitz-Britain.-We-sure-as-hell-cant-lick-terrorism-on-our-own.html

4 Interview transcript, *Australian Financial Review*, 8 May 2014

5 'Who is winning the Crypto War?', Gordon Corera, BBC News, 16 March 2014, http://www.bbc.co.uk/news/magazine-26581130, interview with Ladar Levison, 2014

6 'Snowden's first move against the NSA was a party in Hawaii', *Wired* magazine, 21 May 2014

7 'Job title key to inner access held by Snowden', *New York Times*, 30 June 2013. Snowden may also have turned down a job with the NSA's TAO because of stricter vetting.

8 'Out of Control' (declassified NSA document), 3 September 1991 (author's name redacted)

9 NSA letter to Senator Charles E. Grassley, 11 September 2013

10 Edward Snowden interview with German TV, https://www.tagesschau.de/snowden-interview-englisch100.pdf

11 National Archives, POST 56/57

12 Sir David Omand, 'Intelligence in the Digital Age', RUSI lecture, 6 October 2014

13 Greenwald's *No Place to Hide* provides a detailed explanation of this viewpoint.

14 'NSA growth fueled by need to target terrorists', *Washington Post*, 22 July 2013

15 'Claim on "attacks thwarted" by NSA spreads despite lack of evidence', ProPublica, 23 October 2013

16 Some analysts have claimed there is evidence that Al Qaeda increased its efforts to spread use of its own encryption tools (one called Mujahadeen Secrets has been circulating since at least 2007, but one analyst claims a new version was extensively upgraded in the wake of Snowden's revelations; critics point out that these systems are likely to be less professional and so easier to break). 'Qaeda plot leak has undermined US intelligence', *New York Times*, 29 September 2013, and see Bruce Schneier in http://www.theregister. co.uk/2014/08/13/snowden_effect_terrorist_crypto_debate/

17 Interview with Sir David Omand for BBC *Newsnight*, 2013

18 http://www.technologyreview.com/news/530116/qa-former-nsa-deputy-director-john-c-inglis/

19 'How America spies on Europe and the UN', *Der Spiegel*, 26 August 2013

20 See Wright, *Spycatcher*, on the early instances, Richard Tomlinson's *The Big Breach: From Top Secret to Maximum Security* (Moscow, Narodny Variant Publishers, 2000) on later MI6 cases.

21 http://www.standard.co.uk/lifestyle/london-life/tim-bernerslee-some-people-will-always-try-to-control-the-internet-we-have-to-keep-fighting-for-it-9757371.html 26 September 2014

22 http://www.lemonde.fr/international/article/2014/03/20/dgse-orange-des-liaisons-incestueuses_4386264_3210.html and 'Disclosures about France's big brother', *Le Monde*, 4 July 2013

23 'Industrial espionage: the American advantage', *Le Monde*, 31 October 2013

24 'Memories of Stasi color Germans' view of US surveillance programs', *McClatchy DC*, 26 June 2013

25 'Germany is both a partner to and a target of NSA surveillance', *Der Spiegel*, 12 August 2013

26 'As US weights spying changes, officials say data sweeps must continue', *New York Times*, 4 November 2013

27 Other European countries have also been identified as likely to be providing this kind of raw data, including Denmark – a useful partner since traffic from the data centres of Nordic countries flows through its pipes.

28 'Technically backward and helpless', *Die Welt*, 25 October 2013

29 http://www.washingtonpost.com/world/national-security/2013/10/30/bb9979c8-4192-11e3-8b74-d89d714ca4dd_story.html

30 Greenwald, *No Place to Hide*, p. 125

31 'Germany demands top US intelligence officer be expelled', *New York Times*, 10 July 2014

32 Interview with Chris Inglis, 2013

33 Interview with Ross Anderson for BBC *Newsnight*, 2013

34 Interview with Bruce Schneier, 2014, and 'The US government has betrayed the internet. We need to take it back', Bruce Schneier, *Guardian*, 5 September 2013

35 A series of breakthroughs appears to have been made in performing large-scale internet decryption of fibre-optic cable traffic by the NSA around 2010 – 'those not already briefed were gobsmacked', noted a GCHQ document; 'Revealed: how US and UK spy agencies defeat internet privacy and security', *Guardian*, 6 September 2013

36 Susan Landau, 'Under the Radar: NSA's efforts to secure private-sector telecommunications infrastructure', *Journal of National Security Law and Policy*, vol. 7, 2014, http://jnslp.com/wp-content/uploads/2015/03/NSA's-Efforts-to-Secure-Private-Sector-Telecommunications-Infrastructure_2.pdf

37 'The NSA's new code breakers' by Matthew Aid, *Foreign Policy*, 15 October 2013

38 https://edwardsnowden.com/2015/01/18/the-roc-nsas-epicenter-for-computer-network-operations/; https://edwardsnowden.com/2015/01/18/interview-with-a-sid-hacker-part-1-how-does-tao-do-its-work/ and https://freesnowden.is/2015/01/18/expanding-endpoint-operations/

39 'Inside the NSA's ultra-secret China hacking group' by Matthew Aid, *Foreign Policy*, 12 June 2013, http://www.foreignpolicy.com/articles/2013/06/10/inside_the_nsa_s_ultra_secret_china_hacking_group

40 https://edwardsnowden.com/2015/01/18/interview-with-a-sid-hacker-part-1-how-does-tao-do-its-work/

41 'Unit offers spy gadgets for every need', *Der Spiegel*, 30 December 2013

42 'NSA surveillance: a guide to staying secure', Bruce Schneier, *Guardian*, 6 September 2013

43 'Documents reveal top NSA hacking unit', *Der Spiegel*, 29 December 2013

44 'How the NSA plans to infect "millions" of computers with malware', The Intercept, 12 March 2014

45 Greenwald, *No Place to Hide*, p. 192

46 http://www.nbcnews.com/feature/edward-snowden-interview/exclusive-snowden-docs-show-uk-spies-attacked-anonymous-hackers-n21361

47 'MI6 attacks Al Qaeda in operation cupcake', *Daily Telegraph*, 2 June 2011, http://www.telegraph.co.uk/news/uknews/terrorism-in-the-uk/8553366/MI6-attacks-al-Qaeda-in-Operation-Cupcake.html

48 Greenwald, *No Place to Hide*, p. 187

49 Ibid., p. 191

50 http://www.theguardian.com/technology/2013/sep/11/yahoo-ceo-mayer-jail-nsa-surveillance

51 'Don't listen to Google and Facebook: the public-private surveillance partnership is still going strong', *The Atlantic*, 25 March 2014, http://www.theatlantic.com/technology/archive/2014/03/don-t-listen-to-google-and-facebook-the-public-private-surveillance-partnership-is-still-going-strong/284612/Power in the age of the feudal internet speech by Bruce Schneier

52 http://www.dailydot.com/technology/eric-schmidt-google-nsa-spying-attack/

53 'NSA bombshell shocks former spooks: "Why in the World Would We Burn Google?"', *Foreign Policy*, 30 October 2013

54 'NSA breached Chinese servers seen as security threat', *New York Times*, 22 March 2014

55 'Putin calls internet a "CIA project" renewing fears of web breakup', *Guardian*, 24 April 2014

56 'NSA is no match for FSB', Agentura.ru website

57 Looking at NSA and GCHQ as role models: German intelligence plans their own mass spying programme, 4 June 2013 https://www.accessnow.org/blog/2014/06/04/nsa-gchq-as-role-models-german-intelligence-plans-their-own-mass-spying

58 Analysts at TeleGeography say that in 1999, 47 per cent of all international internet traffic touched the United States and 31 per cent the United Kingdom (with overlap between those two). By 2014 those figures had slipped to 27 per cent for the US and 18 per cent for the UK. Other figures are cited in the US IG Report. Nigel Inkster in 'China in Cyberspace', *Survival*, July 2010, quotes figures that 91 per cent of Asian internet traffic transited the US in 1994 compared to 54 per cent in 2010. It is not clear that these are all using the same definitions.

59 'Mastering the Internet: how GCHQ set out to spy on the world wide web', *Guardian*, 21 June 2013

60 Remarks by Michael Hayden at George Washington University, Federal News Service, 19 February 2013

61 'The world's most wanted cyber-jihadist' by Gordon Corera, BBC News, 16 January 2008, http://news.bbc.co.uk/1/hi/world/americas/7191248.stm

62 *Financial Times*, 4 November 2014

63 Speech by Rachel Whetstone. Munich Security Conference, 13 February 2015

64 http://cpj.org/blog/2014/10/doj-resistance-to-encryption-jeopardizes-journalis.php

Epilogue: To Infinity and Beyond

1 'GCHQ: Cracking the Code', BBC Radio 4 broadcast, 30 March 2010

2 Ibid.

3 http://www.cray.com/Programs/Cascade.aspx

4 Tony Sale, who rebuilt Colossus at the National Museum of Computer History, tested this out.

5 http://m.nextgov.com/emerging-tech/2014/12/intelligence-communitys-research-arm-funding-develop-even-more-super-supercomputer/100409/?oref=ng-HPriver

6 'NSA able to foil basic safeguards of privacy on the web', *New York Times*, 5 September 2013

7 'The NSA is building the country's biggest spy center', *Wired* magazine, 15 March 2012, and William Binney Court Declaration, 2012

8 'Advertising and Technology', *Economist* Special Report, 13 September 2014

9 http://www.independent.co.uk/news/uk/home-news/exclusive-abusers-using-spyware-apps-to-monitor-partners-reaches-epidemic-proportions-9945881.html

10 Mayer-Schönberger and Cukier, *Big Data*, pp. 153 and 92

11 http://geer.tinho.net/geer.uncc.9x13.txt

12 http://www.businessinsider.com/cia-presentation-on-big-data-2013-3?op=1

13 'Edward Snowden, The Dark Prophet', *Time* magazine, 11 December 2013, http://poy.time.com/2013/12/11/runner-up-edward-snowden-the-dark-prophet/